OXFORD MONOGRAPHS IN INTERNATIONAL HUMANITARIAN AND CRIMINAL LAW

The Morality of the Laws of War

OXFORD MONOGRAPHS IN INTERNATIONAL HUMANITARIAN AND CRIMINAL LAW

This series emerged through the vision of the late Antonio Cassese. Since then the series has published original and innovative works on challenging issues in international humanitarian law and international criminal justice, primarily by emerging authors. Given the ever-growing intersection of these areas with other fields of international law such as the jus ad bellum, and disciplines such as criminology, sociology, and history, the scope of the series also incorporates these perspectives. Fresh approaches and new ideas for improving conceptual and practical challenges are welcomed. The series editors encourage submissions from around the world, and will consider edited collections, although the primary focus is on authored works.

ALSO PUBLISHED IN THIS SERIES

Accountability for Mass Starvation
Testing the Limits of the Law
Edited by Bridget Conley, Alex de Waal, Catriona Murdoch, and Wayne Jordash KC

The Defence of Mistake of Law in International Criminal Law
A Study on Ignorance and Blame
Antonio Coco

The Use of Force against Individuals in War under International Law
A Social Ontological Approach
Ka Lok Yip

Non-Binding Norms in International Humanitarian Law
Efficacy, Legitimacy, and Legality
Emily Crawford

Doing Justice to History
Confronting the Past in International Criminal Courts
Barrie Sander

The Morality of the Laws of War

War, Law, and Murder

MARCELA PRIETO RUDOLPHY

OXFORD
UNIVERSITY PRESS

OXFORD
UNIVERSITY PRESS

Great Clarendon Street, Oxford, OX2 6DP,
United Kingdom

Oxford University Press is a department of the University of Oxford.
It furthers the University's objective of excellence in research, scholarship,
and education by publishing worldwide. Oxford is a registered trade mark of
Oxford University Press in the UK and in certain other countries

First Edition published in 2023

Published in the United States of America by Oxford University Press
198 Madison Avenue, New York, NY 10016, United States of America

British Library Cataloguing in Publication Data

Data available

Library of Congress Control Number: 2023935537

ISBN 978-0-19-285547-3

DOI: 10.1093/oso/9780192855473.001.0001

Printed and bound by CPI Group (UK) Ltd, Croydon, CR0 4YY

To Bautista

Series Editors' Preface

The present book, *The Morality of the Laws of War* by Dr Marcela Prieto Rudolphy, revolves around a theme which remains topical over the decades through the changing faces of armed conflict. It poses fundamental questions to both lawyers and philosophers trying to understand this extreme violence, find justifications for why humans take up arms, how to make them less inhuman, and what ultimately justifies (or not) killing in war. In Dr Prieto Rudolphy's analysis, these questions revolve around the potential of projecting justifications for waging wars into the conceptualization of justifications for killing in war.

The philosophical approach which shapes the work of Dr Prieto Rudolphy highlights the complexities of the issues underlying the realities discussed in the book (and even more those underpinning what happens on the battlefield, one should add). Framing the discourse on a theoretical level allows the author to avoid the traps of technicalities which from time to time can make it difficult to see what is essential, as can be the case in black-letter law approaches.

The author unveils the fundamental contradiction of the contemporary system of international law, which outlaws war, while continuing to try to aim at making it less brutal. On the one hand, she emphasizes the additional layer of intricacy given by the fundamental value of peace as a cornerstone of post-Second World War international legal order, and hence the expected (and yet hardly implemented) ban on war. On the other hand, she lucidly demonstrates the reality that we simply moved into an era in which war never ends—a sort of shift from an ideal of perpetual peace to a status of perpetual war. She denounces the hypocrisy of there being decades of wars which have not become less brutal through the laws of war, and the failure to turn towards the direction of ending all wars.

The book is premised on the idea of emphasizing such a contradiction by discussing the nature and extent of 'the privilege to kill' in war, the distinction between individuals possessing different combatant statuses. The legality of targeting some individuals and not others. In so doing, this book first deals with combatants and the privilege to kill; it then discusses the function and purposes of 'regulating' murder and some of the possible inspirational goals of such an approach. Subsequently, it tackles the various aspects of the role of morality in premising the regulations of killing under the laws of war and the necessary links with the motives of a conflict as well as the potential effects (or lack thereof) on the position of individuals and just or unjust combatants of a distinction between just and unjust wars.

The analysis includes arguments relating to possible justifications under the self-defence paradigm, including by looking at the differing positions between those who fight lawful wars (with a potential overlap between self-defence from the perspective of states as well as individual combatants).

Finally, the book closes the circle by highlighting the impact of the contradiction of prohibiting war and yet continuing to try to regulate it. Dr Prieto Rudolphy offers to the reader the sharp (and bitter) reflections of Tamara Stepanovna Umnyagina, a woman interviewed by the Nobel Prize winner Svetlana Alexievich in her well-known work *The Unwomanly Faces of War* (which also opens the conclusions of the book). Tamara notes that '[p]eople still hate each other. They go on killing. That's the most incomprehensible thing to me....' It is even more incomprehensible (or less perhaps) if one looks at it through the prism of the famous poem by Bertolt Brecht, 'The war which is coming':

> The war which is coming
> Is not the first one. There were
> Other wars before it.
> When the last one came to an end
> There were conquerors and conquered.
> Among the conquered the common people
> Starved. Among the conquerors
> The common people starved too.

Dr Prieto Rudolphy has successfully uncovered the roots of the modern regulation of the laws of war, teased out the ambiguities, and convincingly demonstrated the fundamental contradiction of regulating war while declaring that war as such is outlawed. But what is the alternative to trying to regulate war and aim at reducing its impact on ordinary people?

As all clever books, *The Morality of the Laws of War* will certainly promote and provoke discussions. As Series Editors, we are very pleased to present this fine work as part of the Oxford Monographs on International Criminal and Humanitarian Law Series.

The Editors

Acknowledgements

This book started as my dissertation at the Doctor of Juridical Science (JSD) programme in New York University. It takes a village to write a dissertation—and to write a book. This is mine.

First, I would like to thank my JSD advisor, Jeremy Waldron. This book would not exist without him. As a graduate student, he taught me what makes a good argument, and, more importantly, what makes a good and generous scholar. He read and commented on multiple drafts, to the point where it would be impossible to thank him for individual thoughts or ideas. After I graduated, just at the beginning of the pandemic, he took the time to advise me on the publication process, gently pushing me to think of my dissertation as a future book. I have said this before but I truly could not have wished for a better advisor.

I am also indebted to my JSD committee members, Moshe Halbertal and Liam Murphy. Moshe Halbertal provided invaluable comments and suggestions and pushed me to develop the idea of permissibility when it was little more than two lines in a paragraph. Liam Murphy has also provided invaluable feedback on what was then my dissertation and other related work, as well as generous advice and support on professional matters.

During my time at NYU, I was fortunate to attend classes that pretty much defined the way I approach legal and moral philosophy. For that, I thank, again, Jeremy Waldron, as well as Lewis Kornhauser and Liam Murphy.

The influence of Lewis Kornhauser's Legal Theory seminar on my scholarship simply cannot be overstated; I would be a significantly worse scholar without it. He was also the director of the JSD programme during the first years at the programme and provided detailed feedback on the first drafts of the dissertation.

I would also like to thank Joseph Weiler and Gráinne de Búrca for their feedback on previous versions of this book, their constant support and advice on professional matters, and their generosity in making me part of the I.CON community. It is always a pleasure to work with them.

There are many others who have read my work at different stages of development. I thank, in particular, Alejandro Chehtman, Scott Altman, Jeesoo Nam, and Erin Miller for helpful feedback and conversations on the subject;

an anonymous reviewer for Oxford University Press; David Cruz, for his detailed feedback on Chapter II; Marcia Baron for her thoughtful feedback on Chapter V; Wojciech Sadurski, Sergio Verdugo, Michaela Hailbronner, Meir Yarom, Teng Li, the NYU JSD forum participants; and Samuel Scheffler and Gary Bass, who kindly agreed to be commentators in my JSD forums. I am also indebted to the participants of the faculty workshop at ITAM, kindly organized by Joyce Sadka and Jorge Cerdio, and the faculties of USC Gould School of Law, Indiana University Maurer School of Law, and Villanova University who provided helpful comments on what was then my job talk paper.

I am grateful to Dean Andrew Guzman and to the directors of the Center for Law and Philosophy, Greg Keating, Felipe Jiménez, and Jonathan Quong for organizing a book workshop, reading the manuscript, and providing helpful feedback. I'm particularly indebted to Jonathan Quong, who read and commented on a previous version of Chapter VI at the USC Gould Faculty workshop and later reread the entire manuscript for the book workshop and provided detailed feedback as well as helpful conversations on the subject. I also thank Mira Dalpe and Angie Houff for their incredible logistical support.

I would like to thank the workshop attendants, as well as Susanne Burri, Michelle Dempsey, Adil Haque, and Samuel Moyn, who generously agreed to comment on my book manuscript and provided invaluable and gracious comments. It was a privilege that they took a day to discuss and think about my work. Susanne Burri followed up with extensive written notes, which were incredibly helpful when revising the book. Michelle Dempsey helped me immensely to fine tune the discussion regarding exclusionary reasons. Adil Haque provided detailed feedback on both the international law and the moral philosophy aspects of the book and helped me tremendously with Chapter II—not once but twice. Samuel Moyn pushed me to think of the project in a slightly different light, which made more sense of its normative commitments. I am incredibly grateful to each of them.

A special thanks goes to my Oxford University Press editor, Paulina dos Santos Major, for her constant and gracious help during this process, and to my research assistant, Jessica Murphy.

I must also thank my parents and sisters for their long-distance emotional support, as well as much needed and appreciated help with childcare whenever they were able (willing, I should say; they were always).

Almost finally, I thank Michele Krech, Gabi Wadlig, and Naama Goldberg for their constant support and encouragement, hilariously accurate gifs and stickers, feedback, and advice, but above all, for their friendship.

Last, but certainly not least, I thank my husband, Felipe Jiménez, for his constant support and encouragement. He has read and provided feedback on every single chapter, in its many different versions. I have forced him to talk about the morality of war over lunch and dinner (and breakfast) on more than one occasion, and he has graciously endured the vast array of little crises that come with writing. I've been incredibly lucky to be able to count on his emotional and intellectual support, and to share the adventure that is parenthood with him.

To Bautista, I thank him for the joy of watching him grow and discover the world together. It isn't a good or just one, but he makes it a happy one. I hope that when he grows up, he'll make good on his promise to 'buy all the books about the laws of war that money can buy'.

Contents

I. Introduction: The Problem with the Legal Equality of Combatants 1
- I. Just War and Just Combatants 14
- II. The Laws on Killing in War 15
  - A. Who Is a Privileged Combatant? 16
  - B. Who Can Be Targeted? 17
  - C. How to Kill? 20
  - D. When Does the Privilege to Kill Apply? 22

PART 1. THE REVISIONIST MORALITY AND THE LAWS OF WAR

II. Combatants and the Privilege to Kill 27
- I. Introduction 27
- II. War and Hohfeld 30
- III. Privileged Killings and Privileged Combatants 36
- IV. A Mere Immunity from Prosecution? 40
- V. Conclusion 54

III. Regulating Murder: Instrumentalism and the Revisionist Morality 55
- I. Introduction 55
- II. Instrumentalism and the Humanitarian View 58
  - A. The Humanitarian View 61
- III. Instrumentalism: The Appeal and the Objections 63
- IV. War and Perpetual Peace 74
- V. Conclusion 85

PART 2. ATTENUATING INSTRUMENTALISM'S FRAGILITY

IV. Non-Instrumentalism and the Revisionist Morality 93
- I. Introduction 93
- II. The Moral Division of Labour and Role Morality 94
- III. Democratic Authority 105
- IV. Conclusion 110

V. Just Wars, Just Combatants, Just Killings? 111
I. Introduction 111
II. Revisionism 113
A. Methodological Revisionism 114
B. Substantive Revisionism: Liability and Self-Defence 118
1. Liability to harm 119
2. The moral responsibility account of liability 122
III. Liability, Self-Defence, and their Limitations 123
A. Killing Aggressors: Liability, Combatants, and the Villainous Aggressor 125
1. Unjust combatants and the threats they pose 127
2. Unjust combatants and moral responsibility 128
B. Killing Enemy Combatants 133
C. Killing the Many: Unjust Combatants and the Number of Aggressors 139
1. Unjust combatants and narrow proportionality 140
2. Unjust combatants and necessity 143
IV. Conclusion 148

VI. The Non-liability of Unjust Combatants 149
I. Introduction 149
II. Unjust Combatants and the Problem with Forfeiture 150
III. A Lesser-Evil Justification 160
A. What is a Lesser-Evil Justification? 161
IV. Lesser-Evil Justifications, Killing in War, and the Combined Justification 163
A. Lesser Evil instead of Liability 164
B. Lesser Evil *with* Liability: The Combined Justification 169
V. Conclusion 176

VII. Beyond Self-Defence 179
I. Introduction 179
II. Recap and the Chain of Command Issue 181
III. Just War and Killing 186
A. A Just War 190
B. Just War and Unjust Combatants 194
IV. Permissibility in Defensive Harm 199
A. Moral Residue: Acting Against Important Values 201
B. Small Differences in the Weight of Reasons 203
C. Wronging the Non-Liable 205
D. Back to Small Differences in the Weight of Reasons 209
V. Permissibility: Implications and Advantages 216
VI. Permissibility: Objections 227
VII. Conclusion 234

VIII. Unjust Combatants 237
I. Introduction 237
II. A Background of Duress and Uncertainty 241
III. Collective Enterprises and Their Moral Heterogeneity 243
A. Just Wars: Fighting Unjustly 243
B. The Problem of Small Contributions 245
1. It makes no difference whether or not I fight 245
2. It makes no difference whether or not I surrender 247
IV. Risking Your Life 250
A. Unjust Combatants, Counter-Defence, and Defence of Others 253
B. It's You or Me: Doing Your Share and Deflecting Harm 257
V. Conclusion: The Small Moral World of Combatants 262

Conclusion: The Peace That Was Promised 269

Bibliography 279
Index 295

I

Introduction

The Problem with the Legal Equality of Combatants

War is an ever-present feature of human society. It has a long history, and although diverse in its causes and manifestations, it does not seem to be in danger of disappearing in the near future. As of June 2022, armed conflicts rage in Ukraine, Syria, Afghanistan, Libya, and South Sudan, to name just a few.[1] Although ever-present, war is a human—and also a human-made—tragedy. What makes war tragic is, obviously, the destruction of human life. Thousands of civilians die and are wounded as collateral damage; entire populations are displaced from their homes. But the tragedy also lies—even if sometimes we forget this—in the thousands of combatants who die and kill in wars and under circumstances that have not been chosen by them,[2] and for which they are usually not responsible—or, maybe, not responsible enough.

War is also, and perhaps slightly surprisingly, a legally regulated tragedy. And it is legally regulated in a particular way. Although the protection of the right to life and the prohibitions on killing and assault are among the most basic rules of all legal systems,[3] the laws of armed conflict seem to circumvent these commitments: they permit the intentional killing of combatants.

Indeed, the international laws of armed conflict have several outstanding features. First, they distinguish between civilians and combatants: while intentionally targeting civilians is a war crime, intentionally targeting combatants is not. Second, the laws that regulate the resort to force (the *jus ad bellum*) are independent from the laws that regulate conduct in war (the *jus in bello*). This independence entails that a war that is illegal according to the *jus ad bellum* may be conducted in legal ways according to the *jus in bello*, and a war that is

[1] *See* 'Global Conflict Tracker' (*Council on Foreign Relations*) <https://www.cfr.org/interactive/global-conflict-tracker/?category=us> accessed 22 August 2022.

[2] *See* Cheyney C Ryan, 'Self-Defense and the Obligations to Kill and to Die' (2004) 18 Ethics & International Affairs 69, 70–71.

[3] *See* David K Chan, *Beyond Just War: A Virtue Ethics Approach* (Palgrave Macmillan 2012); HLA Hart, *The Concept of Law* (Penelope A Bulloch and Joseph Raz eds, 2nd edn, Clarendon Press 1998) 171–72, 193–200.

The Morality of the Laws of War. Marcela Prieto Rudolphy, Oxford University Press. © Marcela Prieto Rudolphy 2023.
DOI: 10.1093/oso/9780192855473.003.0001

legal according to the *jus ad bellum* may be conducted in ways that contravene the *jus in bello*. Finally, all combatants, regardless of the nature of the war in which they are participating, have the same legal rights, duties, and privileges so long as they comply with the legal requirements to be considered privileged combatants.[4]

These three features—the principle of distinction, the independence of *jus ad bellum* from *jus in bello*, and the symmetry thesis—are basic to the legal regulation of war. Yet their moral standing, once widely accepted, is now dubious. Although these features of the legal regime were defended by Michael Walzer in *Just and Unjust Wars*[5]—what is now known as the conventionalist or traditionalist view—they have been the target of attack in the past years by a group of philosophers known as revisionists.

Revisionists[6] think that these basic features of our legal regime are at odds with the morality of war. Neither the principle of distinction nor the symmetry thesis finds any correlate in morality. Whether someone can be killed defensively depends, revisionists argue, on whether they have moral responsibility for an objectively unjust threat of harm.[7] Applied to war, this entails that *only* those waging a just war—just combatants—are morally justified in killing

[4] This is true of international armed conflicts. In non-international armed conflicts, targeting combatants is not a war crime, but combatants fighting against state armies are not privileged: they can be prosecuted for their participation in the hostilities. *See* Adam Roberts, 'The Principle of Equal Application of the Laws of War' in David Rodin and Henry Shue (eds), *Just and Unjust Warriors: The Moral and Legal Status of Soldiers* (Oxford University Press 2008); Christopher Kutz, 'The Difference Uniforms Make: Collective Violence in Criminal Law and War' (2005) 33 Philosophy & Public Affairs 148; Michael Walzer, *Just and Unjust Wars: A Moral Argument with Historical Illustrations* (Basic Books 2006); Yoram Dinstein, *The Conduct of Hostilities under the Law of International Armed Conflict* (Cambridge University Press 2004); Yoram Dinstein, *Non-International Armed Conflicts in International Law* (Cambridge University Press 2014).

[5] *See* Walzer (n 4).

[6] They are called revisionists because they challenge Walzer's 'traditionalist' view, which asserts the independence of *jus in bello* from *jus ad bellum* and the symmetry thesis (that combatants have the same rights, duties, and privileges regardless of the justice of the war they are fighting). That is, they are revisionists in relation to Walzer's view. In terms of the historical evolution of just war theory, the position is not revisionist. What revisionism says about the connection between *jus in bello* and *jus ad bellum* and against the symmetry thesis has been said before, by most medieval just war theorists, like Vitoria. *See* Uwe Steinhoff, 'Rights, Liability, and the Moral Equality of Combatants' (2012) 16 The Journal of Ethics 339, 339–46. A more detailed definition of the revisionist view and its commitments can be found in Chapter V. The most prominent revisionists are Jeff McMahan, Helen Frowe, Victor Tadros, and Cécile Fabre. *See* Helen Frowe, *Defensive Killing* (1st edn, Oxford University Press 2014); Jeff McMahan, *Killing in War* (Oxford University Press 2009); Victor Tadros, 'The Moral Distinction Between Combatants and Noncombatants: Vulnerable and Defenceless' [2018] Law and Philosophy 1; Victor Tadros, *The Ends of Harm: The Moral Foundations of Criminal Law* (Oxford University Press 2011); Victor Tadros, *To Do, to Die, to Reason Why: Individual Ethics in War* (Oxford University Press 2020); Cécile Fabre, *Cosmopolitan War* (1st edn, Oxford University Press 2012).

[7] McMahan, *Killing in War* (n 6); Adil Ahmad Haque, *Law and Morality at War* (Oxford University Press 2017); Frowe (n 6); Victor Tadros, 'Duty and Liability' (2012) 24 Utilitas 259.

unjust combatants.[8] Unjust combatants, revisionists argue, not only lack a moral justification but are also forbidden from killing just combatants, because killing just combatants is tantamount to killing the innocent (or worse).[9] Revisionists also think that unjust combatants (and some civilians from the unjust side) forfeit their moral rights against harm, and that they are not allowed to defend themselves.[10] These ideas are sometimes referred to as 'the deep morality of war'.[11]

If revisionists are right about the morality of war, the challenge posed to the international laws of armed conflict could hardly be overstated. Our laws of war do not distinguish between just and unjust combatants: all combatants are fair targets, and, in an international armed conflict, all combatants can be privileged in their killings if they comply with certain legal requirements. Yet, according to the revisionist account,[12] our laws of war permit what is equivalent to murdering the innocent.

We thus find ourselves with legal rules that are generally assumed to be significantly divergent from morality. They are not just over- or underinclusive; it is not just a matter of the need to model morality into law.[13] The current laws of war—if revisionists are right—forbid what is morally permitted (targeting some civilians from the unjust side of the war) and permit what is morally forbidden (killing just combatants). The revisionist account leaves us then with

[8] *See* Victor Tadros, 'Orwell's Battle with Brittain: Vicarious Liability for Unjust Aggression' (2014) 42 Philosophy & Public Affairs 42; Frowe (n 6); Haque (n 7); Fabre (n 6); McMahan, *Killing in War* (n 6).

[9] For instance, Tadros has suggested that killing just combatants might be morally worse than killing civilians since, unlike civilians, just combatants are contributing to the achievement of justice. *See* Tadros, 'The Moral Distinction Between Combatants and Noncombatants' (n 6) 290.

[10] Indeed, Tadros says that 'it is uncontroversial that combatants on the unjust side are liable to be harmed'. Tadros, 'Orwell's Battle with Brittain' (n 8) 55. Whether civilians are morally permissible targets will depend on whether they are morally responsible for an unjust threat of harm. On this discussion, *see* McMahan, *Killing in War* (n 6); Tadros, 'Orwell's Battle with Brittain' (n 8); Frowe (n 6); Seth Lazar, *Sparing Civilians* (1st edn, Oxford University Press 2015); Seth Lazar, 'The Responsibility Dilemma for Killing in War: A Review Essay' (2010) 38 Philosophy & Public Affairs 180.

[11] *See* Jeremy Waldron, 'Deep Morality and The Laws of War' in Seth Lazar and Helen Frowe (eds), *The Oxford Handbook of Ethics of War* (2015); Jeff McMahan, 'The Morality of War and The Law of War' in David Rodin and Henry Shue (eds), *Just and Unjust Warriors: The Moral and Legal Status of Soldiers* (Oxford University Press 2008).

[12] And, certainly, according to most forms of pacifism. On pacifism, *see* Craig K Ihara, 'In Defense of a Version of Pacifism' (1978) 88 Ethics 369; Jan Narveson, 'Pacifism: A Philosophical Analysis' (1965) 75 Ethics 259; Andrew Alexandra, 'Political Pacifism' (2003) 29 Social Theory and Practice 589; Cécile Fabre, 'On Jan Narveson's "Pacifism: A Philosophical Analysis"' (2015) 125 Ethics 823; Larry May, 'Contingent Pacifism and the Moral Risks of Participating in War' (2011) 25 Public Affairs Quarterly 95; Jan Narveson, 'Is Pacifism Consistent?' (1968) 78 Ethics 148; M Jay Whitman, 'Is Pacifism Self-Contradictory?' (1966) 76 Ethics 307.

[13] *See* John Finnis, *Natural Law and Natural Rights* (2nd edn, Oxford University Press 2011) 268–86; Jeremy Waldron, *Torture, Terror, and Trade-Offs: Philosophy for the White House* (Oxford University Press 2010) 96; Jeremy Waldron, 'Torture, Suicide, and Determinatio' (2010) 55 American Journal of Jurisprudence 10.

the daunting task of providing a moral justification for our current legal rules, something to which relatively little attention has been paid.[14]

As mentioned above, the laws of war have three basic features: the principle of distinction, the independence of *jus ad bellum* from *jus in bello*, and the symmetry thesis. Each demands a moral justification. This book, however, is concerned with only one dimension of our laws of armed conflict: the legal permission of combatants to target and kill each other during war. This dimension is directly related, of course, to the problem of the symmetry thesis and the independence of *jus ad bellum* and *jus in bello*. It is also indirectly related to the prohibition on intentionally targeting civilians. Nonetheless, the focus of this book is on combatants. I don't intend to provide a justification for international humanitarian law (IHL) as a whole, with its many disparate provisions, nor for the prohibition on the intentional targeting of civilians during war. The question about civilian immunity is undoubtedly important, but it is also the question on which most of the academic debate has focused.[15]

The question in this book is thus how to morally justify a set of laws that permit combatants to kill each other during armed conflict. Or, put differently, the question is what moral justification there might be for the fact that our laws of war make all combatants permissible targets during war. Although the scope of the book is limited, the answer to this justificatory question might have revisionary implications for the assessment of IHL. After all, the permissibility of targeting combatants is central—but not necessarily essential, as I will argue—to the legal regime.

Until revisionism became the predominant moral theory of war, traditional just war theory—as reflected in Michael Walzer's influential book[16]—argued that the basic features of the laws of war were mostly in harmony with the

[14] Except for Lazar, *Sparing Civilians* (n 10); Waldron, *Torture, Terror, and Trade-Offs* (n 13); Haque (n 7); Arthur Ripstein, *Kant and the Law of War* (Oxford University Press 2021).

[15] *See* Gabriella Blum, 'The Dispensable Lives of Soldiers' (2010) 2 Journal of Legal Analysis 115. She notes this precise void in the literature. On the issue of civilian immunity, *see* Richard J Arneson, 'Just Warfare Theory and Noncombatant Immunity' (2006) 39 Cornell International Law Journal 663; Whitley Kaufman, 'What Is the Scope of Civilian Immunity in Wartime?' (2003) 2 Journal of Military Ethics 186; Lawrence A Alexander, 'Self-Defense and the Killing of Noncombatants: A Reply to Fullinwider' (1976) 5 Philosophy & Public Affairs 408; Robert K Fullinwider, 'War and Innocence' (1975) 5 Philosophy & Public Affairs 90.

[16] *See* Walzer (n 4). Other scholars have also tried to defend some version of Walzer's view and can be considered 'traditionalists', but the classification is more complicated than this. I will address this in Chapter V. In the meantime, basic notions of what revisionism holds are enough. Some traditionalists are Waldron, 'Deep Morality and The Laws of War' (n 11); Waldron, *Torture, Terror, and Trade-Offs* (n 13); Lazar, *Sparing Civilians* (n 10); Henry Shue, 'Laws of War' in Samantha Besson and John Tasioulas (eds), *The Philosophy of International Law* (Oxford University Press 2010); Janina Dill and Henry Shue, 'Limiting the Killing in War: Military Necessity and the St. Petersburg Assumption' (2012) 26 Ethics & International Affairs 311.

morality of war, and thus the legal permissibility of targeting combatants was more or less consistent with morality. But revisionism has offered a compelling and complete account of the morality of war which has put into question precisely this basic feature of the legal regime. Revisionists have argued that not all combatants are morally permissible targets and that all combatants are not, in fact, morally equal. They are, on the contrary, decidedly unequal: those who kill just combatants in pursuit of an unjust war are engaged in what is the moral equivalent of murdering the innocent.

The justificatory challenge against the laws of war arises, then, in large part due to the content of the revisionist morality. If revisionists are right, and killing just combatants is tantamount to killing the innocent, justifying our laws of war will involve providing a justification for laws that permit murder on a vast scale. Of course, an obvious response to the challenge is simply to contest the revisionist morality itself. We might think that the moral analogy between unjust combatants who kill just combatants and people who murder the innocent is incorrect. As will become clear later, I am sympathetic to this position, and in Part 2 of the book, the search for an attractive answer to the question of justification of our laws will lead me to contest the morality of war that revisionists endorse.

Nevertheless, in Part 1 of the book, I will leave the revisionist morality in place. I will try to come up with a moral justification of our laws of war that takes for granted that what revisionists think about the morality of war is correct. This strategy may seem odd at first sight. Why not simply contest the revisionist morality, rather than wonder whether we can find a way to reconcile it with our laws? A first reason for doing so is simply that revisionism has become the predominant view about the morality of killing in war in the philosophical literature. A more important reason is that any plausible contestation of the revisionist morality is going to have significant limits. And while I do think that the revisionist morality is, at the very least, partly mistaken, and that we should re-examine it and deny that unjust combatants are *always* and *just like* those who murder the innocent, we should be careful not to exaggerate the extent to which this can be done. Even if not all unjust combatants are engaged in what is the moral equivalent of murder, the truth is that at least *some* killing done by the unjust side of a war will, in fact, be very close to, and sometimes exactly like, murdering the innocent. After all, an unjust war is a catastrophe that is regulated by our laws of war in a (mostly) permissive way.

Because a plausible contestation of the revisionist view will only go so far, and at least some killing permitted by our laws of war will remain the moral equivalent of murder, the question of justifying legal rules that permit the murder of the innocent—even if not *everything* they do is to permit the murder of the innocent—demands an answer. We cannot leave this question unresolved if we want to provide a successful moral justification of the legal permissibility of targeting combatants. Thus, Part 1 of this book will keep the revisionist morality in place and try to provide an attractive moral justification for our laws of war that takes seriously what it means to have laws that effectively allow for murder. In Part 2, I will move on to contest the revisionist morality of war itself.

To complicate things further, the answer to the question about the justification of the legal permission of combatants to kill each other during war could take different forms. In Part 1 of this book, the range of answers which can be provided is constrained by the revisionist morality (ie by the fact that what needs justification is a set of laws that permits the murder of the innocent).

A first alternative is simply to deny that such a legal permission exists. If the laws of war do not grant any strong legal permission for combatants to kill each other, then the problem of justifying a regime that allows for the murder of the innocent loses its urgency. This is, in fact, the route that Adil Haque has taken.[17] Of course, this is not providing a moral justification for the legal privilege of unjust combatants but side-stepping the problem. What Haque argues is that the laws of armed conflict do not provide unjust combatants with a strong legal permission to kill just ones. In other words, Haque claims, the law does not grant a permission to the non-innocent to kill the innocent; it merely provides an immunity from prosecution. If Haque is right, this particular problem of justification—justifying the permissibility of targeting combatants or the symmetry thesis—goes away. Other dimensions of the regime might still demand a justification (such as the prohibition on intentionally targeting civilians), and indeed, it is to these ones that Haque turns his attention. However, in Chapter II, I will argue that this way out of the justificatory problem is not available. The most plausible interpretation of the laws of armed conflict supports the existence of a symmetrical legal permission—and not just of an immunity—for combatants to kill each other during war. We are, thus, where we began: with a legal permission in need of moral justification.

A second possibility is to justify our laws of war instrumentally, that is, on account of their consequences. Precisely because of the difficulties associated with providing a justification for laws that so wildly diverge from our moral commitments by allowing for the murder of the innocent, a consequences-based answer becomes attractive. This is indeed the route that some revisionists have taken. They argue that our laws of war, though unable to reflect morality,

[17] *See* Haque (n 7).

are justified because they can bring about less suffering than laws that tracked the revisionist morality of war. This position is usually known as the 'humanitarian view'. But is an instrumentalist justification of our laws of war successful if revisionists are right about the morality of war?

In Chapter III I will argue that the humanitarian view—that is, the instrumentalist justification provided by some revisionists—fails as a moral justification of the permission to target combatants. I also believe it stands in irreconcilable contradiction with the revisionist morality of war. However, this failure does not provide an argument against instrumentalism itself. A richer and more attractive instrumentalist justification, even if premised on the revisionist morality of war, may be able to morally account for our legal rules. Thus, I will provide an alternative, contingent, instrumentalist justification—based on what I call 'the perpetual peace ideal'—that fares a bit better than the humanitarian view. This account explains how the *adoption* of the symmetry thesis could be justified, but it cannot sanitize its content.

As a result, the instrumentalist justification of the legal privilege of combatants remains incredibly fragile when premised on the revisionist morality of war. In many ways, moral fragility is inescapable for both just war theory and the legal regulation of war. As I will argue in Chapter III, even the pursuit of a just war is second-best to the ideal of peace. Nonetheless, any moral justification of the legal privilege of combatants that starts from the assumption that our laws of war permit the murder of innocents will struggle greatly. Can this fragility be attenuated, and if so, how? This is precisely what I will attempt in Part 2 of this book and which will inevitably lead to contesting the revisionist morality.

In Chapter IV, that attempt will first involve discussing non-instrumentalist justifications of the symmetry thesis, which rely on role morality and democratic authority. I will argue that these justifications are of limited application, and cannot, on their own, justify the symmetry thesis if revisionists are right about the morality of war.

But the attempt to attenuate the fragility of the instrumentalist rationale is also related to a second way in which we can understand legal rules as morally justified. This type of moral justification demands a greater fit between morality and law than an instrumentalist justification: our legal rules must be good not only at achieving some consequences, they must *also* track some moral demands, particularly if they concern important rights (such as the right to life). Thus, for example, legal rules on murder must be good (or at least not very bad) at not encouraging murder *and* they must also track the moral prohibition on murder.

Now we already know that the laws of war are not actually tracking what revisionists take their corresponding moral content to be. On the contrary, our laws of war diverge significantly from it. They permit, under the revisionist account, unjust combatants to murder the innocent. Thus, as should be clear by now, attenuating the fragility of the instrumentalist justification can only be done to the extent that we are able to challenge the revisionist morality and propose an alternative one. In particular, we ought to deny that what unjust combatants do is *just like* killing the innocent.

In doing so, I do not mean to imply that just and unjust combatants are morally equal—they are not. Just combatants are fighting a just war, and so they are contributing to something that is (or should be) objectively good. Unjust combatants are simply not. What I aim to highlight is that individuals participating in war differ significantly and in morally significant ways: some of them we might judge favourably, while others we might not. Thus, that individuals are fighting a just or unjust war is not the end of the moral story we might want to tell—it doesn't tell us everything about the individual combatants. War tends to erase differences between individuals; it is part of what makes it tragic.

The first step in establishing the lack of analogy between what unjust combatants do and murdering the innocent will involve looking closely at the revisionist morality of war and what I take to be its shortcomings. One of the main problems with the revisionist morality, in the version that allows it to explain all killing in war, is that it leaves very little room for the moral tragedy there is in killing and fighting in a just war. If we are to provide an attractive morality of war, the resulting moral view should be able to reflect this conflict.

The argument against the plausibility of the revisionist view has two different, but related, parts. In Chapter V, I will describe the revisionist view in more detail and argue that a self-defence view committed to reductivist individualism fails to account for most killing in war. In order to do so—that is, in order to expand its justificatory reach—it would have to let go of its methodological commitments (which is something that it sometimes inadvertently does.) In Chapter VI, I will argue that a plausible forfeiture-based view cannot make many unjust combatants liable to significant or lethal harm.

If my arguments in Chapters V and VI are successful, the revisionist morality of war leaves a vast space of killing in war morally unaccounted for—it is unable to support the conclusion that many unjust combatants are liable to be killed. It is within this vast space that we can begin to deconstruct the analogy between what unjust combatants do and the murder of the innocent by, obviously, proposing an alternative morality of war. This will be the focus of Chapter VII.

The account of the morality of war I will propose stems, in many ways, from the reasons, attitudes, and experiences that I have found in books about war, written by journalists who have reported on wars, either independently or embedded within the US army, by soldiers and combatants themselves, and by historians who have interviewed former combatants.

In this way, the book parts ways with many works in analytical moral philosophy, which rely primarily on stylized examples against which intuitions and arguments are tested. Although I resort to some examples in this way, I avoid doing so when possible. Part of the reason is that I often do not have strong intuitions about stylized examples. But I do have strong intuitions of a more general nature. I think dying and killing in war is often a tragedy, and that what happens to combatants in war is often a moral wrong, in light of which certain reactive attitudes are appropriate.

These intuitions seem as defensible and plausible to me as a starting point as intuitions about who is liable to defensive force in stylized examples. They are also consistent with the feelings and attitudes of people who have gone to war (though I do not claim, nor do I believe, that whether these attitudes are the most common or are the only ones has any significance).

None of this means that the intuitions can be left untested. On the contrary, the method I and those who use stylized examples rely on is largely the same. As Jonathan Quong explains, the point is to go 'back and forth between particular judgments and more general principles, along with more general theories, with the aspiration of reaching a position where particular judgements cohere with more general principles and theories'.[18] Thus, although the starting point and the materials I will be working with are somewhat different from those of mainstream moral analytical philosophy,[19] the method I will employ is substantially similar.

An important part of the moral account I will propose thus arises out of the very reasons that combatants themselves have expressed as their reasons for fighting. I suppose some may think that this counts against them. Undoubtedly human beings tend to try and justify the things that they do. Yet, if we are going to talk about the morality of war, we have to start somewhere. Morality does not provide any clear roadmap telling us where to begin. We might then begin by listening to those who have been to war and by taking seriously what they say. This doesn't mean that they *must* be right, but we cannot assume from the

[18] Jonathan Quong, *The Morality of Defensive Force* (Oxford University Press 2020) 15.
[19] There are exceptions. *See*, for example, Lazar, *Sparing Civilians* (n 10); Fabre (n 6).

start that they are not.[20] This is, thus, a morality that arises from the little history of war,[21] not from the big history, the big picture. It is about the reasons combatants have for fighting, the reasons they act against, and the regrets that weigh them down.

The alternative morality of war I will propose starts, as well, from the limitations of the revisionist view. If many unjust combatants are *not* liable—as I will conclude in Chapters V and VI—we require an argument that makes those killings permissible, or, alternatively, we must embrace pacifism. This is a difficult position to be in, morally speaking.

Although pacifism about war is attractive, it is also a troubling position to embrace. After all, not resisting collective threats of violence can have significant moral costs. And it is hard to accept that there is nothing morally permissible to do when faced with collective violence against the innocent. I will thus sketch an account that can make sense of the moral conflict that exists in killing in war and that is able, at the same time, to preserve a morally permissible course of action and capture the distinction between combatants and civilians. But if many unjust combatants are not liable, why can just combatants kill them?

I will suggest that just combatants are often only *permitted* to kill non-liable unjust combatants—not justified in doing so. This claim strays away from reductivism, and requires developing a *sui generis* account of moral permissibility, to which most of Chapter VII is devoted. The idea of moral permissibility is meant to capture the moral fragility of certain instances of harming the non-liable, both in the case of agent-relative prerogatives to defend one's life and in the case of killing non-liable unjust combatants in pursuit of a just war. This account focuses on three features about these impositions of harm: killing the non-liable wrongs them grievously; killing the non-liable leaves a significant moral residue; and reasons in favour of killing the non-liable, both in war and in the context of agent-relative prerogatives to defend one's life, are barely weightier than the reasons against doing so.

The account of the morality of war that I will develop in Chapter VII—if successful—allows us to move beyond pacifism and to provide a

[20] In his postscript to *Just and Unjust Wars* (5th edn), Michael Walzer makes a similar point and argues about the importance of this sort of material when working, even if within philosophy, on the morality of war. *See* Walzer (n 4) 336–37.

[21] The idea of 'the little history of war' can be found in Alexsievič's work. For example: 'From a conversation with the censor—This is a lie! This is slander against our soldiers, who liberated half of Europe. Against our partisans. Against our heroic people. We don't need your little history, we need the big history. The history of the Victory. You don't love our heroes! You don't love our great ideas.' Svetlana Aleksievič, *The Unwomanly Face of War* (Random House Trade Paperbacks 2018) xxxvii.

non-instrumentalist justification for the just combatant's legal privilege to kill in war. This justification relies partly on those aspects of the revisionist morality of war that have not been displaced by my account and partly on the idea of permissibility that I will develop.

With this account of moral permissibility in place, we can begin to understand the reasons that unjust combatants might have for participating in an unjust war. In other words, the fact that many unjust combatants are not liable and the idea of moral permissibility open a path where we can continue to deconstruct the analogy between what unjust combatants do and murdering the innocent. If these reasons are powerful enough, we can attenuate the fragility of the instrumentalist justification in accounting for the symmetry thesis. This will occupy Chapter VIII.

The reasons I will propose are fundamentally determined by certain features of war and our institutional arrangements. Unjust combatants face considerable uncertainty about the justness of the wars they are called to fight, and refusing to fight is often exceedingly costly for agents and achieves very little. As a result, sometimes, it will be permissible for unjust combatants to participate in unjust wars. Further, given that many unjust combatants are not liable, each might be allowed to defend themselves in the exercise of agent-relative prerogatives. Finally, some unjust combatants might have evidence relative permissions to participate in war.

Some of the reasons that some (non-liable) unjust combatants have for participating and defending themselves are agent-relative ones. That is, they are reasons for action only for a particular agent to do something.[22] Thus, if we think that there is no such thing as agent-relative reasons, and that all reasons are agent-neutral, then some of the arguments in Chapter VIII will obviously fail to generate permissions (although they might be enough to constitute excuses).

The deconstruction of the revisionist analogy between what unjust combatants do and the murder of the innocent will be, however, inevitably limited. Although there might be plausible moral reasons for participating in an unjust war and even for killing in self-defence and defence of others during it, participation in an unjust war will also call for actions that will cause great harm—harm which will not be morally justified nor permitted. This will be particularly true of those higher up in the chain of command. The highest the position of power an unjust combatant finds herself in, the more important

[22] Christine M Korsgaard, 'The Reasons We Can Share: An Attack on the Distinction between Agent-Relative and Agent-Neutral Values' (1993) 10 Social Philosophy and Policy 24.

the strategic value of the military objective she is defending or trying to gain control over, the harder it will be to judge these actions as permissible in any relevant sense.

Yet, besides criminalizing aggressive war at the highest levels of command,[23] international law does not make those distinctions. Our laws grant a legal privilege to all unjust combatants, in circumstances in which not *every* unjust combatant will have a moral permission to do what she is doing. This divergence between law and morality might not be as deep as the revisionist account suggests, if the arguments of Chapters V to VIII are successful. But the divergence is still quite significant and troubling, for there is a legal permission in place for many individuals in high and mid-level positions of command to pursue unjust goals.

This remaining gap between our laws of war and morality might, of course, be justified under the instrumentalist rationale that I will provide in Chapter III. And if the lack of analogy between what some unjust combatants do and murder can be constructed, even if partially, the moral fragility of this instrumentalist justification is then successfully attenuated, to a certain extent.

This book, then, puts forward a pluralist moral justification for the legal privilege of combatants to kill in war. While the symmetrical legal privilege of combatants might be contingently justified by the instrumentalist justification I will provide in Chapter III, the morality of war I will defend in Chapters VII and VIII will attenuate the fragility of the instrumentalist explanation by denying, at least partially, that what unjust combatants do is just like murdering the innocent.

Yet, as the conclusion will argue, this justification should not give us much comfort about the moral standing of IHL. On the contrary, it might have revisionary implications.

First, if the account developed in Chapter VII is successful, a lot of killing in war, including the killings of non-liable unjust combatants, is morally fragile or very close to killing the innocent. And, in fact, the idea of permissibility I will develop is subject to important objections as an account of killing in war, which might make it untenable. If that is true, we should embrace pacifism.

Whatever the case, the instrumentalist justification I defend in Chapter III must bear a lot of the justificatory burden regarding the symmetry thesis. But this justification can only explain the *adoption* of the symmetry thesis, it

[23] Rome Statute of the International Criminal Court (hereinafter, Rome Statute) ISBN No 92-9227-227-6, 17 July 1998, art 8*bis*; ICC, *Amendments to the Elements of Crimes*, 11 June 2010, Res. RC/Res.6/Annex 2.

cannot sanitize its content. And it is also contingent: it succeeds only to the extent that certain conditions are true. It remains unclear, however, that they were ever true, or that, if they were, they are true now.

Second, this contingent justification has a limited scope: it explains why the legal permissibility of targeting combatants might be (contingently) justified, but it says nothing about other dimensions of the legal regime, such as the fact that inflicting collateral harm on civilians, when proportionate, is not a war crime but permitted (or immunized) behaviour.

Ultimately, there is no avoiding the fact that in various ways, we have regulated murder on a vast scale. Perhaps the contingent justification I will provide for the permissibility to target combatants might be extended to other areas of the legal regime—perhaps not. But even if the book can contingently justify the adoption of the symmetry thesis, the content of these legal permissions of combatants to target each other is often very close to being morally abhorrent. I say this with some trepidation, as I do think, and Chapter III will argue, that what we have is better than other alternatives (eg total war, the absence of any regulation). But what we have is still a significant moral compromise. Because of that, we must relentlessly strive for the regime's reform, to whatever extent possible. I will, in fact, point out several areas where our laws should be modified. And, going further, one of the upshots of the account this book defends is that there is urgent moral pressure to improve our laws—to bring them closer to an ideal whereby wars are not necessary; whereby war does not exist—and to create a world where there is no need for a 'morality of war' at all.

Finally, some of the arguments and conclusions I will put forward in this book have a tentative character. Though this may not be entirely satisfactory, it is intentional. Jonathan Glover once wrote that '[h]uge problems sometimes produce an irrational paralysis of the imagination'.[24] I hope that the hesitant nature of some of these conclusions does not read as a sign of an imagination paralysed by the problems posed by war, but as a reasonable—and perhaps urgently needed—degree of ambivalence towards what I consider to be one of the most difficult moral questions.

Before beginning to provide an answer to the question of justification, some descriptions and explanations are necessary. Hence, the next sections set out to clarify first, what a just combatant and a just war are; and second, what the laws of armed conflict say about killing in wartime. Readers familiar with these terms and with international law may wish to move on to the next chapter.

[24] Jonathan Glover, 'It Makes No Difference Whether or Not I Do It' (1975) 49 Proceedings of the Aristotelian Society 171, 172.

I. Just War and Just Combatants

The term 'just combatant' refers to combatants fighting a morally just war, and most usually, it refers to the combatant fighting a war with a just *cause*.[25] Additionally, in most of the literature, a combatant is deemed 'just' if the war he is fighting is *objectively* just.[26] The war must be just from a God's-eye point of view, that is, due to facts that are independent of the just combatant's beliefs.[27] Because the justness of the war is objective, a war cannot be just on both sides,[28] but it can be unjust on both. Hence, there cannot be just combatants on both sides of a war.

Following most of the literature, I will use the term 'just combatant' to refer to combatants fighting an objectively just war. This is not because I think there is nothing morally significant about the just combatant's beliefs. Presumably, the just combatant's beliefs should be aligned with the justness of the war, in the sense that she would reasonably[29] believe she is fighting a just war.[30] This is a complex moral question, and I think that it would be wrong, in at least *some* sense, for a combatant to kill in an objectively just war without reasonably believing the war to be just.[31]

I will also sometimes use the terms 'evidence-relative' and 'fact-relative' permissibility in conformity with their standard use in ethics. Evidence-relative permissible acts are those acts that would be right if we believed what the available evidence gives us decisive reasons to believe, and these beliefs were true.[32] Or, in simpler terms, evidence-relative permissibility refers to what is permissible for an agent to do given the evidence available to them. Evidence-relative permissible acts might be impermissible in a fact-relative sense, that is, impermissible given all the morally relevant facts.[33]

It is important to note at the outset that a just war is not necessarily a legal one.[34] War in self-defence is the paradigmatic instantiation of a

[25] See Jeff McMahan, 'The Ethics of Killing in War' (2004) 114 Ethics 693; Yitzhak Benbaji, 'A Defense of the Traditional War Convention' (2008) 118 Ethics 464; Yitzhak Benbaji, 'The War Convention and the Moral Division of Labour' (2009) 59 The Philosophical Quarterly 593; Haque (n 7).

[26] McMahan, *Killing in War* (n 6) 43.

[27] ibid.

[28] But both sides may be engaged in a partly just and partly unjust war.

[29] In an evidence-related sense. *See* Derek Parfit, *On What Matters*, vol 1 (Oxford University Press 2013) 151–64; Tadros, *The Ends of Harm* (n 6) 219–26.

[30] McMahan thinks so, but only in reference to the just cause. He does not, however, take a stand on the question of whether it is sufficient that the act be objectively justified for it to be permitted. McMahan, *Killing in War* (n 6) 1–43.

[31] On different moral senses of wrongness, *see* Parfit (n 29) 151–64.

[32] ibid 151.

[33] ibid.

[34] Legal uses of force, according to the UN Charter, are only those in self-defence and collective self-defence, and enforcement actions authorized by the UN Security Council. *Charter of the United Nations* (hereinafter, UN Charter) 24 October 1945, 1 UNTS XVI, arts 2 and 51, and Chapter VII.

just[35] and a legal cause,[36] but the possibility of a legal war being an unjust one remains. For example, if the United Nations (UN) Security Council authorized the use of force under circumstances which would not meet the just war criteria, or if a war in self-defence failed to meet strict requirements of necessity or proportionality. On the other hand, an illegal war might be just, NATO's intervention in Kosovo being the preferred historical example.[37] At the other extreme, aggression is not only banned by the UN Charter, and a crime under the Rome Statute,[38] but is also the paradigmatic example of an unjust war.

In summary, though there is some overlap between what makes war just and what makes war legal, there is also significant dissonance. This issue will arise numerous times throughout the book, as we grapple with the question of justification.

II. The Laws on Killing in War

This section is a survey of the current legal framework on killing in war. It provides a description of the laws on targeting—killing or wounding—enemy combatants.

Killing is not the only activity privileged by the laws of war. Killing is but one of them, along with others, such as the destruction of property. Indeed, the laws of war grant immunity for acts of violence and destruction that are otherwise forbidden by the legal framework in times of peace.[39] Here I only provide a detailed description of the rules on targeting, namely killing and wounding.

[35] *See* Thomas Hurka, 'Proportionality and Necessity' in Larry May and Emily Crookston (eds), *War: Essays in Political Philosophy* (Cambridge University Press 2008); Walzer (n 4); Larry May, 'Humanity, Necessity, and the Rights of Soldiers' in Jens David Ohlin, Larry May, and Claire Finkelstein (eds), *Weighing Lives in War* (Oxford University Press 2017).

[36] UN Charter, art. 51.

[37] *See* Martti Koskenniemi, '"The Lady Doth Protest Too Much" Kosovo, and the Turn to Ethics in International Law' (2002) 65 The Modern Law Review 159.

[38] Given the ban on aggressive war by the UN Charter and its criminalization by the Rome Statute, the political leadership of the state and its top military commanders may be prosecuted for the crime of aggression. Thus, at the very least some combatants of high military rank, who most likely would be privileged combatants, may be prosecuted under the current regime for their initiation of an aggressive war. Given that in theory the ICC could consider factors such as the amount of damage caused and others of the sort in sentencing those guilty of aggression, these provisions could potentially chip away at the symmetry thesis. ICC, Amendments to the Rome Statute of the International Criminal Court on the crime of aggression, 11 June 2010, Res. RC/Res.6/Annex 1.

[39] David Luban, 'War Crimes. The Law of Hell' in Larry May and Emily Crookston (eds), *War: Essays in Political Philosophy* (Cambridge University Press 2008) 277.

Most of the current rules on war have been consolidated over the decades in customary international law, and many of them have been incorporated in international treaties as well.[40] Yet no single treaty nor any cluster of treaties covers the whole span of IHL, and both sources of international law interact with each other on different levels.[41]

A. Who Is a Privileged Combatant?

In an international armed conflict (IAC), the privilege to participate in hostilities (and thus, to engage in acts of war such as killing) belongs to all combatants who fulfil the requisites to be considered privileged combatants (privileged precisely because they cannot be criminally punished for their acts of war).[42] A more technical statement of the possessor of the privilege would be the following: combatants who are entitled to prisoner of war status have a right to participate in hostilities and cannot be prosecuted for taking direct part in them once they fall into the hands of the adverse party.[43]

'Unlawful combatants'—a category of doubtful legal standing[44]—that is, those who engage in violence but do not fulfil the requisites established by the law to be privileged combatants, can be prosecuted for the acts committed during hostilities.[45] Note, however, that combatants waging an unjust or illegal war, as long as they fulfil the requirements, may also be privileged combatants. This is due to the legal independence of *jus ad bellum* from *jus in bello*. It is also related to the 'symmetry thesis' or to what is somewhat misleadingly called 'the

[40] Dinstein, *The Conduct of Hostilities under the Law of International Armed Conflict* (n 4) 5–7.

[41] ibid.

[42] *Geneva Convention III Relative to the Treatment of Prisoners of War* (hereinafter, GCIII), 12 August 1949, art 4; *Convention (IV) respecting the Laws and Customs of War on Land and its annex: Regulations concerning the Laws and Customs of War on Land* (hereinafter Hague Convention IV), The Hague, 18 October 1907, reg 1; API, art 43.2.

[43] API art 43.2 and *see* Prisoners of war and detainees protected under international humanitarian law—(ICRC) <https://www.icrc.org/eng/war-and-law/protected-persons/prisoners-war/overview-detainees-protected-persons.htm> accessed 13 November 2015.

[44] *See* Blum (n 15) 129; Michael N Schmitt, 'Asymmetrical Warfare and International Humanitarian Law' in WH von Heinegg and V Epping (eds), *International Humanitarian Law Facing New Challenges* (Springer 2007); Mary Ellen O'Connell, 'The Legal Case against the Global War on Terror' (2004) 36 Case Western Reserve Journal of International Law 349; George P Fletcher and Jens David Ohlin, *Defending Humanity: When Force Is Justified and Why* (Oxford University Press 2008) xiii.

[45] *See Protocol Additional to the Geneva Conventions of 12 August 1949, and Relating to the Protection of Victims of International Armed Conflicts (Protocol I)* (hereinafter, API) 8 June 1977, 1125 UNTS 3, art 75, which enumerates the fundamental guarantees of persons who fall in the power of a party to the conflict but do not have prisoner of war status. Note as well that article 44 of Additional Protocol I complicates this picture. However, it has not been ratified by the United States and its standing as customary international law is highly dubious. *See* Dinstein, *The Conduct of Hostilities under the Law of International Armed Conflict* (n 4) 45–47.

moral equality of combatants', which is meant to convey that lawful combatants on both sides have the same rights, duties, and privileges.

Civilians (ie anyone who is not a combatant) are not privileged combatants, except for the case of a *levée en masse*, in which inhabitants of a non-occupied territory, on the approach of the enemy, spontaneously take up arms to resist invasion, only if they carry their arms openly and respect IHL.[46] Prisoners of war, though entitled to attempt escape, do not retain the privilege to kill to do so.[47] Finally, neither spies nor mercenaries have the status of prisoners of war, and thus are subject to prosecution for their conduct during hostilities.[48]

In armed conflicts of non-international character (NIAC), there is no immunity from prosecution. Belligerent immunity is reserved by the Geneva Conventions to states and their armies.[49] There will usually be no legal privilege to kill except for the government's army; rebel forces are dealt with under domestic criminal law. The only exception to this regime is stated in Additional Protocol I, which expands the concept of IAC to include those in which peoples are fighting against colonial domination and alien occupation, and against racist regimes in the exercise of their right of self-determination.[50]

In spite of this *ex ante* lack of belligerent immunity for rebel forces in NIAC, Additional Protocol II mandates the authorities in power, at the end of the hostilities, to endeavour to grant the 'broadest possible amnesty' to persons who participated in the armed conflict and to those deprived of their liberty for reasons related to it.[51] This rule, however, provides a weaker protection than the IAC privileged combatant paradigm.[52]

B. Who Can Be Targeted?

The oversimplified answer is the following: combatants can be targeted, civilians (non-combatants) cannot. However, the legal rules are more complex, both in their letter and application. Protected persons that are not to be made

[46] GCIII, art 4.6; Hague Convention IV, reg 2.

[47] GCIII, arts 91–93.

[48] *See Geneva Convention Relative to the Protection of Civilian Persons in Time of War (Fourth Geneva Convention)* (hereinafter, GCIV) 12 August 1949, 75 UNTS 287, art 5; API, art 46(1); Hague Convention IV, reg 30; API, art 47(1); and UNGA, *International Convention against the Recruitment, Use, Financing and Training of Mercenaries*, 4 December 1989, A/RES/44/34.

[49] Luban (n 39) 278.

[50] API, art 1.4.

[51] *Protocol Additional to the Geneva Conventions of 12 August 1949, and Relating to the Protection of Victims of Non-International Armed Conflicts (Protocol II)* (hereinafter, APII) 8 June 1977, 1125 UNTS 609, art 6.5.

[52] Luban (n 39) 278.

the object of attack include some members of the armed forces. And telling combatants from civilians is not an easy task, especially in today's conflicts, where a significant number of combatants belong to non-state actors and the armed forces of states are increasingly '*civilianized*' so that there is a greater intertwining of combatants and civilians.[53] Nevertheless, the basic legal framework is as follows.

In IAC, only combatants are lawful targets, whether engaging in a just or unjust war, privileged or not. The killing of combatants may be intentional or the result of an attack on another legitimate target. However, there are exceptions to this general rule: prisoners of war (ie persons who have fallen into the power of the enemy and that belong to one of the categories specified in article 4 of Geneva Convention III), unless they attempt to escape and only as an extreme measure;[54] those persons that are or should be recognized as *hors de combat*;[55] persons parachuting from an aircraft in distress, only during the time of their descent;[56] members of the armed forces and other persons who are shipwrecked;[57] and members of the armed forces and military units assigned to civil defence organizations, unless their protection ceases for the reasons established in article 65 and provided that they comply with the criteria established in article 67 of Additional Protocol I.[58]

Civilians, on the contrary, are not permissible targets. They shall not be the object of attack unless and for such time as they take direct part in hostilities.[59] Attacks that are not discriminatory, and thus are of a nature to strike military objectives and civilians or civilian objects without discrimination, are forbidden.[60] The killing of civilians as collateral damage[61] is better understood as part of the so-called belligerent immunity for acts of war, since, arguably, there is no intention to kill those civilians. Nevertheless, civilian casualties are allowed if in the attack there has been compliance with the precautionary

[53] Blum (n 15) 140.

[54] GCIII, arts 4, 13, and 42.

[55] API, art 41. According to article 41, a person is *hors de combat* if 'a) he is in the power of an adverse Party; b) he clearly expresses an intention to surrender; or c) he has been rendered unconscious or is otherwise incapacitated by wounds or sickness, and therefore is incapable of defending himself, provided that in any of these cases he abstains from any hostile act and does not attempt to escape'.

[56] API I, art 42. Note that when reaching the ground in territory controlled by an adverse Party, they must be given a chance to surrender before being made the object of attack. According to Blum, this constraint was born out of the early admiration of pilots. *See* Blum (n 15) 126.

[57] *Geneva Convention for the Amelioration of the Condition of the Wounded, Sick and Shipwrecked Members of Armed Forces at Sea (Second Geneva Convention)* (hereinafter, GCII) 12 August 1949, 75 UNTS 85, art 12.

[58] API, arts 62–67.

[59] API, art 51.

[60] API, art 51.4.

[61] API, art 57.

measures established in Additional Protocol I, and the loss of civilian life is not excessive in relation to the concrete and direct military advantage anticipated.[62] Furthermore, the wilful killing of protected persons, the intentional targeting of civilians, and the intentional launching of an attack in the knowledge that it will cause incidental loss of life or injury to civilians which would be clearly excessive in relation to the concrete and overall direct military advantage anticipated constitute war crimes.[63]

In NIAC, only the killings of non-government forces, and only for such time as they take direct part in hostilities,[64] will not constitute a domestic crime nor a war crime. Persons who do not, or have ceased to take direct part in hostilities, including members of the armed forces who have laid down their arms,[65] shall not be victims of acts of violence to their life, their health, and their physical and mental well-being, in particular regarding murder as well as cruel treatment.[66] Those who do not fit this description, but whose liberty has been restricted for reasons related to the conflict, are also protected against murder and cruel treatment.[67] The wounded, the sick, and the shipwrecked are also to be protected,[68] as well as persons engaged in medical activities and religious personnel.[69] Killing the government's armed forces will not be an international crime, but it will be a domestic one. So too will be the participation in hostilities, usually as sedition, insurrection, rebellion, or treason.

Finally, the targeted killing of individual enemy combatants is not forbidden by IHL, though it is constrained by its general norms.[70] Top echelons of the

[62] API, art 57.2.iii.

[63] Rome Statute, arts 8.2.(a)(i), 8.2.(b)(i), and 8.2.(b)(iv).

[64] APII, art 13.3. Note that the ICRC's Interpretive Guidance on the Notion of Direct Participation of Hostilities has reserved the category of civilians in NIAC for those who are not members of the state armed forces or organized armed groups of a party to the conflict. Organized armed groups, for this purpose, constitute the armed forces of the non-state party to the conflict and consist only of individuals whose continuous combat function is to take direct part in the hostilities. *See International Committee of the Red Cross, Interpretive Guidance on the Notion of Direct Participation in Hostilities under International Humanitarian Law* (Nils Melzer ed, ICRC 2009).

[65] GCIII, art 3.1.

[66] APII, art 4.2.

[67] APII, art 5.3.

[68] APII, art 7.

[69] APII, arts 9–12.

[70] The general requisites for the targeting of individual enemy combatants to be permissible are: the existence of an armed conflict; the target must be a legitimate one; there must be a concrete military advantage gained by the killing; possible collateral damage must be assessed; the killing cannot be a perfidious or treacherous one; quarter must be provided, among others. An example of targeted killing that is usually claimed to have been considered legitimate at the time (1943) was the United States targeting of the Commander in Chief of the Japanese Fleet, Admiral Yamamoto, whose plane was ambushed and shot down. *See* Dinstein, *The Conduct of Hostilities under the Law of International Armed Conflict* (n 4) 103; Nils Melzer, *Targeted Killing in International Law* (Oxford University Press 2008) 56–57, 394–420. However, Jeremy Waldron has put into question this narrative and in fact records suggest that the decision was controversial in 1943 not only due to strategic reasons but also due to

chain of command—even a civilian head of state who serves as commander in chief[71]—are legitimate targets, though as a matter of comity or tradition, the killing of leaders has been limited.[72]

C. How to Kill?

The rights of the parties to a conflict (whether IAC or NIAC) to choose methods or means of warfare is limited.[73] One of those limits restricts the way in which the privilege to kill may be exercised: it is forbidden to use weapons, projectiles, materials, and methods of warfare of a nature to cause superfluous injury or unnecessary suffering.[74] This doesn't mean that the weapons cannot be lethal[75] but that they cannot needlessly aggravate the suffering of combatants by causing a harm greater than that unavoidable to achieve legitimate military objectives.[76] The employment of poison or poisoned weapons is forbidden,[77] and there are prohibitions on particular weapons as well, contained in different treaties.[78]

moral concerns about assassination and retribution. *See* Tamar Meisels and Jeremy Waldron, *Debating Targeted Killing: Counter-terrorism or Extrajudicial Execution?* (Oxford University Press 2020) 210–13.

[71] This is particularly relevant in modern democracies, where the principle of civilian control of the military practically ensures that the top position of command in the army will belong to a civilian—and even more, to the head of state.

[72] *See* Markus Gunneflo, *Targeted Killing: A Legal and Political History* (Cambridge University Press 2016) 196–97; Dinstein, *The Conduct of Hostilities under the Law of International Armed Conflict* (n 4) 107–08. Note that there was a period when targeted killing (or wartime assassination) was equated with treachery. The issue of assassination of political and military leaders of a nation at war was addressed by Alberico Gentili, Hugo Grotius, and Emmerich de Vattel, among others. *See* Alberico Gentili, *De Iure Belli Libri Tres* (John C Rolfe tr, The Clarendon Press 1933) 530–34; Hugo Grotius, *Hugo Grotius on the Law of War and Peace* (Stephen C Neff ed, Cambridge University Press 2012) 354–56; Emer de Vattel, *The Law of Nations, or, Principles of the Law of Nature, Applied to the Conduct and Affairs of Nations and Sovereigns, with Three Early Essays on the Origin and Nature of Natural Law and on Luxury* (Bela Kapossy and Richard Whatmore eds, Thomas Nugent tr, Liberty Fund 2008) 155–59.

[73] API, art 35.1.

[74] API, art 35.2; Hague Convention IV, reg 23.e; *Declaration Renouncing the Use, in Time of War, of Explosive Projectiles under 400 Grammes Weight* (hereinafter, 1868 St Petersburg Convention) 11 December 1868, preamble. The other restriction regarding weapons is related to the principle of distinction. Thus, weapons that by their very nature or design cannot possibly distinguish between civilians and combatants, are forbidden. *See* Dinstein, *The Conduct of Hostilities under the Law of International Armed Conflict* (n 4) 62–63.

[75] Blinding lasers are not lethal, and yet, forbidden. *See* ibid 65.

[76] ICJ, *On Legality of the Threat or Use of Nuclear Weapons*, Advisory Opinion, 8 July 1996, para 78. A better account of what counts as excessive suffering is the one proposed by O'Donovan, which considers several factors: the intensity and duration of pain, hope of recovery from injury, and the factors of indignity and humiliation. *See* Oliver O'Donovan, *The Just War Revisited* (Cambridge University Press 2003) 78–94.

[77] Hague Convention IV, reg 23.a.

[78] Some of the forbidden weapons and projectiles regulated by treaties are explosive bullets, expanding bullets, non-detectable fragments, some types of booby-traps and landmines, some types of

Killing by resort to perfidy[79] is also prohibited, but ruses of war, such as the use of camouflage, decoys, mock operations, and misinformation are allowed as long as they do not rise to the level of perfidy.[80]

Finally, quarter must be provided both in IAC and NIAC. It is forbidden to order that there shall be no survivors, to threaten an adversary with this, and to conduct hostilities on this basis.[81]

As can be seen, other than the prohibition on unnecessary suffering and superfluous injury and the prohibition on the use of particular weapons, the regulatory framework on war is quite thin concerning the killing and wounding of enemy combatants. Indeed, the closest we get to a thick principle is the preamble of the 1868 St Petersburg Declaration, which says that the only legitimate object which states should endeavour to accomplish during war is to weaken the military forces of the enemy. For this purpose, the Declaration states, it is sufficient to disable the greatest possible number of combatants, a limit that would be exceeded by the employment of arms that needlessly aggravate the suffering of disabled men or render their death inevitable.[82]

The above means that war *involves* killing combatants, but only as a means. Killing is not supposed to be the sole goal of war.[83] Here we can see how military necessity and the principle of humanity operate both by limiting unnecessary violence against enemy combatants and also justifying the use of force against them.[84] Although the International Committee of the Red Cross (ICRC) has asserted that the unnecessary or disproportionate killing of combatants violates these principles,[85] the directive has been harshly criticized[86]

torpedoes, cluster munitions, chemical and biological weapons. For the details of the regulations, *see* Dinstein, *The Conduct of Hostilities under the Law of International Armed Conflict* (n 4) 61–77. Rome Statute, art 8.2(b)(xviii)–(xx).

[79] Perfidy is defined in art 37 of API, as '[a]cts inviting the confidence of an adversary to lead him to believe that he is entitled to, or is obliged to accord, protection under the rules of international law applicable in armed conflict, with intent to betray that confidence'. Examples of such acts, also in art 37 of API, are feigning intent to surrender, feigning incapacitation by wounds, feigning of non-combatant status, among others.

[80] API, art 37.1; Hague Convention IV, regs 23 and 24.

[81] API, art 40; APII, art 4.1.

[82] 1868 St Petersburg Convention, preamble.

[83] Eyal Benvenisti, 'The Legal Battle to Define the Law on Transnational Asymmetric Warfare' (2009) 20 Duke Journal of Comparative & International Law 339, 352.

[84] *See* Michael N Schmitt, 'Wound, Capture, or Kill: A Reply to Ryan Goodman's "The Power to Kill or Capture Enemy Combatants"' (2013) 24 European Journal of International Law 855, 857; Burrus M Carnahan, 'Lincoln, Lieber and the Laws of War: The Origins and Limits of the Principle of Military Necessity' [1998] American Journal of International Law 213; David Luban, 'Military Necessity and the Cultures of Military Law' (2013) 26 Leiden Journal of International Law 315.

[85] *See* Melzer (n 64) 78–82.

[86] *See* Ryan Goodman, 'The Power to Kill or Capture Enemy Combatants' (2013) 24 European Journal of International Law 819, 820; Michael N Schmitt, 'The Interpretive Guidance on the Notion of

and is not legally binding. Hence, IHL tells us nothing or little about whether states have an obligation to capture rather than to kill[87] or whether they ought to choose military strategies that spare the lives of the highest number of enemy combatants.

D. When Does the Privilege to Kill Apply?

This may seem an easy question to which the answer must surely be, only during an ongoing armed conflict. However, the nature of most of today's conflicts makes it difficult to draw a sharp line between times of war and times of peace (or the cessation of hostilities), and between law enforcement and war. Moreover, the question of when violence amounts to war is a complex one, particularly when distinguishing insurgency from war.

The legal regime for IAC applies to all cases of 'declared war or of any other armed conflict' between two or more Parties to the Geneva Conventions, even if the state of war is not recognized by one of them.[88] Thus, the legal regime for IAC will apply when a declaration of war has been made, and when such a declaration is lacking but there is war in a material sense.[89] Both Kahn and O'Connell argue that a declaration of war is currently meaningless and serves no purpose under international law. It can have no bearing on the underlying legal situation; armed conflict is determined by facts of fighting, not by mere declarations.[90] This would leave only war in the material sense as the situation triggering the application of IHL, a result I take to be the correct one because it is consistent with the legal regime set up by the Rome Statute.[91]

Once an armed conflict has begun, combatants can be killed at any time during the conflict, regardless of the threatening or non-threatening nature of their activities.

Direct Participation in Hostilities: A Critical Analysis' in Michael N Schmitt,, *Essays on Law and War at the Fault Lines* (Springer 2011).

[87] *See* the discussion between Goodman (n 86); Schmitt, 'Wound, Capture, or Kill' (n 84).

[88] GCIII, art 2.

[89] For the difference between technical and material war, *see* Yoram Dinstein, *War, Aggression, and Self-Defence* (5th edn, Cambridge University Press 2012) 9–10.

[90] *See* Paul W Kahn, 'War Powers and the Millennium' (2000) 34 Loyola of Los Angeles Law Review 11, 16–19; Mary Ellen O'Connell, 'Combatants and the Combat Zone' (2008) 43 University of Richmond Law Review 845, 854–57.

[91] See ICC, *Amendments to the Rome Statute of the International Criminal Court on the Crime of Aggression*, 11 June 2010, Res. RC/Res.6/Annex 1.

In NIAC, there is a negative threshold, which states that IHL does not apply 'to situations of internal disturbances and tensions, such as riots, isolated and sporadic acts of violence and other acts of a similar nature, as not being armed conflicts'.[92] In its positive form, the threshold is that the legal regime for NIAC will apply to all armed conflicts not covered by article 1 of Additional Protocol I, and which take place in the territory of a High Contracting Party between its armed forces and dissident armed forces, or other organized armed groups that meet certain requirements.[93]

Despite these threshold provisions for NIAC, for a long time there was no internationally accepted definition of armed conflict.[94] However, the issue has been arguably settled by the International Criminal Tribunal for the former Yugoslavia (ICTY) in *Prosecutor v Tadić*.[95] The judgment sets a threshold which draws a contrast, at the internal level, between law enforcement and war. Thus, there is a demarcation question about war (when does violence become war?) and a question of whether there is a place in inter-state relations for the distinction between law enforcement and war.[96] More importantly, the demarcation question about war is directly related to the issue of killing in war, for outside of armed conflict the term 'combatant' has no legal meaning:[97] there are no combatants in times of peace.

Legal threshold aside, persons who are not members of state armed forces or organized armed groups of a party to the conflict can only be killed when they are taking direct part in hostilities. Members of organized armed groups whose continuous combat function is to take direct part in hostilities are considered combatants by the ICRC Interpretive Guidance.[98] As such, they lose

[92] APII, art 1.2.

[93] That is, dissident armed forces or other organized armed groups that are under responsible command and exercise such control over a part of its territory as to enable them to carry out sustained and concerted military operations and to implement Additional Protocol II. APII, art 1.1.

[94] *See* Lindsay Moir, *The Law of Internal Armed Conflict* (Cambridge University Press 2002) 31–42.

[95] '[A]n armed conflict exists whenever there is a resort to armed force between States or protracted armed violence between governmental authorities and organized armed groups or between such groups within a State. International humanitarian law applies from the initiation of such conflicts and extends beyond the cessation of hostilities until a general conclusion of peace is reached; or, in the case of internal conflicts, a peaceful settlement is achieved. Until that moment, international humanitarian law continues to apply in the whole territory of the warring States or, in the case of internal conflicts, the whole territory under the control of a party, whether or not actual combat takes place there.' ICTY, *Prosecutor v Dusko Tadić*, IT-94-1-A, Appeals Chamber, Decision on the Defense Motion for Interlocutory Appeal on Jurisdiction, 2 October 1995, para 70.

[96] *See* Eric Gregory, 'What Do We Want from the Just War Tradition? New Challenges of Surveillance and the Security State' (2014) 27 Studies in Christian Ethics 50, 57.

[97] O'Connell (n 90) 854.

[98] *See* Melzer (n 64).

their civilian immunity and can be targeted for as long as they retain membership in the armed group by virtue of their continuous combat function.[99]

Finally, the regulatory differences of the privilege to kill for IAC and NIAC are complicated by the fact that some armed conflicts may incorporate elements of both intra-state and inter-state hostilities, and others may start out as internal conflicts and evolve into IAC.[100]

[99] This is far more complicated. *See* Melzer (n 64).
[100] Dinstein, *The Conduct of Hostilities under the Law of International Armed Conflict* (n 4) 14–15.

PART 1

THE REVISIONIST MORALITY AND THE LAWS OF WAR

II

Combatants and the Privilege to Kill

I. Introduction

If revisionists are right about the morality of war, providing a moral justification for the legal privilege of combatants to kill each other during armed conflict could hardly be a more daunting task. This is because one of the basic features of the legal regime—the symmetrical permission to target combatants—is, under the revisionist account, the moral equivalent of a law that permits people to murder the innocent.

The task of justifying the symmetry thesis would be presumably easier if we could deny that a legal permission to target combatants is part of the international law of armed conflict. This is the route that Adil Haque has taken.[1] He argues that the international law of armed conflict is prohibitory in nature and does not grant any strong permissions. What combatants have, he says, is only an immunity from prosecution. If Haque is right, at least part of the justificatory problem disappears: we no longer need to justify a legal rule that *permits* people to murder the innocent. We only need to justify a legal rule that makes people *immune from prosecution* when they murder the innocent. And according to Haque, justifying an immunity is easier than justifying a permission.

In this chapter, I argue that this way out of the justificatory problem is not available. Haque's position is incorrect in two ways. First, the most plausible interpretation of our laws of armed conflict is that combatants are *privileged* to participate in hostilities, not *only* immune from prosecution for doing so. Second, even if combatants have nothing but an immunity from prosecution, the justificatory challenge persists almost unchanged.

Regarding the legal situation of combatants, we usually say that according to international law, combatants *may* kill other combatants during war.[2] So far,

[1] *See* Adil Ahmad Haque, *Law and Morality at War* (Oxford University Press 2017).

[2] The answer to the question of who is a permissible target during armed conflict is not exactly the same as the answer to the question of who is entitled to participate in hostilities. Combatants are permissible targets during armed conflict, but so are civilians who directly participate in hostilities (and only for such a time). And, of course, combatants are entitled to participate in hostilities (provided that they meet the legal requirements for lawful combatant status), but so are civilians who participate in a *leveé en masse*. In other words, although all privileged combatants are permissible targets, there are

The Morality of the Laws of War. Marcela Prieto Rudolphy, Oxford University Press. © Marcela Prieto Rudolphy 2023.
DOI: 10.1093/oso/9780192855473.003.0002

I have characterized that auxiliary verb 'may' as a legal privilege. I have said that killing combatants during wartime is a *privileged* activity, or that combatants have a *legal privilege* to kill.

Saying that someone has a legal privilege can mean different things. Sometimes, having a privilege means enjoying a permission to do something that would be impermissible under different circumstances.[3] At other times, it means that you or you and others of your class have been granted a special favour.[4] Both these meanings convey the existence of a legal advantage. Finally, 'legal privilege' sometimes means a 'Hohfeldian' privilege.[5]

Contrary to Haque's position, I will argue that combatants are privileged in these three different senses. They enjoy an exceptional permission to do something which is usually forbidden—which I characterize as a Kelsenian positive permission[6]—they belong to a favoured class of people, and they have a Hohfeldian privilege.[7] It is important to note that the fact that these three senses apply to the combatant's normative situation is only a coincidence. Having a Hohfeldian privilege is not the same as having an exceptional permission. And, indeed, some Hohfeldian privileges are not positive or exceptional permissions (but all legal permissions, negative and positive, can be characterized as Hohfeldian privileges). As we will see later, it is correct to state that I have a Hohfeldian legal privilege to breathe, yet it would be ludicrous to describe this Hohfeldian privilege as a privilege if by that we mean an exceptional permission: breathing is not something we are generally forbidden from doing. And being privileged in the sense of belonging to a favoured class of people can take different legal forms: the favour may consist of a right, an immunity, a privilege, or a power. Thus, being privileged in this latter sense tells us little

many individuals who do not possess the combatant's privilege to participate in hostilities—they can be criminally prosecuted under the domestic law of the enemy—yet are still permissible targets under the laws of war. *See* Chapter I.

[3] *See* David M Adams, 'Hohfeld on Rights and Privileges' (1985) 71 ARSP: Archiv für Rechts-und Sozialphilosophie/Archives for Philosophy of Law and Social Philosophy 84, 90–93.

[4] *See* Glanville Williams, 'The Concept of Legal Liberty' (1956) 56 Columbia Law Review 1129, 1132; *Black's Law Dictionary* (Bryan A. Garner & Henry Campbell Black eds, 1999).

[5] *See* Wesley Newcomb Hohfeld, 'Some Fundamental Legal Conceptions as Applied in Judicial Reasoning' (1913) 23 Yale Law Journal 16; Wesley Newcomb Hohfeld, 'Fundamental Legal Conceptions as Applied in Judicial Reasoning' (1917) 26 Yale Law Journal 710.

[6] *See* Hans Kelsen, *General Theory of Norms* (Clarendon Press 1991) 98–99; Hans Kelsen, *Pure Theory of Law* (University of California Press 1967) 15–16. These two first claims contradict Haque's position, but the third is entirely compatible with his analysis of international law.

[7] Janina Dill has also argued against Haque's interpretation and thinks that IHL is not only prohibitory in nature. Janina Dill, 'Towards a Moral Division of Labour between IHL and IHRL during the Conduct of Hostilities' in Ziv Bohrer, Helen Duffy, and Janina Dill (eds), *Law Applicable to Armed Conflict* (Cambridge University Press 2020) 216–220.

about whether we enjoy a Hohfeldian privilege or about whether we have an exceptional permission.

I will also argue that Haque's view is mistaken in the moral significance it ascribes to combatants' having a mere immunity from prosecution. He claims that there is no need to insist on the sharp divergence between law and morality observed by revisionists because the law does not confer a legal permission on combatants to fight for an unjust cause;[8] it does not confer a legal permission on combatants to murder the innocent. It only provides an immunity from prosecution, and immunities from prosecution are easier to justify.[9] Yet this cannot be true. Surely the existence of a rule that grants a blanket immunity to people who have murdered the innocent on a vast scale is not easy to justify at all, even if it is somewhat easier to justify than a legal permission. As a result, the divergence between the revisionist morality and the laws of war is still quite troubling—much more troubling than Haque acknowledges.

I will proceed as follows. In section II, I argue that combatants enjoy a legal Hohfeldian privilege. In section III, I argue that combatants belong to a legally favoured class of people and that they enjoy a legal permission to do what is usually forbidden. The latter I characterize as a positive Kelsenian permission.[10] In section IV, I explicitly argue against Haque's understanding of the laws of war as providing *only* an immunity from prosecution,[11] and against the idea that justifying immunities from prosecution is significantly easier than justifying a privilege. Section V concludes.

Before moving on, I am aware that speaking of combatants as having a legal *privilege* to kill, whilst analytically correct, does not convey the terrible experience of war fairly. Combatants often see killing in wartime as a terrible duty, and their immunity from prosecution is usually purchased at the cost of putting themselves in mortal and moral danger.[12] Thus, though the term 'privilege' is unfortunate, I only use it in a restricted technical sense; I do not attach to it any positive significance.

Finally, talking about a privilege to *kill* is not entirely precise. The verb 'to kill' is an action with a specific result: someone's death.[13] What the laws of war provide is a privilege to *target* enemy combatants. Targeting sometimes will result in killing, but not always: legally, the result of death is secondary. And

[8] Haque, *Law and Morality at War* (n 1) 55.
[9] ibid.
[10] *See* Kelsen, *General Theory of Norms* (n 6) 98–99; Kelsen, *Pure Theory of Law* (n 6) 15–16.
[11] My argument does not deny that combatants have an immunity from prosecution.
[12] I thank Wojciech Sadurski and Gary Bass for their suggestions on this.
[13] *See* Oxford Dictionaries. <https://en.oxforddictionaries.com/definition/kill> accessed 18 May 2018.

Table II.i Hohfeld's table of jural relations

Jural Opposites	Right No-right	Privilege Duty	Power Disability	Immunity Liability
Jural Correlatives	Right Duty	Privilege No-right	Power Liability	Immunity Disability

yet I have deliberately chosen to speak of a privilege to *kill* rather than to *target* because the latter gives an impression of a more technical and clinical affair. It obscures what is precisely the most troubling aspect of war and the laws that regulate it: that people's lives are at stake during armed conflict—that soldiers kill each other during battle.

II. War and Hohfeld

In disentangling the complexity of discussions about killing, the right to life, and war, Hohfeld's attempt at reducing juridical or legal relations to their lowest common denominators can be helpful.[14] My purpose in engaging with Hohfeld's language is only of an explanatory nature. I use Hohfeld's terminology in order to clarify conceptually one way in which we say that combatants *may* kill other combatants in war.

Hohfeld's classification of juridical concepts stems from the need to clarify the pervading confusion he saw in legal thought, in which often there is a tacit assumption that all legal relations may be reduced to rights and duties,[15] and particularly, in which the notions of right, privilege, and power are conflated.[16] In order to dispel this confusion, Hohfeld distinguished between eight legal categories in terms of which all legal problems can be stated and clarified.[17] Due to Hohfeld's lack of faith in formal definitions, he chose to exhibit them in a scheme of 'opposites' and 'correlatives' (Table II.i).[18]

As this table shows, a privilege is the opposite (or negation)[19] of a legal duty and the correlative of a no-right.[20] However, due to Hohfeld's resistance to

[14] *See* Hohfeld, 'Some Fundamental Legal Conceptions as Applied in Judicial Reasoning' (n 5); Hohfeld, 'Fundamental Legal Conceptions as Applied in Judicial Reasoning' (n 5).

[15] Hohfeld, 'Some Fundamental Legal Conceptions as Applied in Judicial Reasoning' (n 5) 28.

[16] ibid 34.

[17] *See* Walter Wheeler Cook, 'Hohfeld's Contributions to the Science of Law' (1919) 28 Yale Law Journal 721, 724; Hohfeld, 'Some Fundamental Legal Conceptions as Applied in Judicial Reasoning' (n 5) 30.

[18] Hohfeld, 'Some Fundamental Legal Conceptions as Applied in Judicial Reasoning' (n 5) 30.

[19] ibid 38.

[20] ibid 32.

define the term itself, his conception of privilege has been the subject of much controversy.[21] Probably only adding to the confusion is the fact that Hohfeld's paradigmatic example of a legal privilege is that of the landowner to walk on his own land, which in most legal systems is not merely a tolerated liberty but rather a liberty surrounded by a perimeter of supporting claim-rights of the landowner against attempted interference with the exercise of his liberty.[22]

The privilege to kill combatants looks different from the privilege of the landowner, but leaving those differences aside, let us provisionally say that one party A has a legal privilege in relation to some second party B to perform some action X, if and only if A has no (legal) duty to refrain from performing said action X.[23] Thus, A will have a privilege to jump in relation to B, if and only if A has no duty to B to refrain from jumping.

[21] *See*, eg, Albert Kocourek, 'Hohfeld System of Fundamental Legal Concepts' (1920) 15 Illinois Law Review 24; Williams (n 4); Andrew KW Halpin, 'Hohfeld's Conceptions: From Eight to Two' (1985) 44 The Cambridge Law Journal 435.

[22] *See* John Finnis, 'Some Professorial Fallacies about Rights' (1971) 4 Adelaide Law Review 377, 378.

[23] Carl Wellman, 'Legal Rights' in Carl Wellman (ed), *Rights and Duties* (Routledge 2002) 171. There is also the issue of what it means to say that a duty is owed to someone. In the case of the privilege to kill in war, the point is precisely that A owes *no* duty to enemy combatant B not to kill him, but in order to understand that assertion, we need to have an idea of what it would mean for A to owe a duty to B not to kill him. According to Hohfeld's table, a duty is the opposite of a privilege and the correlative of a claim-right. (Hohfeld, 'Some Fundamental Legal Conceptions as Applied in Judicial Reasoning' (n 5) 30.) Thus, A's duty not to kill B would be correlative to B's claim-right not to be killed by A. The issue gets complicated when we look at this legal relation outside of Hohfeld's schema—that is, when we step outside of describing the form of rights and duties, and start wondering about their function—and ask questions such as: what is it that makes any legal duty a duty *to* some second party? Is the duty owed to the party who can take legal action to enforce that duty or to the person in whose interest the duty is established by the law? Is B (the person who has the correlative right) at liberty to choose whether to exercise that right or not? Can B release A from his duty? If B is not at liberty to do such things, is the duty really *owed to him* or is the duty owed to someone or something else? Is it that, for instance (as could be the case when claim-rights are backed up with criminal sanctions) A has a duty of obedience to the law from which no-one, not even B, can release him? Here, we are asking what claim-rights do for those who hold them (or, in different words, we try to explain the direction of duties) and thus, we venture into the debate between the will theory of rights and the interest theory of rights. I will not take sides on these issues here. I will only say that there is nothing in the description of legal relations in a Hohfeldian fashion which will solve the debate between these two positions. Note, however, that the will theory, unlike the interest theory, has trouble explaining rights we cannot waive (the interest theory has issues of its own as well. *See*, eg, HLA Hart, *Essays on Bentham. Jurisprudence and Political Philosophy* (Oxford University Press 1982) 180–192). This, of course, is problematic when we are talking about the right to life or the right not to be killed (which is precisely what we talk about when we discuss war), because we think of them as rights that we cannot normally waive. Nevertheless, a clarification needs to be made: when we speak of the 'right to life' we may be referring to particular Hohfeldian atomic components of that right (ie the legal privilege of defending one's life when attacked; the legal claim upon some second party that they provide one with the minimal means of sustaining one's life, etc) or we may be referring to all of the components of the right to life at once (and in that case, we use the term 'right' in a non-Hohfeldian way to encompass different Hohfeldian legal relations). When I say that A has a Hohfeldian privilege to kill enemy combatant B, I am only saying that B has no legal right that A does not kill him; that A breaches no legal duty when he kills B. But there is nothing in this description that would allow us to conclude something about B's right to life in general. There is also nothing here regarding the issue of whether B has no right that he is not killed by A because B has waived that right when he entered the army (and whether he may do such thing); or whether there is something about his status as

Before addressing the privilege of the combatant, a few caveats about Hohfeld's language are necessary. When analysing a juridical relation under Hohfeld's system, it is important to note first, that each Hohfeldian relation concerns only *one* activity of *one* person; second, that each legal relation regards an activity with respect to *only two* persons; and finally, that the relevance of legal remedies was left entirely undetermined by Hohfeld.[24] Therefore, analysing a legal relation in a Hohfeldian fashion demands that complex legal relations—where there are multiple issues between two parties, issues involving three or more parties, and other legal structures that surround those relations—are understood in terms of the relation existing between two parties holding opposite ends of the same stick.[25] Thus, when analysing killing combatants in war as a Hohfeldian privilege, we are only looking at the legal relation between *one* combatant A and *one* enemy combatant B, only pertaining to the activity of A killing B.

Having this in mind, we can now say that killing in war is a privileged action if combatant A is under no legal duty to enemy combatant B not to kill him. But what does it mean for A to lack a legal duty not to kill enemy combatant B? In other words, when do we say that someone lacks a legal duty (or has one)? This question, of course, is a matter of interpretation of legal sources, and it might be a very easy question to answer—if, for instance, the law explicitly states that A has a legal duty not to X—or a very difficult one.

In determining whether the legal relation between combatant A and enemy combatant B pertaining to A killing B is a Hohfeldian privilege, we could ask: what may A do, without a penalty assessed for the benefit of B?[26] If we

a combatant that makes the duty not to kill him disappear or his right not to be killed to be forfeited; or whether there is something about the collective and political nature of the violence that does so. On rights and duties and the debate between the will and the interest theories of rights, *see* Carl Wellman, *An Approach to Rights: Studies in the Philosophy of Law and Morals* (Kluwer Academic Publishers 1997); Joel Feinberg and Jan Narveson, 'The Nature and Value of Rights' (1970) 4 The Journal of Value Inquiry 243; Leif Wenar, 'Rights' in Edward N Zalta (ed), *The Stanford Encyclopedia of Philosophy* (Fall, 2015) <http://plato.stanford.edu/archives/fall2015/entries/rights/> accessed 6 June 2016; Joseph Raz, 'On the Nature of Rights' (1984) 93 Mind 194; HLA Hart, 'Are There Any Natural Rights?' (1955) 64 The Philosophical Review 175; HLA Hart, *The Concept of Law* (Penelope A Bulloch and Joseph Raz eds, 2nd edn, Clarendon Press 1998) 82–91; Matthew H Kramer and Hillel Steiner, 'Theories of Rights: Is There a Third Way?' (2007) 27 Oxford Journal of Legal Studies 281; David Lyons, 'The Correlativity of Rights and Duties' [1970] Nous 45.

24 Finnis (n 22) 379–80; Halpin (n 21) 439; David Rodin, *War and Self-Defense* (Clarendon Press 2002) 18.

25 *See* Curtis Nyquist, 'Teaching Wesley Hohfeld's Theory of Legal Relations' (2002) 52 Journal of Legal Education 238, 240–49.

26 Corbin's formulation obviously applies more readily to private law. It is doubtful whether sanctions in the criminal law exist for the benefit of victims. Arthur L Corbin, 'Legal Analysis and Terminology' (1919) 29 The Yale Law Journal 163, 165.

determine that A may kill B without a penalty, then we could conclude that A has a privilege with respect to B. A, knowing that he enjoys a particular privilege, can answer the question 'what *may* I do without a court preventing me or penalizing me?'[27] Posed like this, the answer is obvious: combatants cannot be prosecuted by their participation in hostilities,[28] and therefore we may conclude that they enjoy a Hohfeldian privilege to participate in hostilities.

But formulating the questions in this way is not entirely correct because the existence of a remedy is not necessary for establishing the existence of a duty. Affirming this position would imply adopting a distinct view about the relationship between duties and remedies (something which, as mentioned above, was left undetermined by Hohfeld himself). Thus, in determining what counts as a legal privilege, it is important to keep in mind that the lack of penalty might not necessarily entail the absence of a legal duty. It is possible that legal duties are not enforceable or lack remedies (as we will see in section IV, this is the case of diplomatic immunity: diplomats have a legal duty to obey the laws of the receiving state but, at the same time, they enjoy a jurisdictional immunity). In any case, solving the debate about the relationship between rights and remedies is not necessary to determine the situation of combatants.

The question about the kind of legal relationship that exists between enemy combatants must be answered by analysing the laws of armed conflict. Unfortunately, it doesn't help that these laws are a patchwork of treaties and customary international law. As Haque correctly acknowledges, there is no rule of armed conflict whose content or application is not disputed, and even the purposes of the regime and its relationship to other areas of international law are the subject of deep disagreements.[29]

A reasonable place to start is article 43 of Additional Protocol I, which explicitly states that combatants have 'a right to participate directly in hostilities,'[30] and that '[a]ny combatant, as defined in Article 43, who falls into the power of an adverse Party shall be a prisoner of war.'[31] Combatants entitled to prisoner of war status cannot be criminally prosecuted for their participation in

[27] ibid 167.

[28] *See* Commentary of 2020 to Convention (III) Relative to the Treatment of Prisoners of War (Geneva, 12 August 1949) (hereinafter, 2020 Commentary to GCIII), Article 85: Offences Committed Before Capture, para 3634. <https://ihl-databases.icrc.org/applic/ihl/ihl.nsf/Comment.xsp?action=openDocument&documentId=3A23C27AF17D2326C1258585005366D3> accessed 23 July 2022.

[29] Adil Ahmad Haque, 'Indeterminacy in the Law of Armed Conflict' (2019) 95 International Law Studies 119.

[30] *Protocol Additional to the Geneva Conventions of 12 August 1949, and relating to the Protection of Victims of International Armed Conflicts (Protocol I)* (hereinafter, API) 8 June 1977, 1125 UNTS 3, art 43.2.

[31] API, art 44.

hostilities unless they have committed genocide, crimes against humanity, or war crimes. Targeting and killing other combatants are none of these.

However, the language of Additional Protocol I is somewhat imprecise. Combatants do not have a right to participate in hostilities if we understand 'right' as a Hohfeldian claim-right,[32] which would correlate to a duty: combatant A does not have a right to kill enemy combatant B, because enemy combatant B does not have a legal duty to let himself be killed. Thus, the correct interpretation is that combatants *may* participate directly in hostilities, which is closer to the language utilized by the 1987 Commentary to Additional Protocol I.[33] This document is more careful in pointing out that members of the armed forces 'can participate directly in hostilities i.e., attack and be attacked', and that they are 'entitled to do so'.[34]

In other words, we can now say that combatants have a Hohfeldian privilege to participate in hostilities. Participation in hostilities, of course, includes killing. This is the equivalent of saying that, under international law, combatants lack a legal duty not to kill enemy combatants, or, in more precise terms, that any given combatant A generally has a Hohfeldian privilege to kill enemy combatant B because he lacks a legal duty not to kill B. This more precise formulation allows for the possibility that combatant A has a legal duty to kill B that is owed not to B, but to someone else (most likely his commanding officer or his state).[35]

Note that for a legal advantage to be a privilege, it is not logically necessary that the privilege be accompanied with a right not to be interfered with in the exercise of said privilege.[36] The mere fact of having a privilege says nothing about whether those rights exist. Thus, the privilege of combatant A to kill combatant B does not logically entail that A has a right—against B or third parties—not to be interfered with in the exercise of his privilege to kill B.

In the case of killing in war, International Humanitarian Law (IHL) does not include rights of non-interference alongside the privilege to kill. Combatant

[32] In this sense, Hohfeld was correct when pointing out that the word 'right' is used generally and indiscriminately to denote any sort of legal advantage, whether a claim, a privilege, a power, or an immunity. *See* Hohfeld, 'Fundamental Legal Conceptions as Applied in Judicial Reasoning' (n 5) 717.

[33] Commentary of 1987 to Protocol Additional to the Geneva Conventions of 12 August 1949, and relating to the Protection of Victims of International Armed Conflicts (Protocol I), 8 June 1977 (hereinafter, 1987 Commentary to API). <https://ihl-databases.icrc.org/applic/ihl/ihl.nsf/Treaty.xsp?documentId=D9E6B6264D7723C3C12563CD002D6CE4&action=openDocument> accessed 31 July 2022.

[34] 1987 Commentary to API, Article 43: Armed Forces, para 1677.

[35] This aspect of war is emphasized by Christopher J Finlay, 'Justification and Legitimacy at War: On the Sources of Moral Guidance for Soldiers' (2019) 129 Ethics 576, 580.

[36] On this conception of privilege *see* Williams (n 4) 1143–45; Alan D Cullison, 'Review of Hohfeld's Fundamental Legal Concepts, A' (1967) 16 Cleveland-Marshall Law Review 559, 567; Adams (n 3) 93. Against, *see* Halpin (n 21) 441–50.

A does have a privilege to kill B, but A has no right that B—or other enemy combatants—does not interfere in the exercise of A's privilege. Adding to this complexity is the fact that under the current rules of international armed conflict, enemy combatant B also has a privilege in relation to combatant A to kill him, as he is under no duty not to do so.[37] This means two things: first, that both A and B are under no duty not to kill the other or, in simpler terms, they both have a privilege to kill each other; and second, that both A and B do not have a right against each other that the other does not interfere with the other's privilege to kill him. Though this may seem puzzling, the liberty to trade works similarly:[38] by trading and attempting to trade you commit no breach of duty, but this does not mean that another trader is forbidden from interfering with your trading by, say, selling better products or lowering her prices. This permissible interference, of course, does not logically entail that anything goes in trading. The same applies for killing in war.[39]

As I said earlier, Hohfeld's scheme calls for focusing on the legal relation between combatant A and enemy combatant B, while leaving out the complex legal reality where that relation exists. If we step outside the relation between combatant A and enemy combatant B regarding the killing of B, we can introduce three distinct roles at play in the privilege: the role of the second party against whom the privilege holds (combatant B), the role of the possessor of the privilege (combatant A), and the role of third parties in a position to intervene in any confrontation between the privilege holder and some second party.[40] The reason for bringing in these complications is that the privilege to kill in war, as any other privilege, does not exist in a legal vacuum where only A, B, and the conduct of killing take place but is part of a complex system of rules. Within this system, the privilege to kill stands in relation to other claim-rights

[37] As explained in the introduction, in a non-international armed conflict, there is no symmetrical legal privilege to kill. Rebel forces are usually dealt with under domestic criminal law. The only exception to this regime is stated in Additional Protocol I, which expanded the concept of international armed conflict to include those in which peoples are fighting against colonial domination and alien occupation, and against racist régimes in the exercise of their right of self-determination. API, art 1. Regarding the normative and legal issues posed by asymmetric warfare, *see* Alejandro Chehtman, *A Theory of Asymmetric Warfare* (Oxford University Press, forthcoming 2023).

[38] I take this example from Williams (n 4) 1146. And the comparison, from Carl von Clausewitz, *On War* (Michael Eliot Howard and Peter Paret eds, Princeton University Press 1989) 149.

[39] This is only a limited formal analogy, and I acknowledge that the comparison between war and trading is problematic on several levels but, particularly, because traders do not have a right to economic success, whereas combatants do have a right to life—unless one would be willing to say that during war, combatants' right to life is somehow lost. Revisionists (and sometimes Walzer) do talk about the forfeiture of combatants' rights. *See* Michael Walzer, *Just and Unjust Wars: A Moral Argument with Historical Illustrations* (5th edn, Basic Books 2006) 146; Haque, *Law and Morality at War* (n 1); Jeff McMahan, *Killing in War* (Oxford University Press 2009); Helen Frowe, *Defensive Killing* (1st edn, Oxford University Press 2014); Victor Tadros, 'Duty and Liability' (2012) 24 Utilitas 259.

[40] *See* Wellman, 'Legal Rights' (n 23) 212–13.

(such as the right to life), privileges, and duties, and the roles of third parties and state officials also need to be accounted for.

Introducing these three roles allows us to say a few more things about the privilege to kill in war. First, from the perspective of enemy combatant B, and consistent with Hohfeld's table of juridical or legal correlations, A's privilege to kill is the lack of a legal claim-right of B not to be killed by A. Second, to say that combatant A has a privilege to kill enemy combatant B does not mean that combatant A is necessarily at liberty to kill enemy combatant B, that is, that he can choose whether to kill B.[41] It may be that he is under a duty to do so because of orders from his superior. This fact does not change the nature of the relationship between both combatants as it pertains to a third party. Finally, from the perspective of official third parties, the privilege can be understood as an immunity from prosecution. In the case of war, these third parties are the authorities of the opposite party to the conflict, particularly those under whose power a privileged enemy combatant has fallen. When a privileged combatant is taken prisoner by authorities belonging to the adversary, the combatant cannot be prosecuted for his participation in the hostilities, including acts that constituted killing.[42] The point of the combatant's detention is not punishment but incapacitation: to prevent them from going back to the battlefield for the duration of the conflict.[43]

III. Privileged Killings and Privileged Combatants

I have already explained that combatants have a Hohfeldian privilege to kill: they are under no legal duty not to kill enemy combatants. But under Hohfeld's analysis, we can also generally describe individuals as having a Hohfeldian privilege to breathe: I am under no legal duty *not* to breathe in relation to you. Thus, having a Hohfeldian privilege is, in a sense, a somewhat weak assertion: we enjoy Hohfeldian privileges to do all those things the law doesn't directly concern itself with. So, does the law concern itself with killing in war? Or does killing in war enjoy the same legal status as breathing?

[41] For the relation between liberty, privilege, and choice, *see* Adams (n 3) 87–88; HLA Hart, 'Bentham on Legal Rights' in Carl Wellman (ed), *Rights and duties* (Routledge 2002) 95–97; JC Smith, 'Liberties and Choice' (1974) 19 American Journal of Jurisprudence 87, 87.

[42] *See* Yoram Dinstein, *The Conduct of Hostilities under the Law of International Armed Conflict* (Cambridge University Press 2004) 35–37. *Geneva Convention Relative to the Treatment of Prisoners of War (Third Geneva Convention)* (hereinafter, GCIII) 12 August 1949, 75 UNTS 135, arts 87 and 102.

[43] Dinstein (n 42) 35.

Killing in war is not *only* like breathing. The law of war is not silent on killing as the law is on breathing. We can correctly speak of combatants as privileged in their killings not just in a Hohfeldian sense but also in the sense of them enjoying an exceptional permission to do what is usually forbidden. However, let me be upfront about this: there is no individual provision codified in IHL that explicitly states something like 'combatants have a privilege to kill' or 'combatants have a permission to kill enemy combatants'.[44] The closest IHL gets to an explicit permission is in its consistent use of language such as 'lawful acts of war',[45] acts that 'are not unlawful under humanitarian law',[46] and 'lawful targets'.[47] All these expressions are a reference to the intentional targeting and killing of combatants (among other acts).[48] Similar language is used in the International Committee of the Red Cross's (ICRC) formulation of rule 1 of customary international law, which reads as follows: 'The parties to the conflict must at all times distinguish between civilians and combatants. Attacks *may only* be directed against combatants. Attacks must not be directed against civilians.'[49]

These formulations suggest the existence of an exceptional permission to kill—exceptional in relation to times of peace—although they are, perhaps, inconclusive of the existence of an explicit legal permission (ie 'you *may* kill'). In any case, the absence of an explicit rule does not entail that the laws of war do not provide an exceptional permission to kill. That, I believe, would be a mistaken interpretation of our legal sources.

In order to explain why combatants have an exceptional legal permission to kill enemy combatants, it is useful to consider Kelsen's distinction between negative and positive permissions. An action is negatively permitted when there is no valid legal norm on the subject.[50] By contrast, an action is positively permitted when the validity of a norm prohibiting a certain behaviour is repealed or restricted by another, derogating norm, so that this behaviour is no longer forbidden or the prohibition is partially lifted.[51] As can be seen from this definition, what we ordinarily call an exceptional permission is a Kelsenian

[44] For a brief description of the laws of armed conflict on killing, *see* Chapter I.

[45] 2020 Commentary to GCIII, Article 2: Application of the Convention, para 272; Article 5: Beginning and End of Application, para 1114, 3634.

[46] 2020 Commentary to GCIII, Article 3: Conflicts Not of an International Character, para 565.

[47] 2020 Commentary to GCIII, Article 3: Conflicts Not of an International Character para 784.

[48] 1987 Commentary to API, Article 48: Basic Rule, para 1874. ('As regards military objectives, these include the armed forces and their installations and transports.')

[49] ICRC. IHL Database, Customary IHL, Rule 1. <https://ihl-databases.icrc.org/customary-ihl/eng/docs/v1_rul_rule1> accessed 28 July 2022 (emphasis added).

[50] *See* Kelsen, *General Theory of Norms* (n 6) 98; Kelsen, *Pure Theory of Law* (n 6) 16.

[51] Kelsen, *General Theory of Norms* (n 6) 98–99; Kelsen, *Pure Theory of Law* (n 6) 15–16.

positive permission: permitting positively can be characterized as *making* something free, unlike permitting negatively, whereby behaviour is permitted or free, but not *made* free.[52]

The absence of an explicit provision permitting a given behaviour—as is the case with killing enemy combatants under the laws of war—doesn't necessarily mean that the action is negatively permitted.[53] Usually, the absence of an explicit norm that states the exceptionally permitted conduct is a matter of legislative technique. The law often limits itself to establishing the sanctions associated with particular behaviours so that the enactment of a norm explicitly prescribing the behaviour that avoids the sanction is rendered superfluous: such prescription is presupposed in the sanction-decreeing norm.[54]

If we are to conclude that killing in war is a positive permission, then there must be first, a prohibition on killing, and second, a norm that partially lifts that prohibition. But remember, the norm that partially lifts the prohibition does not need to contain an explicit permission to kill. The permission, as Kelsen explains, can be presupposed by a sanction-decreeing norm.

That there is a general legal prohibition on killing does not need much of an argument. Suffice it to say that IHL is not an isolated and completely independent legal system: it is embedded within international law, and it interacts with domestic law and other areas of international law (though sometimes, that interaction is uneasy in its fit). In the case of international law, the protection of the right to life is central to all human rights treaties, which protect against arbitrary deprivations of life.[55] In domestic law, the protection of the right to life manifests itself in the prohibition on murder.

We already know that the existence of an explicit norm permitting combatants to kill other combatants is dubious, but we do find a sanction-decreeing norm from which we can conclude a presupposed permission: intentionally targeting civilians and other protected persons during armed conflict constitutes a war crime.[56] Thus, even though the laws of war do not explicitly state

[52] Kelsen, *General Theory of Norms* (n 6) 98–99.

[53] The fact that law can contain norms other than explicit ones is not novel. Famously, Ronald Dworkin argued that law can include principles that are implicit. *See* Ronald Dworkin, *Taking Rights Seriously* (Harvard Univ Press 1978) 14–81; Ronald Dworkin, *Law's Empire* (Hart 2010).

[54] *See* Kelsen, *General Theory of Norms* (n 6) 142. However, the point about legislative technique must be distinguished from the logical operation we engage in to ascertain that the permission to kill is presupposed in the sanction-decreeing norm.

[55] *See*, eg, *International Covenant on Civil and Political Rights*, 1966, 999 UNTS 171 (entered into force 23 March 1976), art 6; *European Convention on Human Rights. Convention for the Protection of Human Rights and Fundamental Freedoms*, 4 November 1950, 213 UNTS 222, art 2; *African Charter on Human and Peoples' Rights*, 21 June 1986, OAU Doc. CAB/LEG/67/3 rev.5, 21 ILM 58 (entered into force 21 October 1986), art 4.

[56] *Rome Statute of the International Criminal Court* (hereinafter, Rome Statute) ISBN No. 92-9227-227-6, 17 July 1998, art 8.2.(a)(i); art 8.2(b)(i); and art 8.2(b)(iv).

that combatants are allowed to kill other combatants, the sanction-decreeing norm—the prohibition on targeting *civilians*—allows us to conclude that intentionally targeting *combatants* is permitted, and that this permission is exceptional to the circumstances of war (there are simply *no* combatants outside war). Indeed, IHL is considered *lex specialis* precisely because one of its consequences is to determine what an arbitrary deprivation of life during armed hostilities comprises.[57] At the domestic level, this might take the form of a criminal law defence for soldiers who kill in battle.[58]

In summary, it would be a mistake to think that the laws of war are silent regarding killing enemy combatants. They are not. A careful reading of the relevant treaties, as well as of customary international law, strongly supports the conclusion that the laws of armed conflict provide a Kelsenian positive (exceptional) permission to kill during wartime. In other words, combatants enjoy a privilege to kill enemy combatants if by that we mean that they are permitted to do something which is usually forbidden: killing.[59]

Additionally, combatants are privileged in a third sense: they enjoy something that others do not. As we saw above, *anyone* who kills a combatant during wartime *does not* commit an international crime, yet civilians and non-privileged combatants who kill combatants during armed conflict *do* commit a domestic crime: murder.[60] Lawful combatants, on the other hand, do not. Unlike civilians and unlawful combatants, lawful combatants cannot be prosecuted for their participation in hostilities if they fall under the power of the adversary.[61] As can be seen, lawful combatants—usually referred to, precisely,

[57] *See* David Luban, 'Military Necessity and the Cultures of Military Law' (2013) 26 Leiden Journal of International Law 315, 328. ICJ, *Legality of the Threat or Use of Nuclear Weapons*, Advisory Opinion, ICJ Reports 1996, para 25. Note that this is independent of the discussion about the application of human rights treaties during times of war. *See* Theodor Meron, 'The Humanization of Humanitarian Law' (2000) 94 American Journal of International Law 239.

[58] *See* Gerald D Robin, 'Justifiable Homicide by Police Officers' (1963) 54 The Journal of Criminal Law, Criminology, and Police Science 225. Model Penal Code art 3 s 3.03; United States Manual for Courts Martial, Rule 916(c); Uniform Code of Military Justice 802.2(a)(9), 802.2(a)13, 805.5.

[59] In other work, Kelsen uses different definitions of negative and positive permissions and agrees with Haque's position. Kelsen argues that acts of warfare committed within an illegal war cannot be considered permitted, neither in a negative nor in a positive sense of the term. Because the war itself is forbidden, each act is forbidden as well. States cannot punish other combatants: doing so would violate international law, but only because it involves making an individual responsible for an act of another state and violates jurisdictional norms. *See* Hans Kelsen, 'Collective and Individual Responsibility in International Law with Particular Regard to the Punishment of War Criminals' (1942) 31 California Law Review 530, 547–51. I'm indebted to Adil Haque for pointing this out to me.

[60] *See* Dinstein (n 42) 44–61.

[61] GCIII, arts 4, 87, 99, and 102; *Convention (IV) Respecting the Laws and Customs of War on Land and Its Annex: Regulations Concerning the Laws and Customs of War on Land (Hague Convention IV)* (hereinafter, Hague Convention IV) The Hague, 18 October 1907, reg 1; and Prisoners of War and Detainees Protected Under International Humanitarian Law *in* ICRC <https://www.icrc.org/eng/war-and-law/protected-persons/prisoners-war/overview-detainees-protected-persons.htm> accessed 28 July 2022.

as *privileged* combatants—belong to a favoured class of people, as opposed to civilians and unprivileged combatants.

IV. A Mere Immunity from Prosecution?

In sections II and III, I have characterized combatants as enjoying a privilege in three different senses. I have said that they enjoy a Hohfeldian privilege to kill, that they enjoy an exceptional permission to kill, and that they belong to a legally favoured class of people. The fact that we can describe combatants as privileged in these three different ways is only a coincidence: the existence of one of them does not logically entail the existence of the others. More precisely, it is perfectly possible to enjoy a Hohfeldian privilege that is not a positive or exceptional permission—but both positive and negative legal permissions can be understood as Hohfeldian privileges—and to enjoy a special favour that is neither a positive permission nor a Hohfeldian privilege. Therefore, even if one were to disagree with my argument that the laws of war contain a positive permission to kill,[62] in and of itself, that disagreement would not provide an argument against the characterization of the combatants' permission to kill as a Hohfeldian privilege, nor against combatants belonging to a favoured class of people. Unfortunately, this is not an uncommon mistake.[63]

At the end of section II, I said that the Hohfeldian privilege of the combatant exists alongside an immunity from prosecution. This characterization of the laws of armed conflict is correct and is not inconsistent with the existence of a Hohfeldian privilege.[64] As already explained, the same legal advantage can

[62] As Haque seems to think is the case. *See* Haque, *Law and Morality at War* (n 1) 32. 'Put another way, the law of targeting and attack confers no *strong* permissions. As Joseph Raz explains, "an act is strongly permitted only if its being permitted is entailed by a norm. It is permitted in the weak sense if the permission . . . is simply a consequence of there being no norms prohibiting the performance of the action." '

[63] *See*, eg: 'For many years, when teaching Jurisprudence in the University of London, I used Hohfeld's term, abandoning it only when I had become convinced that it created an excessive amount of difficulty. The term "privilege" carries the strong popular and etymological meaning of a special favour given to the individual or a narrow class (*privilegium*, a private law), and the tendency is for this to fight down the general meaning that Hohfeld chose to assign to it. I found that, however clearly I tried to explain to students that "privilege" in jurisprudence was to mean merely what Salmond called a liberty, some seemed to be unable to hold this explanation in the mind. Hence I was constantly faced with the objection to particular illustrations: "That is not a privilege, because everybody has it." ' Williams (n 4) 1132.

[64] It is also consistent with the language in some parts of the 2020 Commentary to GCIII. For example, Article 4: Prisoners of War, para 1048, speaks of civilians as not having 'a right to participate directly in hostilities *with* immunity from prosecution for such acts' (emphasis added); and the ICRC's Interpretive Guidance on Direct Participation in the Hostilities (2009), fn 52, states that 'combatant privilege, namely the right to directly participate in hostilities *with* immunity from prosecution for lawful acts of war' (emphasis added). I will go back to this later on. *See International Committee of the*

be seen both as a privilege, in relation to the enemy combatant and the action of killing, and as an immunity, from the point of view of legal officials (in relation to the power to prosecute) and enemy combatants (in relation to their power to initiate legal action). These relations can also be understood in terms of Hohfeld's notions of immunity and liability.[65] However, the view that combatants have a privilege or that they are permitted to kill enemy combatants has some detractors. The most pre-eminent one is Adil Haque.[66]

According to Haque, international law does not grant a legal privilege to *all* lawful combatants: what all lawful combatants have is merely an immunity from criminal prosecution.[67] By *all* combatants, Haque means combatants who are fighting a just war and those fighting an unjust one. According to him, international law does not grant combatants a legal permission to fight for an unjust cause. Both just and unjust combatants only have a symmetrical immunity from criminal prosecution;[68] it is merely a matter of enforcement. Under this interpretation, saying that combatants have a privilege to kill—either understood as a positive permission or as a Hohfeldian privilege—would be as absurd as saying that someone who has committed a crime whose statute of limitations has expired had a positive permission or a Hohfeldian privilege to commit said crime.

There are three main arguments Haque makes. The first one relates to the nature of international humanitarian law. Haque believes that the law relating to the conduct of hostilities is primarily a law of prohibition. Thus, IHL does not tell combatants what they may do, it only tells them what they are forbidden to do; it does not authorize, it merely prohibits certain things.[69] What I understand as a Kelsenian positive permission to kill is, according to Haque, merely

Red Cross, *Interpretive Guidance on the Notion of Direct Participation in Hostilities under International Humanitarian Law* (Nils Melzer ed, ICRC 2009).

[65] A Hohfeldian immunity is one's freedom from the legal power or control of another as regards some legal relation. Hohfeld, 'Some Fundamental Legal Conceptions as Applied in Judicial Reasoning' (n 5) 55. Another definition, by Wellman, is the following: one party X has a legal immunity against some second party Y from some specified legal consequence C, if and only if Y lacks the legal power to do any action whatsoever that would be recognized by the law as having the consequence C for X. Wellman, 'Legal Rights' (n 23) 171. *See* also Williams (n 4) 1132; Rodin (n 24) 21. Williams provides two examples of immunities: the privilege of members of parliament to be free from arrest and diplomatic immunity. Rodin defines a Hohfeldian immunity as 'not being subject to having one's first-order relations altered by a particular party'.

[66] Recently, Ripstein has agreed with Haque that IHL is prohibitory in nature and does not generate novel permissions. It only imposes distinctive prohibitions which are specific to war. Arthur Ripstein, *Kant and the Law of War* (Oxford University Press 2021) 107; 115–20.

[67] Haque, *Law and Morality at War* (n 1) 24–25.

[68] ibid 55.

[69] ibid 31.

the absence of a prohibition,[70] and acts not prohibited by the laws of war are, according to Haque, merely 'tolerated' rather than 'permitted'.[71]

I have already explained why it is a mistake to think that international humanitarian law is silent regarding whether combatants may kill other combatants during armed conflict, and why the absence of a positive permission does not logically establish the absence of a Hohfeldian privilege. Thus, I will not focus too much on this strand of the argument here, though I cannot avoid it entirely, for Haque thinks that they are related, and he does not distinguish between having a Hohfeldian privilege and having an exceptional or positive permission.

The second aspect of Haque's argument is his denial of the existence of a symmetrical legal privilege for all lawful combatants. This position dovetails with the discussion in the literature concerning the symmetry thesis,[72] that is, the fact that the law of international armed conflict provides the same rights, duties, and privileges to all lawful combatants, regardless of whether the war they are fighting for is just or unjust. According to Haque, what is symmetrical between just and unjust combatants is only an immunity from prosecution, not a legal privilege; international law does not confer on combatants a legal permission to fight for an unjust cause.[73] This is important, as we will see later, if we are concerned with justifying the symmetry thesis while holding on to the revisionist morality, at least insofar as this morality claims that what unjust combatants do is the equivalent of murdering the innocent. For indeed, Haque thinks that if the law provides only an immunity from prosecution to all combatants, that is, if the law does not privilege killing in war—either in a Hohfeldian sense or in the sense of a Kelsenian positive permission—justifying our laws on killing in war presents less of a problem. If our laws do not provide a symmetrical legal privilege, then the 'hard' part of the justificatory question disappears: we only need to justify an immunity, and justifying immunities, Haque says, is easier than justifying a privilege.[74] I will return to this point at the end.

I will start with Haque's arguments against understanding combatants as having a privilege to kill, both in a Hohfeldian sense and as a positive

[70] ibid 31, 33.
[71] ibid 31.
[72] On this discussion, *see*, ie, Walzer (n 39) 34–47; McMahan (n 39) 38–103; Yitzhak Benbaji, 'The War Convention and the Moral Division of Labour' (2009) 59 The Philosophical Quarterly 593; Seth Lazar, 'Method in the Morality of War' in Seth Lazar and Helen Frowe (eds), *The Oxford Handbook of Ethics of War* (Oxford University Press 2015).
[73] Haque, *Law and Morality at War* (n 1) 55.
[74] ibid 28–29, 55.

permission. First, he states that the right to participate directly in hostilities granted to combatants cannot be understood as a legal permission because if it were a permission, it would then follow that combatants are legally permitted to fight, while civilians are legally forbidden to do so. However, Haque continues, international law does not in fact forbid civilians from participating directly in hostilities: civilians do not violate international law by fighting in an armed conflict.[75] Haque thinks this is consistent with the ICRC's guidance on direct participation in hostilities that states that 'international humanitarian law neither prohibits *nor privileges* civilian direct participation in hostilities'.[76]

Haque argues that as a result, the legal right to participate directly in hostilities conferred on combatants by Additional Protocol I, but denied to civilians, cannot be a privilege to participate in hostilities, that is, the absence of a legal duty not to fight. He claims that neither combatants *nor* civilians have a legal duty under international law not to fight.[77] The right to fight, says Haque, is just a limited legal immunity from prosecution for fighting: lawful combatants have specific rights against foreign governments' prosecuting them for acts that violate domestic criminal law but that do not violate international law.[78]

Haque is right that civilians do not commit an international crime if they participate in hostilities during an international armed conflict. Yet, the fact that they do not commit an international crime is not sufficient to ascertain whether they violate international law: not every violation of international law is also an international crime. And although civilians who participate in hostilities do not commit an international crime, they can be prosecuted under the enemy's domestic criminal law for doing so,[79] unlike privileged combatants, who cannot. Thus, there is a good case for thinking that civilians who choose to fight in an armed conflict are indeed violating international law, since the existence of this violation could explain why international law does not grant an immunity from prosecution to civilians. Such a view is also suggested by a footnote to the ICRC's guidance on direct participation in hostilities, which observes that, at least historically, direct participation in hostilities with the authority of a state has 'always been regarded as legitimate, and as such, exempt from domestic prosecution'.[80] In any case, at the very least, we would want to say that international law disincentivizes the participation of civilians in war.

[75] ibid 24.

[76] *International Committee of the Red Cross* (n 64) 83.

[77] Haque, *Law and Morality at War* (n 1) 25.

[78] ibid 24–25.

[79] And, some say, for facts that render their participation in hostilities unlawful (which are international norms). *See* Dinstein (n 42) 41–61.

[80] *International Committee of the Red Cross* (n 64) fn 71.

Second, and more importantly, whether civilians are forbidden to participate in hostilities doesn't say anything about whether combatants have a Hohfeldian legal privilege to participate in hostilities. Both sets of facts are logically independent.

Third, when the ICRC's guidance states that international law 'neither prohibits nor privileges' civilian direct participation in hostilities, it is not using 'privilege' to indicate a Hohfeldian privilege. It is using the term 'privilege' in a different sense: as a special favour granted to a particular class of people. Civilians, the ICRC's guidance tells us, are not privileged by international law regarding their participation in hostilities—unlike combatants, who are. And, the ICRC's guidance adds, international law does not prohibit civilians from fighting either, in the sense that they do not commit an international crime if they do so. Nothing here provides an argument against combatants having a Hohfeldian privilege or against combatants having a positive permission. It only says something about the legal status of civilians who participate in hostilities.

Finally, Haque's assertion that the combatant's right to participate in hostilities cannot be a privilege because 'neither combatants nor civilians have a legal duty under international law not to fight' is puzzling.[81] According to Haque's own interpretation of the norms, just quoted, *both* civilians and combatants have a Hohfeldian legal privilege to fight under international law. Indeed, neither civilians nor combatants are under a duty not to fight, Haque says, and, according to him, they do not have a Hohfeldian claim-right either.[82] But that is precisely what a Hohfeldian privilege is: the absence of a legal duty. What Haque could say is that international law grants both civilians and combatants a Hohfeldian privilege to participate in hostilities and an immunity from prosecution only to combatants (and hence, combatants would be understood as privileged also in the sense of having something that others do not). Yet Haque would probably not embrace this interpretation. If he did, he would then have to admit that the justificatory problem is indeed urgent: not only would combatants have a Hohfeldian privilege to kill combatants, but so would civilians.

Perhaps this is where the distinction between weak and strong legal permissions and Hohfeldian privileges comes into place. Haque could concede that the laws of armed conflict confer a Hohfeldian privilege on both civilians and combatants to kill combatants, but that this privilege is still merely what he refers to as a 'weak permission' or the 'absence of a prohibition'.[83] If that were the

[81] Haque, *Law and Morality at War* (n 1) 25.
[82] ibid.
[83] ibid 31.

case, then much would turn on the distinction between weak and strong permissions. I will come back to this point later.

Now, such an interpretation of the laws of armed conflict—according to which, under international law, both civilians and combatants have a Hohfeldian privilege to target combatants—is not the most satisfying interpretation of our rules. Perhaps it is better to understand civilians as not having a Hohfeldian privilege to kill during armed conflict precisely because they can be criminally prosecuted for their killings during wartime. To my mind, the existence and applicability of the criminal sanction for murder is a powerful reason to support the conclusion that civilians are under the legal duty that all of us generally have of not committing homicide. They remain subject to the legal duties of the domestic jurisdiction in question: they do not have a privilege to kill, not even in a Hohfeldian sense.

Someone could say that this only entails that civilians are under a *domestic* legal duty, but not under an *international* one: under international law they still have a Hohfeldian privilege, as do combatants. Though reasonable, I am unsure if drawing this distinction is necessary, since it is international law itself that impedes the prosecution of lawful combatants, while failing to do the same regarding civilians. Perhaps it is best simply to state that combatants have a Hohfeldian privilege to kill, whereas civilians lack it. They do so because the latter remain subject to the legal duty not to kill which exists in every domestic jurisdiction, and regarding which the international law of armed conflict deliberately chooses not to grant an immunity from prosecution. But it is also correct to say that under international law, the targeting of combatants is not an international crime, regardless of whether it is committed by combatants or civilians. That is, under international law, both civilians and combatants have a Hohfeldian privilege to target combatants. Drawing this distinction between international and domestic law is also useful to describe the legal situation of combatants during non-international armed conflicts: both members of the army and civilians taking direct part in hostilities can be understood as having, under international law, a Hohfeldian privilege to kill other combatants, since they do not commit war crimes for doing so, although domestically, some of them can be prosecuted for their participation in hostilities (those fighting for the government will avail themselves of domestic criminal law justifications, whereas those fighting against the regime will be guilty of sedition, among other criminal offences.) Regardless of the position one takes on this minor issue—which is related to whether one is a monist or a dualist about the relation between international and domestic law—both interpretations support the view that combatants have a Hohfeldian privilege to kill other combatants.

As I said above, perhaps Haque would be happy to concede that international law does grant a Hohfeldian privilege to combatants, insofar as this Hohfeldian privilege is understood merely as a weak permission. This position is not logically incoherent; as I have stated repeatedly, Hohfeldian privileges can be both weak and strong permissions. In section III, I argued that the laws of war contain a positive (Kelsenian) permission, which is closer to a strong permission. Here, I want to look closely at Haque's interpretation of international law as merely containing a 'weak permission' to kill or the 'absence of a prohibition'[84] not to kill. This interpretation allows him to say that combatants only have an immunity from prosecution, a claim to which Haque ascribes great significance when assessing the moral standing of the laws of armed conflict.[85]

Haque's understanding of the right to participate in hostilities as the absence of a prohibition and a mere immunity from prosecution is unsatisfactory for several reasons. According to him, combatants charged with murder under domestic law may raise the defence of lawful combatant immunity which he argues resembles non-exculpatory defences, such as diplomatic immunity, that bar the prosecution irrespective of legal wrongdoing or moral blameworthiness.[86]

Ripstein agrees with Haque in this regard. He argues that those who think that the Geneva Conventions grant novel permissions mistake a principle governing enforcement for a permissive norm of conduct: from the fact that an act—killing enemy combatants—cannot be punished, they infer that the act is not prohibited.[87] I agree that such inference is incorrect—that is why I stated that we cannot conclude the absence of a duty from the lack of remedies. But, as I have argued, there are other reasons that support the existence of a Hohfeldian privilege and a positive permission—reasons that suggest that combatants have more than just an immunity from prosecution.

First, as a rather minor point, the defence that lawful combatants will have in the particular jurisdiction where they are being prosecuted is likely to depend on the military and criminal law of that country. There is nothing in

[84] ibid.

[85] Haque correctly sees the absence of a prohibition and the immunity from prosecution as conceptually distinct (and, I might add, logically independent). However, he thinks they are inextricably linked, because if 'international law prohibited lawful combatants from participating in aggression, then foreign states would be free to prosecute them on that basis', since there is no combatant immunity for acts prohibited by international law. I think an immunity for prosecution could still be provided in such circumstances, and as a result, there is no need to choose between the existence of the prohibition and the existence of the immunity. They are not inextricably linked. Nevertheless, I leave this for next chapter. *See* ibid 30.

[86] Adil Ahmad Haque, 'Law and Morality at War' (2014) 8 Criminal Law and Philosophy 79, 80; Haque, *Law and Morality at War* (n 1) 27.

[87] Ripstein (n 66) 115.

international law that mandates states to comply with their obligation not to prosecute lawful combatants by incorporating a bar to prosecution into their domestic legal systems; they might enact a criminal justification instead.

Second, and more fundamentally, Haque's comparison between diplomatic immunity and the combatant's situation is deeply troubling, and this trouble further showcases that understanding combatants' permission to kill in war as a mere immunity, or as a matter of enforcement, is incorrect.

Diplomatic immunity operates very differently from that of the combatant: it bars the receiving state from exercising its criminal jurisdiction over the diplomatic agent but does not prevent the sending state from exercising its criminal jurisdiction over the matter. Moreover, the sending state can waive that immunity.[88] By contrast, the 'sending' state of the combatant is unlikely to exercise its criminal jurisdiction for killing during an illegal war, considering it was the state itself that ordered him to go to war in the first place.[89] Unlike the diplomat, who may be prosecuted once he gets home, there really is no body which could call the combatant into account for participating in an illegal war while complying with IHL. And here, precisely, we can find the difference between diplomatic immunity and belligerent immunity: the diplomatic agent is not usually sent by her state to the receiving state to commit crimes, whereas the combatant is sent to war by her state precisely to kill other enemy combatants.

Third, the Vienna Convention on Diplomatic Relations explicitly states that it is the duty of all persons enjoying diplomatic privileges and immunities to respect the laws and regulations of the receiving state.[90] When someone is charged with an offence and claims diplomatic immunity, they cannot claim that they were authorized to commit that behaviour or that their conduct was legally permissible. On the contrary, they were under a legal duty not to perform it. Hence, they can only claim that a particular court does not have the authority to call them to account for their conduct.[91] The criminal conduct of the diplomatic agent—over which, granted, the receiving state cannot exercise jurisdiction—remains criminal and worthy of legal and (presumably) moral

[88] *See Vienna Convention on Diplomatic Relations*, 18 April 1961, 500 UNTS 95, arts 29–32.

[89] The sending state could, however, prosecute combatants for failing to comply with the rules of engagement, which are sometimes more demanding than IHL. The point is that the prosecution could not be for the fact that the acts are committed within the context of an illegal war. I'm indebted to Adil Haque for this clarification.

[90] *Vienna Convention on Diplomatic Relations*, art 41.1: 'Without prejudice to their privileges and immunities, it is the duty of all persons enjoying such privileges and immunities to respect the laws and regulations of the receiving State.'

[91] R Anthony Duff, 'I Might Be Guilty, but You Can't Try Me: Estoppel and Other Bars to Trial' (2003) 1 Ohio State Journal of Criminal Law 245, 247.

reproach. Combatants, on the other hand, do not appear to have a legal duty to comply with the enemy state's domestic criminal law during their participation in hostilities or to not fight when the war is illegal: international law does not state one. If they did have such a duty, we could conclude that they only have an immunity from prosecution, yet the absence of that legal duty makes the case for understanding the law as granting something more than a mere immunity for prosecution even stronger.

Haque perhaps would say here that combatants do have a legal duty not to participate in an unjust war. This duty does not exist under the laws of war, but it could be read from other bodies of law, such as human rights law and the laws on the resort to force. Indeed, Haque claims that the laws of war do not authorize or justify the killings that they fail to prohibit and that they cannot permit killings that are prohibited by other bodies of law.[92] Human rights law prohibits the arbitrary deprivation of life,[93] and according to Haque, not every killing that the laws of war fail to prohibit is for that very reason a *non*-arbitrary deprivation of life.[94]

Haque's view about the relationship between the laws of armed conflict and human rights law is controversial, as he acknowledges,[95] but aside from that, a significant problem is that human rights law does not, in fact, give any sort of guidance to individuals as to when it is permissible to fight a war. In other words, human rights law does not say that killings that are not prohibited under the laws of armed conflict constitute arbitrary deprivations of life under human rights law when they are killings undertaken in pursuit of an unjust war. Indeed, the distinction between just and unjust wars simply does not exist in international law—not even in human rights law. Now, my sense is that here is where the laws on the resort to force (the *jus ad bellum*) would come into play. Haque argues that under international law, 'authorization for acts of violence, if any, must come from the *jus ad bellum*'.[96] In other words, any killings of combatants not prohibited by the *jus in bello* may find their justification in the *jus ad bellum* insofar as those killings take place during the course of a use of force that is lawful under international law. Otherwise—regarding killings

[92] Haque, *Law and Morality at War* (n 1) 35.
[93] ibid.
[94] ibid 36.
[95] ibid. However, the Human Rights Committee General Comment No 36 (2018) on art 6 of the International Covenant on Civil and Political Rights, on the right to life, para 70 embraces something close to Haque's position when it states that '[s]tate parties engaged in acts of aggression as defined in international law, resulting in deprivation of life, violate ipso facto article 6 of the Covenant'. The Comment is not binding and somewhat controversial, but it might be indicative of a trend towards that position in international law.
[96] Haque, *Law and Morality at War* (n 1) 31.

undertaken during an unlawful use of force—we can only say that the killings are *immunized* by international law; merely tolerated, but not authorized.

This interpretation is also controversial, of course, as it risks collapsing the independence of *jus in bello* and *jus ad bellum*. But even if it were correct, this interpretation can only claim that killings in pursuit of a *legal* use of force find their authorization in the *jus ad bellum*, whereas killings in pursuit of an *illegal* use of force—most notably, a war of aggression—will be merely immunized by the *jus in bello* and only tolerated by international law. The problem, however, is that what constitutes a *legal* war under international law is not necessarily the same as what makes a war *just*. Most obviously, an armed conflict authorized by the United Nations Security Council under Chapter VII of the UN Charter can perfectly well be an authorization to conduct an unjust war. The same is true of a war in self-defence. Even if we concede that it has a just cause, there are other reasons (eg stricter conceptions of proportionality, lack of chance of success, etc.) why said war might be unjust. Yet in these cases, international law will be not only immunizing or tolerating the murder of just combatants at the hands of unjust combatants but *authorizing* them through the application of the *jus ad bellum*. And this, in fact, is as big a chasm between international law and the revisionist morality as there could be.

As a result, even if we were to accept a reading of the laws of war that incorporates human rights law and the laws on the resort to force, at best we would be able to claim that arbitrary deprivations of life include the deaths of combatants who are fighting *legal* wars at the hands of those fighting *illegal* wars. This claim, however, would still leave open the possibility that the legal war is unjust. Not even these controversial interpretations of international law would be able to support the idea that international law merely fails to prohibit the killings of just combatants at the hands of unjust combatants. Sometimes, under these interpretations, international law would be legitimizing the latter through the *jus ad bellum*.

Fourth, an immunity from prosecution points to a mere matter of enforcement. But this seems problematic in the case of the legal rules on killing combatants in war. Perhaps a helpful way to see why is to consider Dan-Cohen's well-known idea of 'acoustic separation' and his distinction between decision rules and conduct rules: decision rules are rules addressed to officials, whereas conduct rules are those addressed to the general public.[97] In order to classify a rule,

[97] Meir Dan-Cohen, 'Decision Rules and Conduct Rules: On Acoustic Separation in Criminal Law' (1984) 97 Harvard Law Review 625, 627.

> we must conceive of the rule in the imaginary universe characterized by acoustic separation, and then decide—in light of the policies underlying the legal system—whether the rule would in that universe be a directive to the general public, to officials, or to both.[98]

If we understand the current rule as an immunity from prosecution, and as such, as a mere matter of enforcement, it ought to be understood as a decision rule. Therefore, in a world with perfect acoustic separation, combatants would not know such a rule exists. But this will not do. The current rule cannot simply be understood as a decision rule; it is a conduct rule. Without it, soldiers would not know how to behave in the battlefield; commanders would not be able to make targeting decisions. That combatants may kill enemy combatants during wartime is a fundamental rule around which the whole system of IHL articulates itself, by defining permissible and prohibited targets, collateral damage, proportionality, and so on, around it. Thus, understanding the rule as a Hohfeldian privilege and as a Kelsenian positive permission is a better characterization of our current rules of war than understanding them as a mere immunity. Consider, on the other hand, what happens if we undertake the same imaginary exercise with diplomatic immunity (or with any other bar to prosecution). If there were perfect acoustic separation between officials and the targets of the rule, the latter would still know how to behave. Diplomats have, indeed, a legal duty to respect the rules of the receiving state, and this is the rule that they ought to follow; this is their conduct rule. Their immunity, on the other hand, works perfectly well as a decision rule. The same is true of other bars to prosecution, such as the statute of limitations, lack of jurisdiction, and so on.[99]

This distinction can also be appreciated if we think about the fact that armies have manuals of operations and engagement rules, which are directed at telling combatants what they may and may not do during armed conflict. By contrast, we would react with a certain degree of shock if we found out that states had manuals for their diplomats instructing them on how to 'permissibly' violate the laws of the host state.

[98] ibid 631.

[99] 'When a person has a substantive defense in law to an action or prosecution, that is to say a legal defense on the merits, he has a liberty, i.e., the conduct of which complaint is made is not a breach of duty. This is not necessarily true of a merely procedural or adjectival defense. The defense under statutes of limitation, for example, generally does not deny the duty, but alleges that the duty has become unenforceable through lapse of time.' Williams (n 4) 1129.

Finally, I fear that there is no meaningful way in which we can understand the laws of armed conflict as merely failing to prohibit certain things, and thus, I do not think that we can claim that the acts that are not prohibited by the laws of war—such as the killings of combatants—are merely 'tolerated'.[100]

I agree with Haque that the killing of combatants is not prohibited by international law. But international law also consistently speaks of 'lawful acts of war',[101] acts that 'are not unlawful under humanitarian law',[102] and 'lawful targets'.[103] All these 'lawful' acts include the intentional targeting and killing of combatants,[104] and the language here does not suggest merely the absence of a prohibition but a permission exceptional to times of war. Now, I do not think that much should turn on this issue of language. Haque thinks international law merely 'tolerates'[105] the killings of combatants, and he is surely right in suggesting that the *jus in bello* proceeds with certain (moral) reluctance to regulate what in fact it regulates (homicide). But I do not think that from this reluctance we can read the mere absence of a prohibition into the laws of war, nor can we ascribe to it the significance that Haque ascribes to it.

The fact that international law fails to prohibit the targeting of combatants is far from being an oversight of the legal regime. On the contrary, it is decidedly intentional. We cannot just ignore this. It is a main feature of the regime that combatants are permissible—*lawful*—targets. These killings are not international crimes; they cannot be prosecuted domestically. If what we find is only the absence of a prohibition—and I have argued against this interpretation—it is a prohibition that is absent entirely *by design*. It seems implausible that a mere matter of legislative technique could have such a significant impact on the moral assessment of our legal regime.

The above is related to, but not the same as, Lieblich's call to look at the effects of IHL and how, even if we think of it as formally prohibitive, we must concede that at the very least it has a facilitative function.[106] I believe Lieblich is certainly right in focusing on how the laws of armed conflict are facilitative of war in various ways, particularly pertaining to the 'real-world' effects of combatants' immunity from prosecution,[107] and how said immunity is remarkably

[100] Haque, *Law and Morality at War* (n 1) 31.

[101] 2020 Commentary to GCIII, Article 2: Application of the Convention, para 272; Article 5: Beginning and End of Application, para 1114, 3634.

[102] 2020 Commentary to GCIII, Article 3: Conflicts Not of an International Character, para 565.

[103] 2020 Commentary to GCIII, Article 3: Conflicts Not of an International Character, para 784.

[104] 1987 Commentary to API, Article 48: Basic Rule, para 1874. 'As regards military objectives, these include the armed forces and their installations and transports.'

[105] Haque, *Law and Morality at War* (n 1) 31.

[106] Eliav Lieblich, 'The Facilitative Function of Jus in Bello' (2019) 30 European Journal of International Law 321.

[107] ibid 335.

close to a full-fledged right.[108] As Lieblich says, '[a] 'weak permission' is still a real-world permission'.[109] I would add that it cannot even be properly understood as a weak one, for the reasons already explained. Lieblich adds to this list of reasons by emphasizing the language in article 48(1) of API, which states that parties 'shall direct their operations only against military objectives'.[110] As I said before, military objectives include combatants.

In sum, as a matter of legal interpretation, it is doubtful that IHL merely contains the absence of a prohibition. At the very least, it contains a Hohfeldian privilege, and, sometimes, the language is much more suggestive of a permission. The most plausible interpretation supports the existence of what is closer to a strong permission. Even more problematic is that the legal regime cannot even be understood as granting a weak permission in what matters most to Haque (and to revisionists). At best we will be able to claim that international law weakly permits or fails to prohibit the killings of combatants in pursuit of an *illegal* war, while possibly authorizing—through the *jus ad bellum*—the killings of combatants in pursuit of a *legal* war. And a legal war can be an unjust one.

I want to return now to the third point made by Haque, which is about the moral significance of the absence of a strong permission to fight (and kill) in international law. Haque seems to think that proving that combatants are only granted a legal immunity—and not a strong permission—makes the question of justification of the symmetry thesis significantly easier. He thinks that lawful combatants' immunity rests on 'prosaic considerations of treaty and custom, reciprocity and impartiality, marginal incentives and aggregate consequences' and that although such prosaic considerations may be too weak to morally justify a legal permission to kill in pursuit of an unjust cause, they might be strong enough to ground an immunity from prosecution.[111] Bars to prosecution, argues Haque, make relatively modest moral claims, as opposed to justifications and excuses.[112] Haque's position rests on Raz's conceptual view about law and the correspondence that Haque thinks follows between the form of legal and moral norms:[113] if the law is not claiming a symmetrical legal privilege, it is

[108] I would call it a 'privilege', however. ibid.

[109] ibid 330.

[110] ibid.

[111] Haque, *Law and Morality at War* (n 1) 28.

[112] ibid.

[113] ibid 28–29, 55. 'On many jurisprudential views, including my own, law necessarily makes moral claims. On such views, a legal right is a (putative) moral right claimed by law. In some cases, the law purports to identify and protect a pre-legal moral right, such as the right not to be tortured. In other cases, the law purports to create a new moral right, such as the right to vote, by changing the moral reasons that apply to those subject to the laws. On such views, if the law permits some actor to perform some act, then the law claims that actor is morally permitted to perform that act. In particular, if international law

then not claiming a moral privilege, and thus, the symmetry is easier to justify. I disagree.

First, as I have already said, I do not think that the law grants combatants merely a weak permission and an immunity from prosecution. But second, whatever justifies an immunity from prosecution need not be prosaic considerations. The literature on diplomatic immunity readily shows the troubling implications of such immunities and the nature and extent of the injustices they allow.[114] Now, I think it is true that sometimes an immunity from prosecution might be easier to justify than a legal permission. The important point, however, is that although one of them might be slightly easier to justify than the other, *neither* of them might be easy to justify at all. In the case of the laws of war, even if a symmetrical legal immunity from prosecution is a little less problematic than a symmetrical legal privilege, I don't think that prosaic considerations are enough to justify immunities that allow for massive death and destruction to go completely unpunished. Even if we think that many unjust combatants are fully excused, and thus should not be punished, their actions remain wrong, in a fact-relative sense. That is, in a fact-relative sense, they are still killing the innocent. Further, there are many mid-level and higher-level combatants who do not plausibly have full excuses, and who can effectively use (fully excused) individuals to pursue unjust ends. The question of why we would provide an *ex-ante* blanket immunity regarding such behaviour remains. It is thus a mistake to think that if what combatants have is only an immunity from prosecution, the problem of moral justification of the laws of war disappears or loses its urgency.

Although Haque rightfully acknowledges that the legal immunity of combatants carries a moral cost,[115] I think that moral cost is much more significant than Haque allows. This immunity is, in practice, a blanket immunity for what under revisionism amounts to murdering the innocent. It behaves, as Lieblich says, as a 'real-life'[116] permission to commit murder on a vast scale.

confers a symmetrical legal permission to fight then it asserts a symmetrical moral permission to fight. Indeed, if international law did not make such moral claims, then conventionalists would not consider it part of the war convention that conventionalists seek to interpret and defend.

Thankfully, on my view, international law does not claim that killing in pursuit of an unjust cause is morally permissible, that combatants are exempt from ordinary moral constraints, or that combatants are always or even typically justified or excused.' ibid 29.

[114] *See* Veronica L Maginnis, 'Limiting Diplomatic Immunity: Lessons Learned From the 1946 Convention on the Privileges and Immunities of the United Nations' (2002) 28 Brooklyn Journal of International Law 989, 1005–09; Michael B McDonough, 'Privileged Outlaws: Diplomats, Crime and Immunity' (1996) 20 Suffolk Transnational Law Review 475.

[115] Haque, *Law and Morality at War* (n 1) 29.

[116] Lieblich (n 106) 330.

V. Conclusion

In this chapter I hope to have shown that combatants can be understood as privileged in three different ways. First, they have a Hohfeldian privilege to kill enemy combatants. Second, they enjoy a positive legal permission to do so. And, third, they belong to a favoured class of people because, unlike unlawful combatants and civilians, they cannot be prosecuted for their participation in hostilities. I also hope to have shown that understanding them as merely possessing an immunity from prosecution is incorrect.

Indeed, the two most plausible interpretations of our laws of war support the existence of a Hohfeldian privilege to kill combatants. The first does so by arguing that combatants have a Hohfeldian privilege to kill combatants. The second does so by arguing that both civilians *and* combatants have a Hohfeldian privilege to kill combatants. Most importantly, I hope to have shown that even if the laws of war only provided an immunity from prosecution (and a weak permission), it would be an immunity to commit what under revisionism is the equivalent of murdering the innocent on a vast scale.

Therefore, we are, in a way, where we began: with a legal rule—the symmetry thesis—that is, according to revisionists, the moral equivalent of a law that permits the murder of the innocent. If we cannot make the justificatory problem disappear by denying the existence of a legal permission to kill combatants, can we try instead to provide a moral justification for this divergence between law and morality? To this I turn next.

III

Regulating Murder

Instrumentalism and the Revisionist Morality

I. Introduction

In Chapter II, I argued that under international law combatants have a legal privilege to kill and target enemy combatants—not just a mere immunity from prosecution. This symmetrical legal privilege, which according to revisionists is a legal privilege to do what amounts to murdering the innocent, demands a moral justification. We cannot wish away the tension between the revisionist morality and our laws of war by denying the existence of a symmetrical permission to kill.

This tension is not easy to justify, not even if one thinks that what needs justification is only an immunity from prosecution and a weak permission. It is difficult to make moral sense of a legal regime that, by design, allows murder on a vast scale to go unpunished. What kind of justification could possibly be provided for laws that, according to revisionism, allow for the murder of the innocent?

Revisionists think that *only* combatants pursuing a just war have an objective moral justification to kill combatants pursuing an unjust one.[1] They think that unjust combatants not only lack a moral justification but are also forbidden to kill just combatants, for killing just combatants is just like killing the innocent (or worse).[2] Revisionists also think that unjust combatants are (morally) liable to be killed—that is, that they are not wronged when killed[3]—and that they are

[1] There are some exceptions to this generalization, as revisionists accept that combatants with a just cause might be acting wrongly by, for example, torturing others. *See* Victor Tadros, 'Orwell's Battle with Brittain: Vicarious Liability for Unjust Aggression' (2014) 42 Philosophy & Public Affairs 42; Helen Frowe, *Defensive Killing* (1st edn, Oxford University Press 2014); Adil Ahmad Haque, *Law and Morality at War* (Oxford University Press 2017); Cécile Fabre, *Cosmopolitan War* (1st edn, Oxford University Press 2012); Jeff McMahan, *Killing in War* (Oxford University Press 2009).

[2] For instance, as mentioned before, Tadros has suggested that killing just combatants might be morally worse than killing civilians, since, unlike civilians, just combatants are contributing to the achievement of justice. *See* Victor Tadros, 'The Moral Distinction Between Combatants and Noncombatants: Vulnerable and Defenceless' (2018) 37 Law and Philosophy 289.

[3] In most cases, because they think unjust combatants have forfeited their rights.

The Morality of the Laws of War. Marcela Prieto Rudolphy, Oxford University Press. © Marcela Prieto Rudolphy 2023.
DOI: 10.1093/oso/9780192855473.003.0003

not allowed to defend themselves.[4] Some revisionists also think that there is no general moral prohibition on targeting civilians: it depends on whether civilians are morally responsible for an unjust threat of harm.[5]

Our laws of war, as we have seen, do not permit the intentional targeting of civilians: it is an absolute legal prohibition. And they do not distinguish between just and unjust combatants: all combatants are fair targets, and, in an international armed conflict, all combatants can be privileged in their killings if they comply with the legal requirements to be prisoners of war.

Hence, revisionists think that our legal rules are significantly divergent from morality. The current laws of war forbid what is morally permitted (targeting some civilians from the unjust side of the war) and permit what is morally forbidden (killing just combatants).

Despite this divergence between our laws of war and the morality of war, most revisionists don't think that the laws should be modified according to their conception of the morality of war.[6] Instead, some of them have endorsed a justification for our laws of armed conflict that is usually known as 'the humanitarian view',[7] and which finds a correlate in the preamble of many of our international treaties on war.[8] According to the humanitarian view, our laws of war are justified, even if they permit murder, in virtue of their good effects or their ability to achieve certain aims—in this case, in virtue of their ability to reduce more suffering than laws that would track the revisionist morality.

The reasons why some revisionists think that our laws of war must diverge from morality are consequentialist in nature. A law divergent from the morality

[4] Indeed, Tadros says that 'it is uncontroversial that combatants on the unjust side are liable to be harmed'. Tadros, 'Orwell's Battle with Brittain' (n 1) 55.

[5] On this discussion *see* McMahan, *Killing in War* (n 1); Victor Tadros, 'Duty and Liability' (2012) 24 Utilitas 259; Frowe (n 1); Seth Lazar, *Sparing Civilians* (1st edn, Oxford University Press 2015); Seth Lazar, 'The Responsibility Dilemma for Killing in War: A Review Essay' (2010) 38 Philosophy & Public Affairs 180.

[6] A notable exception is Victor Tadros. Victor Tadros, *To Do, to Die, to Reason Why: Individual Ethics in War* (Oxford University Press 2020).

[7] *See*, eg, Haque (n 1) 38.

[8] The St Petersburg Declaration, in its preamble, states that 'the progress of civilization should have the effect of alleviating as much as possible the calamities of war'. The preamble of the Fourth Hague Convention similarly notes that it is important to revise the general laws and customs of war, 'either with a view to defining them with greater precision or to confining them within such limits as would mitigate their severity as far as possible'. *See Declaration Renouncing the Use, in Time of War, of Explosive Projectiles Under 400 Grammes Weight* (hereinafter, St Petersburg Declaration), 29 November/11 December 1868, 1 American Journal of International Law Supp. 95 (1907), preamble; *Hague Convention (IV) Concerning the Laws and Customs of War on Land and the Regulations annexed thereto* (hereinafter, The Hague Convention IV), 18 October 1907, 36 Stat. 2277, 1 Bevans 631 (entered into force 26 January 1910), preamble; *Advisory Serv. on Int'l Humanitarian Law, Int'l Comm. of the Red Cross, What Is International Humanitarian Law?* (2004). <https://www.icrc.org/en/war-and-law> accessed 1 August 2022; Gabriella Blum, 'The Laws of War and the Lesser Evil,' (2010) 35 Yale Journal of International Law 1, 7.

of war, the argument goes, is morally preferable because it achieves better consequences than a law that tracks that morality more closely. I will refer to this justification of our rules as an instrumentalist one. In the ethics of war, instrumentalists are consequentialists about the laws of armed conflict (or about a set of them), but they are not consequentialists about the *morality* of war. The instrumentalist label aims to make this clear.[9]

But is this instrumentalist justification of our laws of war, premised on the revisionist morality of war, successful? In other words, can the humanitarian view justify a legal regime that, if revisionists are right, permits the murder of innocents on a vast scale? And if it can't, could a different instrumentalist justification succeed? These are the questions this chapter will try to answer.[10]

This chapter assumes then that the revisionist morality of war is correct. Both the humanitarian view and the instrumentalist account I will propose must contend with this assumption. In Part 2 of this book, I will move on to argue that the revisionist morality itself is, in many ways, mistaken. But because any plausible contestation of the revisionist morality will have significant limits—at the very least, some killing done by the unjust side of a war will be very close to (and sometimes exactly like) murdering the innocent—the instrumentalist justification I will propose in this chapter will remain necessary and important, even after having contested the revisionist morality. Any justification of the legal privilege of combatants to kill must be able to account for the fact that an unjust war is a moral catastrophe that is regulated by our laws in a permissive way—for the fact that at least some killing done by the unjust side of a war will be the equivalent of murder.

I will proceed as follows. In section II, I will say more about instrumentalism and describe the humanitarian view. In section III, I will argue that the humanitarian view fails as a justification of the symmetry thesis if revisionists are right about the morality of war.[11] However, I will argue that the failure of the humanitarian view does not provide a conclusive argument against

[9] *See* Victor Tadros, *The Ends of Harm: The Moral Foundations of Criminal Law* (Oxford University Press 2011) 13. We could also call them 'pragmatics' about the laws of war, but I find the 'instrumentalist' label more transparent.

[10] Given that an immunity is as hard or only slightly easier to justify than a privilege, both the humanitarian view and the instrumentalist justification I propose can also serve to justify an immunity from prosecution. Because I think that what combatants have is a legal privilege, I will mostly focus on justifying the latter.

[11] As I will explain below, strictly speaking an instrumentalist justification premised on the revisionist morality is only meant to justify the legal privilege granted to unjust combatants to target just ones. The legal privilege granted to just combatants to target unjust ones is non-instrumentally justified, for it mirrors the moral content of the revisionist morality. Because it is quite cumbersome to repeat each time 'the legal privilege granted to unjust combatants to target just ones', I will sometimes simply say 'the symmetry thesis', though this is undoubtedly imprecise.

instrumentalism itself. A stronger defence of instrumentalism can be made if we try to understand war and its regulation from the perspective of ideal and non-ideal theory. Section IV offers that defence. Yet, in section V, I suggest that even this case for instrumentalism, when premised on the revisionist morality, remains incredibly fragile and that this fragility can be attenuated only to the extent that we reconsider the revisionist morality itself.

II. Instrumentalism and the Humanitarian View

Instrumentalist accounts of our laws of war have not been the focus of as much attention as the morality of war. We have careful accounts of the morality of killing in war, but we still lack a similarly developed view on the justification of our legal regime.[12]

The problem for an instrumentalist position—and for the humanitarian view in particular—about the laws of war does not come entirely from instrumentalism itself, that is, from instrumentalism as a general position about the law or certain rules. In fact, I find some version of instrumentalism the most plausible justification of most of our legal rules. The problem doesn't come either just from the law's permission to do what is morally forbidden or from the law's failure to track some moral demands: some morally forbidden behaviours are simply not harmful enough to warrant legal prohibition, and some moral demands are simply not important enough to merit legal enshrinement. The problem is the combination of an instrumentalist justification of our laws of war with the revisionist morality, which tells us that killing just combatants is just like killing the innocent (or worse).[13] In other words, the problem is that an instrumentalist account premised on the revisionist morality of war has to provide a justification for legally permitting the massive infliction of death and grievous harm against individuals who are innocent.

This dissonance between law, which permits murdering the innocent, and morality, which forbids murdering the innocent, should thus trouble anyone who is not a thorough consequentialist. Although there is a wide range of

[12] With the exception of Haque's and Lamb's. *See* Haque (n 1); Antony Lamb, *Ethics and the Laws of War: The Moral Justification of Legal Norms* (Routledge 2013). As Frowe has noted, '[d]espite the recent surge of philosophical interest in the ethics of war, a lacuna persists between the ethics of war and the laws of war. Many moral philosophers defend claims that are, or at least appear to be, at odds with the current law of armed conflict (LOAC), and little has been said about whether or how we should reconcile the laws of war with our best moral theory'. Helen Frowe, 'Introduction' (2019) 129 Ethics 544, 544.

[13] As suggested by Tadros. *See* Tadros, 'The Moral Distinction Between Combatants and Noncombatants' (n 2).

jurisprudential views about the appropriate relationship between moral demands and the law that are non-consequentialist, most of them would accept that laws that are inconsistent with the moral prohibition on killing the innocent pose a justificatory challenge that is not easy to meet. I doubt we would be as cavalier about this dissonance in International Humanitarian Law (IHL) if we weren't so used to it. Most of us would be deeply troubled about leaving similarly important moral prohibitions—like the prohibition on torture—behind in favour of permissive regulatory regimes that would be better at reducing suffering.

Instrumentalists about the laws of war can be partially or completely instrumentalist. Most instrumentalists are instrumentalist about the legal permission of unjust combatants to target just ones, and, sometimes, also about the prohibition on targeting civilians. That is, they think the laws of war are non-instrumentally justified regarding the privilege to kill granted to just combatants, but instrumentally justified in extending that permission—which does not track a moral one—to unjust ones. And it is also instrumentally justified in prohibiting the intentional targeting of civilians who belong to the unjust side of the war. My focus will be on an instrumentally justified permission to target combatants. I think—but will not argue—that instrumentally justifying the prohibition against targeting civilians is easier.

There are three main instrumentalist justifications of the laws of war. First, Haque's service conception, which argues that the laws of war are justified insofar as they aim to get both just and unjust combatants to conform their actions as closely as possible to the constraints imposed by the deep morality of war.[14] Under this view, the law performs a service for combatants: it helps them conform to their moral obligations more closely than they would if they relied on their individual moral judgment.[15] Second, Lamb's and Draper's view, which posits that the laws of war are justified due to their ability at minimizing rights violations[16] or unjust harm.[17] Finally, what has been called the 'humanitarian view',[18] defended in different versions by Jeff McMahan and Henry Shue and Janina Dill, and which aims at reducing suffering.

[14] Jeff McMahan, 'The Ethics of Killing in War' (2004) 114 Ethics 693, 731; Haque (n 1) 53; Jeff McMahan, 'The Morality of War and The Law of War' in David Rodin and Henry Shue (eds), *Just and Unjust Warriors: The Moral and Legal Status of Soldiers* (Oxford University Press 2008) 35.

[15] Haque (n 1) 43.

[16] Lamb (n 12).

[17] Kai Draper, 'Double Effect and the Laws of War' in Saba Bazargan and Samuel C Rickless (eds), *The Ethics of War* (Oxford University Press 2017) 201.

[18] *See*, eg, Haque (n 1) 38.

As can be seen, these accounts disagree on which effects the laws of war should achieve for them to be instrumentally justified. They also differ in scope. While the humanitarian view aims to provide a justification for the prohibition on targeting civilians and the symmetry thesis, the other two only aim to provide a justification for the prohibition on targeting civilians and the rules about collateral damage.[19] Thus, most of my focus in this chapter will be on the humanitarian view and its attempt at justifying the symmetry thesis while also holding that killing just combatants is just like killing the innocent.

In general terms, we might call 'humanitarian' any view whose main aim is to justify IHL (in particular, the prohibition on targeting civilians and the symmetry thesis) on the basis of its ability to reduce suffering or limit the carnage of war.[20] The commitment to reducing suffering is explicitly stated in some of our international treaties on war.[21]

The humanitarian justification of IHL based on the revisionist morality makes something like the following claim: IHL is justified because it is better at reducing suffering than a legal regime more consistent with the content of the revisionist morality. The view is thus instrumentalist: it justifies the laws of war on account of their consequences (ie the reduction of suffering). And it is also a kind of lesser-evil justification: the current laws of war are the lesser evil compared to laws that would track the revisionist morality more closely.

The humanitarian view, although defended in slightly different versions by McMahan and Shue and Dill, has not been fully developed as a justification of IHL. Thus, part of my aim in the first part of this chapter will be to reconstruct and reinterpret it in its best version. This reconstruction is also necessary because the humanitarian view is often posited as a justification of IHL as a whole or as a set, which includes prohibitory norms (eg the prohibition on targeting civilians) and permissive norms (eg the symmetry thesis). But this wholesale justification, I will argue, fails unless we argue that the prohibitory aspects of the regime are connected to its permissive norms. This can only be contingently true.

[19] In the case of Haque, this is because he does not believe that our current laws of war provide a strong legal permission to target combatants. I have argued against his position in Chapter II.

[20] Haque (n 1) 38.

[21] *See* note 8.

A. The Humanitarian View

In some of his work, Jeff McMahan has defended a version of the humanitarian view.[22] He has suggested that the laws of war are conventions designed to limit and repair the breakdown of morality that has led to war and to mitigate its savagery.[23] McMahan thinks we can justify symmetrical rules for both just and unjust combatants in order to achieve those aims.

Regarding the prohibition on killing civilians, McMahan says that since most soldiers will believe that their cause is just,[24] all of them will be strongly disposed to kill civilians if they believe it is permissible to do so. Thus, McMahan argues, there might be a role in the laws of war for the principle of discrimination, even though it is a false criterion of moral liability to attack in war. The point of our laws, he says, is to mitigate or limit the violence of war, not to protect moral rights: having no conventions at all would be worse for everyone.[25]

McMahan also seems to believe that it would be pointless for the rules to be asymmetrical. Since most combatants believe their wars are just, whatever legal advantages are granted only to just combatants will be claimed by unjust ones. Therefore, he says, a rule granting a permission only to just combatants would be ineffective in constraining the unjust.[26] Additionally, McMahan suggests that the legal equality of combatants may fulfil other important aims: discouraging a moralized perception of war with its corresponding crusading mentality and promoting among combatants the view of themselves and their adversaries as professional warriors.[27] Finally, asymmetrical laws of war, especially those that entail criminal punishment for unjust combatants, would be most likely to provide incentives against surrendering and for abandoning all restraint so as to win the war and avoid punishment.[28]

The above leads McMahan to think that it is absolutely clear that the laws of war must diverge significantly from the morality of war. Most of these reasons, McMahan thinks, are largely pragmatic in nature.[29] But neither the symmetry thesis nor the principle of distinction has a foundation in basic morality.[30] It is

[22] In other passages, he seems more inclined towards Haque's service conception, and sometimes, towards reducing unjust harm.

[23] McMahan, 'The Morality of War and The Law of War' (n 14) 35; McMahan, 'The Ethics of Killing in War' (n 14) 730; McMahan, *Killing in War* (n 1) 107.

[24] McMahan, 'The Ethics of Killing in War' (n 14) 730; McMahan, 'The Morality of War and The Law of War' (n 14) 28, 35.

[25] McMahan, 'The Ethics of Killing in War' (n 14) 730; McMahan, *Killing in War* (n 1) 107.

[26] McMahan, 'The Morality of War and The Law of War' (n 14) 28; McMahan, *Killing in War* (n 1) 108.

[27] McMahan, 'The Morality of War and The Law of War' (n 14) 30.

[28] ibid.

[29] ibid 19.

[30] ibid 27.

just that an exclusive concern with consequences has its maximal plausibility in the regulation of war, in which the enormity of the possible consequences may seem to overshadow all other considerations.[31]

A similar view is held by Janina Dill and Henry Shue.[32] Although they characterize their view as 'not some disreputable form of consequentialist reasoning',[33] their view is surely instrumentalist. It aims to minimize all killing as much as possible, though sometimes they seem to be closer to arguing for the aim of protecting rights as much as possible (those of civilians). Importantly, Dill and Shue are not revisionists: they think that an individualized criterion of moral liability to harm is simply impossible to apply in war and that most combatants are not liable to be killed. Nevertheless, their defence of the humanitarian view is still worthy of examination because it incorporates elements different from McMahan's view and provides a defence of a set of laws that effectively permit people to murder the innocent (which in Dill and Shue's view includes both just and unjust combatants).

Dill and Shue argue that the laws of war should limit killing as much as possible.[34] Given that systematically protecting individual rights is not an achievable purpose for the laws of war, we should aim instead at general 'moral damage limitation': not straightforwardly doing what is right and avoiding what is wrong (which is killing only individuals who are liable) but avoiding as much wrong as possible, given the circumstances.[35]

This pragmatic normative framework, they argue, acknowledges two realities on which the laws of war are already premised. First, agents are committed to the pursuit of military success, and second, rules designed to make combat difficult might miss the morally valuable alternative of limiting harm.[36] Although the moral ambition of this approach is limited, Shue and Dill think that these workable and honest limits are preferable to others that might be false, unworkable, or hypocritical.[37]

[31] McMahan, *Killing in War* (n 1) 107.

[32] *See* Henry Shue, 'Laws of War' in Samantha Besson and John Tasioulas (eds), *The Philosophy of International Law* (Oxford University Press 2010); Janina Dill and Henry Shue, 'Limiting the Killing in War: Military Necessity and the St. Petersburg Assumption' (2012) 26 Ethics & International Affairs 311. Janina Dill later adopted a different view, whereby law has two moral tasks: to guide its subjects' conduct toward actions that conform to their moral obligations and to secure the fullest feasible protection of rights in the outcome of conduct. *See* Janina Dill, 'Towards a Moral Division of Labour between IHL and IHRL during the Conduct of Hostilities' in Ziv Bohrer, Helen Duffy, and Janina Dill (eds), *Law Applicable to Armed Conflict* (Cambridge University Press 2020) 198.

[33] Dill and Shue (n 32) 329.

[34] ibid, 319.

[35] ibid.

[36] ibid 323.

[37] ibid 324.

Puzzlingly, as I mentioned before, Shue and Dill think that their approach is not consequentialist.[38] What they advocate, they say, is a type of rule that is effective in protecting rights in the situation for which it is intended: war.[39] These rules, they argue, will produce better consequences than stricter rules: they accomplish the important goal of protecting the rights of civilians, even if they admittedly fail to protect the rights of combatants who are not morally liable to be killed.[40] This is not a consequentialist argument, they say, because the alternative they prefer is not better because it produces the maximum net good.[41] Thus, the claim is not that the rights or lives of civilians protected are worth more than the rights or lives of combatants who are not.[42] Rather, the claim is that protecting the rights of civilians is an achievable aim, whereas the rights of combatants are beyond the reach of a workable normative framework.[43] Thus, the purpose of the laws is to minimize the rights violated and the evils committed, given that war exists.[44] It is not the purpose of these rules to end the practice or to maintain it.[45] The practice of war is simply presupposed.[46]

III. Instrumentalism: The Appeal and the Objections

Instrumentalist justifications of the law are nothing new. They have been a continuous problem for legal philosophy, and attached to them comes the usual set of advantages and challenges.[47] But my main concern is not with the objections against instrumentalism itself but with those that apply more urgently to an instrumentalist justification of our laws of war premised on the existence of a non-consequentialist morality of war that pointedly prohibits what the law is permitting.

Probably one of the greatest advantages of instrumentalism is that it can accommodate moral disagreement better than non-instrumentalist justifications of our laws: it might be easier for us to agree on what consequences our laws should pursue than on what moral content they should mirror. Yet while this

[38] ibid 329.
[39] ibid.
[40] ibid.
[41] ibid.
[42] ibid.
[43] ibid.
[44] Shue (n 32) 29.
[45] ibid.
[46] ibid.
[47] *See* John Gardner, 'Tort Law and Its Theory' in John Tasioulas (ed), *The Cambridge Companion to Philosophy of Law* (Cambridge University Press 2020) 352.

might be true of the humanitarian view (it is easier for all of us to agree that harm is a bad thing that needs to be limited than to agree on a given morality of war), it is not true of instrumentalist views that set the aims of the law at inducing compliance with morality or reducing unjust harm or violations of rights.[48] Moral disagreement slams these views with full force, because we do not actually agree about which morality must be complied with, which harms are unjust, and what counts as rights violations. And we disagree not only about the content of morality but also about its retail application. So any instrumentalist view that requires us to agree on the morality of war will lack the advantage of an instrumentalist view, such as the humanitarian view, whose goals we might think of as 'morally thinner' (in the sense that it is easier to agree that harm is a bad thing that should be prevented than agreeing on the correct morality of war). Still, someone might wonder why we should prevent *all* harm rather than only *unjust* harm. In the end, the problem of moral disagreement is inescapable, both for instrumentalist and non-instrumentalist views. As a result, even if instrumentalist views, when morally thin, have an important advantage, I am not sure it makes their case.

The moral thinness of some instrumentalist views is also important in a second way. It is true that we disagree about morality, but other times, we simply lack clear moral answers to difficult questions, and presumably questions about killing in war are among the hardest. Yet the law needs to settle these questions anyway.[49] In these cases, an instrumentalist view like the humanitarian one can have an advantage over a view that requires us to provide a moral answer that right now we find harder (such as, who is liable to be killed in war). This is not to imply that we must abandon moral inquiry and settle for any legal response. It just means that law sometimes must venture into territories where we have not yet found moral answers. And war may be one of these instances.

Yet despite its appeal, an instrumentalist account that calls for divergence between morality and our laws, if we believe that divergence to entail a legal permission to wrongfully kill innocents (as revisionists do), still suffers from many flaws.

There are usually four main objections against instrumentalism. First, instrumentally justified rules cannot offer categorical support for the categorical rules they aim to justify.[50] Instrumentally justified rules are justified only

[48] Jeremy Waldron makes this argument regarding Haque's view. *See* Jeremy Waldron, 'Responses to Zedner, Haque and Mendus' (2014) 8 Criminal Law and Philosophy 135, 137–39.

[49] *See* Jeremy Waldron, *Torture, Terror, and Trade-Offs: Philosophy for the White House* (Oxford University Press 2010) Chapter IV; Waldron, 'Responses to Zedner, Haque and Mendus' (n 48).

[50] Christopher Kutz, *On War and Democracy* (Princeton University Press 2016) 51.

insofar as it is empirically true that following them achieves their underlying aims to a larger extent than following a different rule.[51] Yet even the best formulated rule must eventually come up against a case where the set of aims it pursues are better achieved by breaking it.[52] Second, if we want the rules to be perceived as categorical, yet the rules are instrumentally justified, we must hide the justification from those who must obey the rules.[53] In other words, the purpose of the rules needs to be opaque, for a combatant who knows what those purposes are will be inclined to engage in first-level reasoning about the optimal course of action rather than on following the rule. Indeed, any moral agent who knows what the purpose of any given rule is would have reasons to ask herself not whether her future conduct is permitted or forbidden by the rule itself but whether following the rule would further the law's moral aims. Presumably because we would not like that to happen, the purposes of the rules should be kept from those at whom the rules are directed—in this case, combatants—so as to make them *think* that the rules are categorical in nature. If we want combatants to be sufficiently motivated to obey these rules, they will have to believe that these rules *do* constitute the deep morality of war; that the prohibition on targeting civilians is an absolute moral prohibition.[54]

These two objections do not pose a major problem for the humanitarian view's defence of the unjust combatant's legal privilege to target just combatants. After all, if unjust combatants know that the purpose of the rules—reducing suffering—will be achieved to a greater extent by not doing what the law allows, and accordingly, decide against killing just combatants, so much the better for the instrumentalist justification of the rule and its underlying morality.[55] The objections are nevertheless unsettling for the humanitarian view as a justification of the legal regime in general, which includes rules that are prohibitory in nature and do not have exceptions, such as the prohibition on killing civilians, *hors de combat*, and prisoners of war. I will not discuss these issues here.

[51] *See* Waldron, *Torture, Terror, and Trade-Offs* (n 49) 51; Kutz (n 50) 52.

[52] Gardner (n 47) 15.

[53] Kutz (n 50).

[54] Both McMahan and Tadros consider this possibility and are favourable to it (though McMahan is more uncertain). *See* Tadros, 'Orwell's Battle with Brittain' (n 1); McMahan, 'The Ethics of Killing in War' (n 14) 732.

[55] This, of course, obviates the fact that in most cases, combatants are under a legal duty to target enemy combatants because they have a legal duty to follow orders. *See* Christopher J Finlay, 'Justification and Legitimacy at War: On the Sources of Moral Guidance for Soldiers' (2019) 129 Ethics 576, 580–81. Discussing this problem in the context of co-belligerency, *see* Marcela Prieto Rudolphy, 'Who is at War? On the Question of Co-belligerency' (forthcoming) Yearbook of International Humanitarian Law.

The third and fourth objections against instrumentalism do apply, however, to the humanitarian view's justification of the unjust combatant's legal privilege. The third objection is that we don't actually know whether the rules we have are the best rules to achieve our chosen aims or whether they are simply *any* good at all. Kahn has suggested that there seems to be no direct correlation between complying with international rules and minimizing suffering.[56] In the end, the strength of instrumentalist justifications lies on empirical proof that we currently lack. We lack proof of how much the current rules can reduce suffering, how they fare compared to other alternatives, how much non-compliance is required for us to move from prohibiting immoral conduct to regulating it, and so on. And this lack of definite proof is also one of their weaknesses, for if we are going to allow thousands of innocents to be killed, we'd better be pretty sure that it is going to reduce suffering or achieve whatever aims we deem worthy. Yet we are not, and it is unclear how—and if—we could ever reach a level of certainty commensurate with the amount of wrongdoing we are permitting.

The final objection against instrumentalist justifications is that they don't take rights—either legal or moral ones—seriously. They are in tension with what underpins most deontological theories, for they fail to 'give full expression to the deontological insight that individual rights are not interchangeable units of agent-neutral value'.[57] The problem, again, is to instrumentally justify a legal rule that permits seriously wrongful conduct—like killing and maiming—against innocent just combatants, that is, against those who have a moral right not to be treated in that way, in order to achieve a set of desired consequences.

This objection has force because the humanitarian view is premised, precisely, on a non-consequentialist morality of war. In other words, the objection about rights wouldn't be a problem if we thought that the morality of war itself was purely consequentialist.[58] But if we are to take seriously the sacredness of human life and the separability of persons, as non-consequentialists do, we should have something more to say to just combatants other than 'you see, we have these rules that permit others to kill and maim you, but not to worry, these

[56] Paul W Kahn, 'Lessons for International Law from the Gulf War' (1992) 45 Stanford Law Review 425, 436.

[57] Jonathan Quong and Rebecca Stone, 'Rules and Rights' in David Sobel, Peter Vallentyne, and Steven Wall (eds), *Oxford Studies in Political Philosophy, Volume 1* (Oxford University Press 2015) 230–31. See also Frances M Kamm, 'Non-Consequentialism, the Person as an End-in-Itself, and the Significance of Status' (1992) 21 Philosophy & Public Affairs 354; Robert Nozick, *Anarchy, Utopia and State* (Basil Blackwell 1974).

[58] Which is Richard Brandt's case. *See* Richard B Brandt, 'Utilitarianism and the Rules of War' (1972) 1 Philosophy & Public Affairs 145.

rules have very good consequences. Not for *you*, perhaps, but for the rest of us, they are, overall, very good.' We should be able to say something that speaks to each of them and their standpoint, and not just to the combination of all our interests.[59]

Of course, some of our current laws of war do benefit just combatants: regulations on the treatment of prisoners of war, prohibitions of certain weapons, and the prohibition on killing *hors de combat* are beneficial to all combatants, just and unjust ones.[60] And, perhaps more importantly, what benefits all of us, and what is indeed a significant achievement, is that there are laws of war at all, and that they aim at things like reducing suffering. But it is difficult to understand how the rules that permit targeting combatants can be of any benefit to just combatants. It sounds very much like saying to a victim of murder that regulating murder rather than prohibiting it makes all murders less cruel, and so it also benefits her. The victim would rightly complain that no-one should be legally allowed to murder her in the first place.

This is just a manifestation of the very old and very difficult problem of having to draw a line in public policy that defines when the end, however valuable, stops justifying the means.[61] Indeed, letting go of our most fundamental moral prohibitions against killing the innocent in order to achieve some consequences, such as the reduction of suffering, is deeply troubling, to say the least. Or, as Lazar says, then 'morality is a truly irrelevant constraint on the legal enterprise'.[62]

And yet there is still a way in which instrumentalism may be defended. Rules of war must be able to function as rules *for war*. As every other rule, rules of war need to assume some non-compliance, but they also need to be administrable during combat, and they need to be learnable by those who will apply them. This includes infantry soldiers, artillery soldiers, and so on, as well as their superiors, from platoon sergeants to those in charge of full companies. It is not just that most laws of war are self-applied. Even if they weren't, legal rules cannot establish standards that are impossible to apply by those who will apply them.[63] They also need to be learnable by those who need to learn them.[64] Thus,

[59] Thomas Nagel, 'Ruthlessness in Public Life' in Thomas Nagel, *Mortal Questions* (Cambridge University Press 2012) 75, 86.

[60] *See* Yitzhak Benbaji, 'The War Convention and the Moral Division of Labour' (2009) 59 The Philosophical Quarterly 593.

[61] Nagel, 'Ruthlessness in Public Life' (n 59) 89.

[62] Seth Lazar, 'The Morality and Law of War' in Andrei Marmor (ed), *The Routledge Companion to Philosophy of Law* (Routledge 2012) 367.

[63] Making these points in much more depth, *see* Jeremy Waldron, 'Deep Morality and The Laws of War' in Seth Lazar and Helen Frowe (eds), *The Oxford Handbook of Ethics of War* (2015) 85–87.

[64] ibid.

even if we agreed on morality's content, this morality might be too fine-grained to be a workable legal rule.[65]

Legal rules have particular standards to meet—what we might call 'considerations of legal technicality'[66]—that moral rules do not. As David Luban points out, law has three idiosyncratic general characteristics: its need for binary, on/off answers to complex qualitative questions (Is this an armed conflict? Is this person a civilian?); its package or bundled character (if it is an armed conflict, then IHL applies; if it is a civilian, then civilian protections apply); and the detachment of legal propositions from the reasoning that justifies them.[67] Thus, any workable legal rule, even one that closely mirrors the morality of war, will stray away from that morality in order to be precisely what is meant to be: a workable legal rule.

Legal rules for war, then, must be a sound instrument of their own moral purpose: they must meet what we can think of as a 'weak instrumentalist requirement'. This entails that laws, even those that mirror some moral content closely, mustn't be directly or indirectly self-defeating.[68] A law that, for example, prohibits murder while also providing perverse incentives that encourage people to commit murder is indirectly self-defeating.[69] And a law that, in aiming to reduce accidents, imposes a restriction on speed that when complied with causes more accidents than before is directly self-defeating.[70]

The humanitarian view's defence of the symmetry thesis could then be understood from the perspective of the weak instrumentalist requirement. It could be interpreted as saying that rules that would track the revisionist morality of war would be a lousy instrument of their own moral purpose; that they would, for example, encourage people to violate the law and thus, would give way to more suffering. To avoid this, law and morality should diverge, so that the law's moral purpose is achieved.

It is probably true that laws of war that tracked the revisionist morality of war without any modifications would be indirectly or directly self-defeating. And they would also be incredibly difficult to apply. This would in turn lead to some—necessary—divergence between the morality of war and the laws of war. However, the problem with the humanitarian view premised on the revisionist

[65] ibid.

[66] ibid 86–87; Brandt (n 58) 163.

[67] David Luban, 'Just War Theory and the Laws of War as Nonidentical Twins' (2017) 31 Ethics & International Affairs 433, 434, 438.

[68] John Gardner, 'What Is Tort Law For? Part 1. The Place of Corrective Justice' (2011) 30 Law and Philosophy 1, 20–26.

[69] ibid 20.

[70] ibid 20–26.

morality of war is that the divergence between law and morality goes well beyond a plausible understanding of the weak instrumentalist requirement.

Think about the law concerning murder that provides incentives for people to commit murder (and that is, then, indirectly self-defeating). This law would certainly fail the weak instrumentalist requirement; we would have to reform it. But we would not change the legal rule that tracks or mirrors the moral prohibition to commit murder. That moral rule is precisely what gives the legal rule its moral purpose. In other words, the moral purpose of the rule stays fixed: that is what makes the instrumentalist requirement a weak one. And so, if in order to be a sound instrument of their own moral purpose (ie in order to meet the weak instrumentalist requirement), the laws of war must be different from the morality of war, those differences cannot extend to the moral purpose itself. Their original moral purpose, given by the moral prohibition of unjust combatants to kill just ones, must be maintained. They cannot leave *that* moral purpose behind. Once they do, we are outside of the scope of the weak instrumentalist requirement.

Revisionists who defend the humanitarian view could perhaps argue that laws that reflected the moral disparity between just and unjust combatants would have no effect in restraining conduct during war, for they would not be complied with. It is not that they would be directly or indirectly self-defeating. They would simply have no effect whatsoever. Thus, they could say, we would be better off with symmetrical rules that would reduce suffering and would be complied with.

The problems related to non-compliance are, of course, extremely important in understanding why law and morality must diverge. And, indeed, part of the reasons why we find ourselves torn between the current laws of war and laws that would track the morality of war is due to non-compliance with the rules on resort to war. But non-compliance is pervasive in our society: people don't even comply with the rule that prohibits murder. In fact, close to half a million people were murdered in 2016.[71] We, however, do not actually regulate murder in a permissive way to make it less cruel: we forbid it.

Perhaps the argument that underlies the humanitarian view is that non-compliance can be so huge as to render the legal prohibition useless or self-defeating. In those cases, it would be better to regulate a conduct rather than to prohibit it. In the case of the rules about murder, we are imagining a world where there would be fewer murders if we chose to regulate murder to make

[71] Max Roser, 'Homicides' *(Our World in Data* 2019) <https://ourworldindata.org/homicides#sources-homicide-data-in-the-post-war-time> accessed 17 August 2022.

it more humane rather than to prohibit it. Yet, moving from prohibition to regulation changes the moral purpose of the legal rule or, at the very least, it catastrophically fails to capture it. In non-consequentialist views—such as the revisionist morality of war—the moral prohibition on murder is not really an agent-neutral rule concerned with the number of murders there are in the world:[72] it is concerned with what agents do to other people.

In a similar vein, Rodin sees in regulating war nothing but an abdication to immorality. [73] He thinks that regulating war to limit its damage simply because we are unable to do what we ought to do—which is to eliminate war altogether—is tempting but cannot be right.[74] This kind of thinking, says Rodin, is morally equivalent to endorsing the regulation of slavery.[75] Although today there are more enslaved persons in absolute terms than before its prohibition, no-one supposes that because slave-holding will probably never be eradicated from Earth, we should implement a permissive regulatory regime that implicitly recognizes its legitimacy.[76] This is true, Rodin says, of most criminal practices: it is most likely that practices such as rape and murder will never be eliminated,[77] yet no-one thinks law should mitigate their savagery by regulating them at least partially through legal permissions. Perhaps, he concedes, the prohibition of these practices protects the rights of victims more effectively than a regime of permissive regulation, which is an expectation we do not have in warfare.[78] But Rodin thinks that this response, which turns on the consequences of prohibition compared to the consequences of permissive regulation, entirely misrepresents the moral objection to a regulatory regime for something like slavery, or even most criminal practices. When fundamental rights are at stake, the content of legal norms cannot be simply determined by consequences: the content of rights sets basic constraints on the content of the law.[79]

[72] As Thomas Nagel explains, both agent-neutral and agent-relative reasons are objective if they can be understood and affirmed outside the viewpoint of the individual who has them. Neutral reasons underlie judgments about what should happen, whereas relative reasons can affect judgments on what people may do. Deontological reasons are a type of agent-relative reason, and they stem from the claims of other persons not to be harmed in certain ways. They limit what we are permitted to do in the service of impersonal or autonomous values. Deontological (and more generally, agent-relative) reasons are objective because they can be affirmed outside the viewpoint of the individual who has them. *See* Thomas Nagel, *The View from Nowhere* (Oxford University Press 1989) 152–85.

[73] David Rodin, 'Morality and Law in War' in Hew Strachan and Sibylle Scheipers (eds), *The Changing Character of War* (Oxford University Press 2011) 459–60.

[74] ibid.

[75] ibid.

[76] ibid.

[77] ibid.

[78] ibid.

[79] ibid 459.

Kant also held a similar position.[80] He argued that there can be no dispute between politics and morality, unless one were to regard morality as a doctrine of prudence, which would be equal to denying that morality exists at all.[81] The rights of humankind, he says, must be held sacred whatever they cost to those in power:

> Here one cannot go halfway, cooking up hybrid, pragmatically-conditioned rights (which are somewhere between the right and the expedient); instead, all politics must bend its knee before morality, and by doing so, it can hope to reach, though but gradually, the stage where it will shine in light perpetual.[82]

Both Rodin and Kant make compelling arguments. If we have moral rights, it is hard to accept that a legal rule that allows for their violation on a massive scale on account of (possibly) achieving some consequences can ever be morally justified. After all, just combatants bear most of the brunt of this arrangement. For, in this case, the humanitarian view aims to justify a legal permission to murder innocents (just combatants) so that we minimize the suffering and damage caused by war. Yet by its own lights, this view aims to minimize everyone's suffering at the expense of the moral rights of just combatants. Its aims are in tension with the very same morality revisionists endorse. This is partly because the goal of reducing suffering seems to be aggregative in nature: it seems to be concerned with the total addition of suffering rather than with who suffers and to what extent. There seems to be no concern about the distribution of that suffering. This is not because instrumentalist views cannot incorporate distributive concerns coherently.[83] It is only because the humanitarian view doesn't, and it remains unclear whether it could incorporate them and remain an attractive justification. After all, to justify the symmetry thesis the humanitarian view must redistribute harm and suffering by burdening those who are morally innocent.

[80] Kant has been read as a committed pacifist, opposed to the regulation of war to the point of not even having a just war theory. However, Orend has argued against this interpretation of Kant, pointing out that he did have a just war theory, although he was opposed to the existence of standing armies. *See* Brian Orend, 'Kant's Just War Theory' (1999) 37 Journal of the History of Philosophy 323.

[81] Immanuel Kant, 'To Perpetual Peace. A Philosophical Sketch' *in Perpetual Peace and Other Essays* (Hackett Publishing Company 1983) 8:370.

[82] ibid 8:380.

[83] On utilitarianism, consequentialism, and distributivism *see* John Rawls, *A Theory of Justice* (Belknap Press of Harvard University Press 1999) 22–27; Samuel Scheffler, *The Rejection of Consequentialism: A Philosophical Investigation of the Considerations Underlying Rival Moral Conceptions* (Clarendon Press 1995) 20–35.

Laws that engage in the trade-off of significant moral rights against harm and death with the aim of reducing suffering pose thus an urgent justificatory challenge for those of us who hold non-consequentialist moral views or who think that moral rights such as the right to life should act as constraints on the legal enterprise. This is true even if we focus on the humanitarian view as a lesser-evil justification, whereby the current laws of war are the lesser evil compared to laws that would track the revisionist morality more closely.

In the ethics of war, lesser-evil justifications are accepted as justifications to impose harm on the innocent when harming the innocent brings about a significantly better outcome than not doing so.[84] Lesser-evil justifications proceed by assessing the badness of two states of affairs, and opting for the less bad one,[85] but they are also sensitive to deontic considerations, such as the disparity between allowing harm and inflicting harm.[86] When a lesser justification is available, the rights of the innocent (ie the non-liable) are infringed, but not violated—they are overridden.

There are some difficulties in thinking about the humanitarian view as a lesser-evil justification. First, lesser-evil justifications are often understood as justifications that certain agents might have to inflict harm on the non-liable, that is, they are concerned with an actor's conduct. They are thus not immediately nor obviously applicable to justify a legal regime, whereby what needs to be justified is not only, nor primarily, its adoption but its content. Second, it is relatively easier to assess the harms at stake in individual actions: they are limited to a given instance such as when one must choose between killing the one to save the many. But a lesser-evil justification for a legal regime is not time-limited in this way: it must justify a legal regime that will allow for multiple instantiations of harm, often for an unknown period of time, until the regime is modified. There is thus significant room for debate as to what the correct application of a lesser-evil justification to the contents of a given legal regime looks like or if it is possible at all to provide one.

One of the most pressing difficulties for the humanitarian view understood as a lesser-evil justification is, thus, as follows. Most views that accept lesser evil as a justification require a significant disparity between the two evils that are balanced in order to prefer one over the other. Some accept a 1:10 ratio, whereby it is justified to intentionally kill one innocent to save 10 others from death.[87] The ratio in the case we are concerned with might be similarly or less

[84] *See* Haque (n 1) 7; David Rodin, 'Justifying Harm' (2011) 122 Ethics 73, 74.

[85] Kerah Gordon-Solmon, 'How (and How Not) to Defend Lesser-Evil Options' [2022] Journal of Moral Philosophy 1, 3.

[86] ibid.

[87] Saba Barzagan, 'Killing Minimally Responsible Threats' (2014) 125 Ethics 114, 126–27.

demanding, given that the laws of war allow harm, rather than inflict it, and allowing harm is generally easier to justify than inflicting harm.[88]

In choosing between two alternative legal regimes (eg IHL as it is and IHL that tracks the revisionist morality (IHLR for short)) we should thus prefer IHL only if something close to the 1:10 ratio holds; that is, IHL will be the lesser evil only if it is true that it brings about 10 times less suffering than IHLR. This seems highly unlikely, and incredibly difficult to confirm empirically. Further, we again face the question of why we would prefer a regime that reduces suffering by burdening those who are innocent.

An even more pressing objection against the humanitarian view is that it remains obscure how it could instrumentally justify a symmetrical legal privilege to kill simply because it is hard to see how a legal *permission to kill* people could reduce *any* suffering at all. It is easy to see how the prohibition on targeting civilians might do so: it introduces constraints on actions that cause immense suffering. But a legal privilege to target combatants, in and of itself, does not introduce any constraints at all: it is a permission to inflict suffering. There is thus a clear disconnect between the aims the view purports to have(reduce suffering) and the things that the view actually justifies (a permission to inflict suffering and death).

Therefore, for the humanitarian view to be successful, it needs to argue that the different dimensions of our laws of war fall or stand together. In particular, it needs to argue that the prohibition on targeting civilians, which might indeed reduce suffering (though whether it meets the 1:10 ratio is dubious), necessarily entails or must be accompanied by a permission to target just combatants. As I will argue later, this is, at the very least, not logically true.

Ultimately, I don't think there can be a proper solution to the tension between the revisionist morality and the symmetry thesis.[89] The humanitarian view cannot be held together with the revisionist morality of war. We have to do away with one of them. But, perhaps, there is a different way to solve this tension between the revisionist morality of war and our laws. A richer account of instrumentalism might be successful. Such an account might not be entirely inconsistent with the humanitarian view, but it must give a more satisfactory answer to the question of how a legal regime that permits killing the innocent can be justified. The perspective I will propose invites us to take a step back and reflect on how we come to think about war in the first place. By doing so, we

[88] ibid 126.

[89] Others who agree with this position are Thorburn and Benbaji and Statman. *See* Malcolm Thorburn, 'Soldiers as Public Officials: A Moral Justification for Combatant Immunity' (2019) 32 Ratio Juris 395, 396–97; Yitzhak Benbaji and Daniel Statman, *War by Agreement: A Contractarian Ethics of War* (1st edn, Oxford University Press 2019) 36.

can achieve a better grasp of the moral nature of war, which can serve as the guiding light when thinking about the content of its moral and legal regulation. Taking this step back also allows us to construct a more plausible instrumentalist account, even assuming that revisionists are right about the morality of war. The goals pursued by this instrumentalist account are different from those of the humanitarian view, but they are meant to enrich it, not contradict it.[90] Further, this instrumentalist view appropriately distinguishes between the permissive and the prohibitory aspects of the rules on targeting, and how providing a justification for the first necessitates the existence of the latter.

Throughout section IV, I will continue to assume that the revisionist morality of war is correct. That is, I will assume that what needs a justification is a set of legal rules that permits the murder of innocents (just combatants). I believe that this instrumentalist account is more plausible than the humanitarian view: it can at least provide a contingent justification for the *adoption* of the symmetry thesis as it is now—but not for the content of its permissions. Yet, as I will argue towards the end, this account might not be entirely successful. It is open to some of the same objections I have discussed in this section, and, particularly, much of whether it succeeds or not depends on empirical certainties that we lack. Thus, this lens does not entirely salvage an instrumentalist position about the symmetry thesis premised on the revisionist morality of war. And it might also put some pressure on that very same morality. I am happy that it does.

IV. War and Perpetual Peace

We all seem to agree that war is a terrible thing. We also seem to agree that war should not be undertaken. Faced with these two facts—agreement that something should not be done and that it is indeed very bad that this thing is done—societies usually respond by enacting a legal prohibition and/or criminalizing the behaviour in question. This is what we have done in the case of war: aggression is a crime under the Rome Statute, and the resort to force is prohibited by the United Nations Charter, except in instances of self and collective defence, as well as United Nations Security Council (UNSC) enforcement actions.[91] Yet even though we have these prohibitions that overlap with some of our moral

[90] And I also read those who propose the humanitarian view—particularly, Shue and Dill—as sometimes hinting at this richer instrumentalist account.

[91] *See Rome Statute of the International Criminal Court* (hereinafter, Rome Statute) 17 July 1998, 2187 UNTS 90, art 5; UN Charter arts 2, 51, and Chapter VII.

commitments, we also have legal and moral rules that tell those engaged in war what they are allowed to do. Though we agree that war is a terrible thing that should not be undertaken, we still think about how to regulate conduct during war.

The first question we are concerned with here is simply *why*. Why are we thinking about regulating war if we agree war should not be undertaken? Why do we begin to think about war in the first place?

In a realistic Utopia[92] of how our world might be, there would be no wars: war only takes place because someone, somewhere, is not doing what they *should* be doing. In more technical terms, there is no place for war in what John Rawls calls 'ideal theory'. If we assume strict compliance with our moral and political principles, and we agree that those principles are that war should not be undertaken, those principles will be complied with.[93] In other words, no-one will wage wars due to self-interest or imperial ambition, and as a result, no-one will have to defend themselves by resorting to war. Let's call this the 'perpetual peace ideal'.[94]

The perpetual peace ideal is not the same as pacifism. Pacifism is a moral or political position by which individual or collective agents may not resort to force, no matter what: not in self-defence, not in defence of others, not during a just war.[95] The perpetual peace ideal is meant to convey that if everyone complied with the principles they ought to comply with, there would be no wars. And because there would be no wars, there would be no need to say anything else about war: war would be outlawed, and all agents and institutions would comply with that principle.

Yet because the world we live in is not a Utopia, there are, in fact, wars. Plenty of them. This is the background we come into, which cannot be avoided. Faced against this reality, the perpetual peace ideal is going to struggle. But it is going to struggle not because it is misguided about how things should be (no wars)

[92] John Rawls, *The Law of Peoples* (Harvard University Press 1999) 126.

[93] Here, I am using the ideal/non-ideal distinction in its full-compliance and non-full (or partial) compliance approach. Other approaches, as explained by Hamlin and Stemplowska, are idealization and abstraction; fact-sensitivity and fact-insensitivity; and perfect justice (or another value) and local improvement in justice (or another value). *See* Alan Hamlin and Zofia Stemplowska, 'Theory, Ideal Theory and the Theory of Ideals' (2012) 10 Political Studies Review 48. Other distinctions are between utopian and realistic theory, and end-state and transitional theory. *See* Laura Valentini, 'Ideal vs. Non-Ideal Theory: A Conceptual Map' (2012) 7 Philosophy Compass 654.

[94] In an obvious hint to Kant. *See* Kant (n 81). However, I am indebted to Jeremy Waldron for the name suggestion.

[95] On pacifism, *see* Jan Narveson, 'Pacifism: A Philosophical Analysis' (1965) 75 Ethics 259; Andrew Alexandra, 'Political Pacifism' (2003) 29 Social Theory and Practice 589; Craig K Ihara, 'In Defense of a Version of Pacifism' (1978) 88 Ethics 369; Larry May, 'Contingent Pacifism and the Moral Risks of Participating in War' (2011) 25 Public Affairs Quarterly 95.

or what principles should regulate the behaviour of states and agents (do not engage in war). It is going to struggle because we don't live in that world, and in the world we live in, agents do resort to war: they don't comply with the principles they ought to comply with.

Now it is perfectly possible that the perpetual peace ideal turns into pacifism. In other words, it is perfectly possible that our non-ideal theory in response to war is pacifism: in the face of aggression, violence is simply not permitted. Yet the perpetual peace ideal need not turn into pacifism. It may turn into another theory—a non-ideal theory—of what principles we should follow given that there will be wars: we usually call this 'just war theory'. Just war theory tells us precisely when it is justified to resort to war and what principles we ought to follow when fighting a just war. Accordingly, the international legal response has been to criminalize aggression, to prohibit the resort to force, and—here is the important part—to establish rules that tell agents how to conduct wars.

Both just war theory and the laws of armed conflict[96] come into place against this reality (a reality of non-compliance), and it would be unhelpful if the perpetual peace ideal insisted on only telling agents that they ought not to pursue war. We would hope for some insight that would have something useful—and I mean *morally* useful—to say about *this* background *right now*. This doesn't mean that the perpetual peace ideal must be discarded, for it does indeed correctly identify what it is that agents must do. There is no incoherence in that ideal: 'one can correctly identify what it is that agents ought to do while being entirely hopeless about the prospect of their doing it'.[97] It is not that the perpetual peace ideal is demanding the impossible: it is feasible for people to comply with their moral obligation of not resorting to war. But it is not realistic as an empirical prediction of their behaviour:[98] we know that they will not comply, and that as a result, there will be wars.

[96] Several international treaties—among them, of course, the UN Charter—identify, in their preambles, peace and the absence of war as the goals to reach. Some examples of this are the preambles to Hague Convention (IV), which states that 'seeing that while seeking means to preserve peace and prevent armed conflicts between nations, it is likewise necessary to bear in mind the case where the appeal to arms has been brought about by events which their care was unable to avert'; and the preamble to Protocol Additional to the Geneva Conventions of 12 August 1949, which states that the High Contracting Parties proclaim 'their earnest wish to see peace prevail among peoples'. *See* The Hague Convention (IV), preamble; *Protocol Additional to the Geneva Conventions of 12 August 1949, and relating to the Protection of Victims of International Armed Conflicts (Protocol I)*, preamble, 8 June 1977, UNTS 3 (hereinafter API). Additionally, Larry May has argued that one way of interpreting the UN's purpose is to understand international law as generally committed to some form of pacifism. *See* Larry May, *Aggression and Crimes Against Peace* (Cambridge University Press 2008).

[97] Zofia Stemplowska and Adam Swift, 'Ideal and Nonideal Theory' in David Estlund (ed), *The Oxford Handbook of Political Philosophy* (Oxford University Press 2012) 373.

[98] *See* ibid, 377; David Estlund, 'Utopophobia' (2014) 42 Philosophy & Public Affairs 113, 116.

Therefore, we need to do something different from simply saying, 'well, you shouldn't be going to war and killing people in the first place'. We already know that. But we also know that people will go to war. The perpetual peace ideal is not enough in the face of massive (and deadly) non-compliance. We need some guidelines on how people ought to act given that they will not act as they really ought to.[99] The perpetual peace ideal simply does not offer any other guidance to people fighting wars except for telling them that they ought to lay down their weapons.[100] Importantly, the perpetual peace ideal tells *all* combatants that they should lay down their weapons, not only unjust ones: fighting a war, even a just war, is only second-best in relation to the ideal. That even a just war is a moral tragedy should never be forgotten—not in the legal regulation of war, not in thinking about the morality of war. As Bao Nihn puts it, after a victorious just war, 'justice may have won, but cruelty, death and inhuman violence ha[ve] also won'.[101]

This is how we come to regulate killing in war; this is *why* we regulate war. Our hand is forced by non-compliance into a kind of damage control. We have a need—a moral need—to put some order amidst chaos. I want to emphasize that I am indeed talking about a moral need. The answer to why we regulate war, both morally and legally, is not a logical consequence of anything. It is a substantive moral position on what we should do given massive non-compliance. An extreme pacifist position is not incoherent.[102] In other words, someone might see in the regulation of war nothing but hypocrisy: regulating killing in war, they would say, is very much like regulating torture. Once one is committing either of these actions, one has already engaged in behaviour so deeply wrongful that no further distinctions can be made. There is simply no point in telling people not to torture but that if and when they torture, it is all right to use electric shock but not all right to pull the victim's nails out. We might think a line has been crossed by the commission of the forbidden act itself, and that once that line is crossed, there is nothing more to say about what agents ought to do. Everything has already been said: thou shall not torture (or kill).

[99] Stemplowska and Swift (n 97) 377.

[100] *See* Lazar making this point about the revisionist morality in Seth Lazar, 'Strengthening Moral Distinction' (2018) 37 Law and Philosophy 327, 332–33.

[101] Bao Nihn, *The Sorrow of War* (Riverhead Books 1996) 195. I first came upon Bao Nihn's novel when reading Shue (n 32).

[102] Although, in all fairness, as Scheffler argues, it is notoriously difficult to provide a rationale for absolutist positions (or agent-centred restrictions) as they do have an air of paradox about them. *See* Scheffler (n 83).

A position like this seems very implausible when talking about killing in war. Most of us seem able to distinguish between forms of violence that are permitted (ie killing in self-defence and perhaps killing in a war of self-defence) and those that are not (cold-blooded murder and torture).[103] And even among some types of wrongful violence, we seem able to distinguish between what is less and more wrongful (ie killing a defenceless child and killing a grown man). Of course, pacifism may be able to recognize that some forms of violence are worse than others. However, that some forms of violence are worse than others does not make some of them permissible: pointedly, to the pacifist, they are all impermissible, no matter what.[104] What I am saying, precisely, is that *it may matter what.*

How we draw these distinctions within just war theory and our laws of war—that is, how we distinguish between violence that is permitted and violence that is not—is a different (though related) question, and I will not say much about it in this chapter. But I will focus on the relation between the perpetual peace ideal and the regulation of war, both in its legal and moral form.

First, the perpetual peace ideal should never be discarded nor put in the background. Although it is frighteningly easy to lose sight of it, it is fundamental to remember that it is *entirely* correct. Its only shortcoming is that it provides an incomplete account of what to do in the face of people not doing what they should be doing in the first place: that space is filled up with non-ideal theory (just war theory and its legal counterpart). But that people will not comply with the perpetual peace ideal is pointedly *not* an objection against it.[105] We need a theory of how to act given that people will not act as they should, but this theory, though morally necessary, does not change the original moral demands:[106] do not initiate war.

Second, the perpetual peace ideal is what provides both just war theory and the legal regulation of war with their ultimate target. What should guide the regulatory enterprise is what that ideal is: that there are no wars (because people comply with the moral principles they ought to comply with).[107] In the literature on ideal and non-ideal theory this is known as the 'target role of ideal theory'.[108] In its target role, ideal theory helps non-ideal theory answer

[103] Thomas Nagel makes a similar point. *See* Thomas Nagel, 'War and Massacre' (1972) 1 Philosophy & Public Affairs 123, 134–36.

[104] *See* Cheyney Ryan, 'The Hard Hand of War' (2018) 37 Law and Philosophy 269, 273–74.

[105] *See* Estlund (n 98) 114.

[106] On the relationship between aspirational theory and concessive normative theory *see* ibid, 121–23.

[107] *See* Rawls, *A Theory of Justice* (n 83) 379.

[108] *See* Stemplowska and Swift (n 97) 376; David Wiens, 'Political Ideals and the Feasibility Frontier' (2015) 31 Economics & Philosophy 447, 448.

the question of what to do now, given what is feasible and what is the ideal end point that we want to reach;[109] it allows us to set up an aim to guide the course of social reform and a landmark to evaluate what we have achieved against a final landmark of where we ought to be.[110] Without identifying that ideal, non-ideal theory lacks an objective by reference to which its questions can be answered.[111] This also means that short-term reforms should be avoided if they set out a path that might be inconsistent with that long-term goal:[112] we have to keep the perpetual peace ideal in place somehow.

Third, that the perpetual peace ideal is going to struggle and that we ought to regulate war are not arguments in favour of the status quo. We cannot be content with the achievements of the past. Our rules, precisely because of what they respond to, and what they target, are contingent; they need to be constantly evaluated against the perpetual peace ideal. I believe this can show the evolution of our rules of war in an illuminating way: we chip away at the permissive rules. First, you may not target civilians. Then, we ban the resort to force; we criminalize aggression. But we keep the rules, so that when hell breaks out, some of the perpetual peace ideal remains in place. After all, questions of non-ideal theory are questions of transition, of how to work from a world containing war to a world in which we all accept and follow the perpetual peace ideal,[113] even if we know that this day might never come.

By providing the regulation of war with its ultimate moral target, the perpetual peace ideal also provides some guidance on the content of our rules.[114] It may do so in several ways. First, it gives us a notion of what a commitment to certain values entails and how to strive for them.[115] Second, it gives us guidance in identifying which wrongs are more grievous and hence, more urgent to correct.[116] Finally, it guards us from making our demands too permissive towards those who are simply unwilling to comply with reasonable limits on the permissibility of their actions.[117]

[109] Stemplowska and Swift (n 97) 376; Hamlin and Stemplowska (n 93) 60.

[110] *See* Rawls, *A Theory of Justice* (n 83) 245; Zofia Stemplowska, 'What's Ideal about Ideal Theory?' (2008) 34 Social Theory and Practice 319, 332.

[111] *See* Rawls, *The Law of Peoples* (n 92) 90; Stemplowska (n 110) 332.

[112] cf Hamlin and Stemplowska (n 93) 59.

[113] *See* Rawls, *The Law of Peoples* (n 92) 90. On ideal/non-ideal theory as end-state vs transitional theory, *see* Valentini (n 93) 60.

[114] Ripstein also thinks that peace, understood as a positive condition in which states accept that disputes will not be resolved through force, restricts the grounds for resorting to force and conduct during war. Means that are inconsistent with the possibility of peace are thus forbidden. However, Ripstein's view is not instrumentalist; peace is an internal rather than external end. *See* Arthur Ripstein, *Kant and the Law of War* (Oxford University Press 2021) 21, 29–32.

[115] Stemplowska (n 110) 334.

[116] John Rawls, *Justice as Fairness: A Restatement* (Erin Kelly ed, Harvard University Press 2001) 13.

[117] Stemplowska (n 110) 333.

And here is where we can go back to the question of divergence between the law and the morality of war. Just war theory, I have said, is part of non-ideal theory. The revisionist morality of war—their non-ideal just war theory—is a non-consequentialist theory that deals with the problem of non-compliance: it can be understood as an attempt to bring non-compliers into compliance (stopping the unjust war) and to remedy the consequences of their non-compliance.[118]

But the problem of non-compliance is also pervasive at a second level. At the first level, people shouldn't go to war, yet they do: this is why we need just war theory. And then, we must ask what this theory ought to look like, so that it keeps the perpetual peace ideal in place. This theory tells just combatants what they are permitted to do during war, and, according to revisionists, it also tells unjust combatants that they are doing what is morally forbidden. But the revisionist morality does not tell unjust combatants anything else other than they are morally prohibited from fighting, yet we know that unjust combatants will not actually comply with that. And here precisely is where we seem to need a second level: because unjust combatants will not comply with the principles they ought to comply with, we tell them something else. We tell them to only target combatants, not to target civilians.[119] People will, of course, not comply with these principles either. And we could concede even more, but we stop here: deliberately targeting civilians is a war crime.[120]

Can we keep the perpetual peace ideal in place while conceding at this second level? How can we justify a law that is significantly divergent from its correlative moral content, and which allows for massive injury and death against the innocent, without the perpetual peace ideal slipping away from us?

In many ways, it does slip away. But there are other ways in which we keep it in place. First, the aims in reference to which we construct this instrumentalist justification arise out of our ordinary moral commitments—the same moral commitments to which the perpetual peace ideal is devoted. If we aim at reducing suffering, we do so because we believe *all* human suffering is bad, regardless of whether people are just or unjust combatants. We protect civilians, *hors de combat*, and medical and religious personnel because we value human life. These values are not in open contradiction with the perpetual peace ideal. Even if some dimensions of the rules—like the permission to target just

[118] Talking about deontological theories that deal with non-compliance, *see* Quong and Stone (n 57) 223–24.

[119] I think these are the principles that ought to regulate the just combatant's conduct as well, but since this is controversial, I leave it aside for now.

[120] On the relationship between aspirational and concessive theory *see* Estlund (n 98).

combatants—are in tension with some of that ideal's content, this tension is not full-blown inconsistency, as it would be if the aims we chose for our rules of war were, say, to make wars cheaper.

What I have said above also provides an argument about how an instrumentalist account should be constructed, and how we should choose what aims such an account should favour. Instrumentalism does not, in and of itself, tell us which consequences matter, which set of aims we ought to choose for the laws of war to realize.[121] The important thing is that those aims need not be completely out of touch with the perpetual peace ideal. On the contrary, they must find a way of letting it shine through. An instrumentalist assessment of our rules can value peace for its own sake. It doesn't just surrender and abandon our most important moral commitments.

The second way in which we keep the perpetual peace ideal in place is given by the fact that our current laws are not completely symmetrical in nature. Aggression is an international crime, and the resort to force is prohibited by the UN Charter, except in instances of self and collective defence and UNSC enforcement actions.[122] At least some unjust combatants then—those in high-command positions—will not actually enjoy the same permissions than just combatants do. They will commit an international crime by resorting to and conducting an aggressive war, which will often be an unjust one.

However, even though our laws are not entirely symmetrical, they are also not entirely consistent with what makes a war just. There is significant dissonance between the legal regime—which will sometimes authorize unjust wars—and morality. And the *jus ad bellum* could certainly—and must—do a better job in providing guidance to individuals regarding the conditions that make war permissible. A true commitment to the perpetual peace ideal would thus require important modifications of our legal regime.

But even if we are able to keep the perpetual peace ideal in place in the above ways, the content of our laws, which includes a permission to kill the innocent, is inconsistent with it. Even in the realm of non-ideal theory, we do not surrender our most important moral demands: non-ideal principles should be morally permissible,[123] yet the content of our laws is not—they contain a permission to kill the innocent.

We might think, however, that in assessing moral permissibility an important distinction should be drawn between the content of IHL and its

[121] On this point about consequentialism in general, *see* Nagel, 'Ruthlessness in Public Life' (n 59) 85.

[122] *See* UN Charter, arts 2, 51, and Chapter VII; Rome Statute, art 8*bis*.

[123] *See* Rawls, *The Law of Peoples* (n 92) 89; A John Simmons, 'Ideal and Nonideal Theory' (2010) 38 Philosophy & Public Affairs 5.

adoption and/or maintenance. And in assessing its adoption and/or maintenance, we might think that moral permissibility ought to be judged in consideration of where we are and where we want to go and the alternatives that we actually have. In fact, at times, those who hold the humanitarian view hint at this notion of constrained choices or impossibility: the impossibility of having laws that track the revisionist morality, the impossibility of eliminating war, or the impossibility of protecting the rights of just combatants.[124]

Suppose our laws of war permitted the execution of prisoners of war. Suppose we could only agree to modify them so that they are imprisoned for life: any further restrictions are currently impossible. Undoubtedly, both laws permit what is morally impermissible. Yet it is less clear whether the second is morally impermissible when we judge it in relation to the first one and against the fact that it is the only possible alternative. That is, the fact that the second law constitutes moral progress in relation to the first one, that it gets us closer to the perpetual peace ideal, might make us judge it as morally permissible in at least some sense. But what is morally permissible, rather than the content of the law itself, is *agreeing on* or *adopting* such a law, given the alternative courses of action available. Sometimes it may be that the options with impermissible content are the *only* options that we actually have.

If the options with impermissible content are the only options we have, we might be able to see the legal permission to target just combatants as instrumentally justified if it constructs moral progress through the law. The aim, then, is to bring the law closer to our moral ideal within the parameters of what is possible—to get closer and closer to the perpetual peace ideal. Sometimes, even though we know what would be morally best to happen—that everyone stops fighting—there is no way we can make *that* happen. But we might be able to make other things happen that, though worse than the ideal, are better than doing nothing at all, like telling unjust combatants that they can target *only* combatants and not civilians, rather than telling them only to lay down their weapons. Thus, we might think that regulating war by constraining some of its unjust conduct (like targeting civilians) while permitting (although at the same time regulating) some of its unjust conduct (like the killing of just combatants) is but a step in this direction; and that this might be morally better than banning war and telling unjust combatants nothing other than 'you must surrender'.

Yet we must take seriously the possibility that certain practices are so abhorrent that seeking to control their use through law only magnifies their

[124] *See* McMahan, 'The Morality of War and The Law of War' (n 14); Dill and Shue (n 32).

horror.[125] Sanford Kadish makes precisely this point concerning the legal regulation of torture: consider, he says, how much respect a book of law would evoke if it contained detailed regulations on the use of torture.[126] Is warfare so abhorrent that seeking to control its use through law only magnifies that horror? In many ways, it is. And indeed, an instrumentalist justification for the permission to target just combatants premised on the revisionist morality of war has the flavour of attempting to carry 'the decorum of sanity'[127] into doing something that seems, by the standards of the revisionist morality, irrationality itself. As Bernard Williams points out, an instrumentalist rationality seems to have no limits.[128] It aims to make the best of a bad job even when saying anything will be very much like saying something about the difference between massacring 7 million and massacring 7 million plus one.[129] But what happens when massacring 7 million and massacring 7 million plus one are the only options we have? What happens when the impermissible options, as I have suggested, are the only options available?

This is a strange question to ask because it matters why the options are constrained in the first place. One reason why we might have only impermissible alternatives available is simply given by non-compliance. But this is not really a matter of impossibility; people are not actually incapable of complying with the perpetual peace ideal. Why should we concede to those who are acting wrongfully? Wouldn't we be conceding too much?[130] Perhaps it is important to notice here that even a concessive solution can be challenging to implement and that it can also be morally demanding.[131] But this doesn't really tell us where to draw the line between conceding and not; it does not make other (permissible) options unavailable.

Another source of impossibility—why we are faced with two morally impermissible options—might be given by what we are able to agree upon, as I suggested above. Perhaps, given the lack of centralized enforcement mechanisms at the international level and the pervasive disagreement among states about what a just war is and how it should be conducted, we can only agree on regulating war symmetrically, but nothing more. And we need to take very seriously the value there is on agreeing on anything that constrains conduct

[125] Sanford H Kadish, 'Torture, the State and the Individual' (1989) 23 Israel Law Review 345, 356.
[126] ibid.
[127] Bernard Williams, 'A Critique of Utilitarianism' in Jack JC Smart and Bernard Williams (eds), *Utilitarianism: For and Against* (Cambridge University Press 1973) 93.
[128] ibid.
[129] ibid.
[130] On the absurdity of what Estlund calls 'complacent realism', *see* Estlund (n 98) 132.
[131] *See* ibid.

in war. So it might be that although some prefer that war is regulated through prohibition, and others through a permissive regime, each side prefers either of the coordinating outcomes to non-coordination.[132] Since we disagree about the morality of war, and since the problem of disagreement is not going to disappear,[133] it might be that we all prefer that these matters be settled even in a way that some of us oppose. This is a particularly pressing consideration if the alternative is no laws of war at all (with everyone who has a view enforcing it as best they can), or a law confined only to those cases where it is uncontroversial in the community that a wrong has been committed.[134]

In sum, even though the content of the symmetry thesis is morally impermissible, it might be morally permissible to adopt it as law, provided that there are no other permissible options available and that the failure to agree on the symmetry thesis would entail that there would be no laws regulating war, not even the prohibition against targeting civilians. But justifying the symmetry thesis like this entails that what we are providing is nothing more than a contingent justification. It is contingent on its ability to lead us closer to peace. And it is also contingent on whether our options are constrained by what we can agree upon, and on whether the failure to agree entails that we would not have prohibitory rules or any rules of war at all.

I am sceptical, however, that our only options are symmetrical rules or no rules at all. Certainly, it is not *logically* true that these are our only alternatives. An asymmetrical law of war need not stay silent on the conduct of war; it can take many forms. One possibility would be simply to state that combatants are permissible targets only in just or legal wars. The existence of such a rule wouldn't necessarily mean that unjust combatants would lack an immunity from prosecution, nor would it include a permission to target civilians. And it clearly would mean the absence of a legal permission for unjust combatants to target just ones. We might think such a rule would be inefficient or pointless, since all combatants would probably claim that their war is just. But it would not bring the kind of disaster that instrumentalists like McMahan seem to be worried about, which is that there would be no rules of war to guide anyone, or that all combatants could then target civilians.[135] So, going back to a point I made

[132] Jeremy Waldron, *Law and Disagreement* (Clarendon Press 2004) 103.

[133] *See* ibid 102. 'We may say, along similar lines, that the felt need among the members of a certain group for a common framework or decision or course of action on some matter, even in the face of disagreement about what that framework, decision or action should be, are *the circumstances of politics*.'

[134] ibid 105.

[135] *See* Lazar, 'The Morality and Law of War' (n 62) 376. He argues that a positive permission to fight is unnecessary from the perspective of the humanitarian view: an immunity from prosecution is enough to restrain conduct in war.

before, it is not really available to the humanitarian view to argue that our laws of war must stand or fall together: at the very least, it is not logically true.

There is, however, a kind of hypocrisy in the solution just described, which simply introduces a prohibition. Is a legal rule that we all know would not be complied with enough to render our laws morally justified? Would a law with nothing other than symbolic value be enough? I am inclined to think that, on its own, it wouldn't.[136] And, certainly, we would still need to justify the lack of punishment against unjust combatants.[137] Given the kind and extension of wrongdoing war entails, this isn't an easy task, though, along with a prohibition on the murder of just combatants, it might be slightly easier than justifying the current regime.[138]

What I have said certainly highlights the value of having laws that regulate war. And if, due to the limits of what we can agree upon, our alternatives are no laws of war or the ones we have, then the ones we have can be justified instrumentally, even if revisionists are right, and these rules allow for what is equal to killing the innocent. But the fact that we might be able to make war more humane through its regulation,[139] or that we can inch closer to the perpetual peace ideal does not render the content of our rules morally acceptable. It is not. Yet as long as wars are fought, and as long as our options are limited, a symmetrical privilege premised on an instrumentalist justification might be all we can achieve. But, and perhaps this is what is most important, we must not forget the ideal of perpetual peace in the process. If what we have is the most we can achieve now, we must still strive relentlessly to keep changing our laws so that they can be brought as close as possible to our ideal: that there are no wars at all.

V. Conclusion

If revisionists are right, and killing just combatants is simply like killing the innocent, the humanitarian view fails as an instrumentalist justification of the symmetrical legal permission to target combatants. Though the instrumentalist view I propose fares a bit better, the fact that it is still premised on the revisionist morality of war makes it incredibly fragile.

136 Though it would be a slight improvement.

137 Ripstein has recently justified combatants' immunity from prosecution on non-consequentialist grounds. He argues that states are equal; they cannot punish other state officials for carrying out their official roles, such as soldiers fighting war. Ripstein (n 114) 155–56.

138 *See* Haque (n 1) 28.

139 *See* Henry Shue, 'Do We Need a "Morality of War"?' in David Rodin and Henry Shue (eds), *Just and Unjust Warriors: The Moral and Legal Status of Soldiers* (Oxford University Press 2008) 106.

It is fragile because its force lies in achieving some aims—constructing moral progress through the law, getting closer to the perpetual peace ideal—we don't actually know it is good at achieving. It is fragile because it depends on there being no other alternatives available (and I am not sure that this was ever true or is true now). It is constantly struggling to keep the perpetual peace ideal in sight. It makes us wonder whether we are conceding way too much. Where do we stop? Why do we stop where we do? I am not only speaking about how contingent it all is (though it certainly is): it is contingent on what our current alternatives are, on what we can agree upon, on whether the aims are achieved and to what extent. As a result, our legal rules must be constantly under revision so that we can bring them as close as possible to our moral ideal. But I am also speaking of how *morally* fragile the compromise is. We might as well remember that '[t]he entire history of social improvement has been a series of transitions by which one custom or institution after another, from being a supposed primary necessity of social existence, has passed into the rank of a universally stigmatized injustice and tyranny'.[140]

In many ways, this moral fragility is an inescapable fact of both just war theory and the legal regulation of war. As I said, even the pursuit of a just war is second-best to our ideal. But there is more to say about what lies behind the symmetrical legal privilege of combatants. An instrumentalist justification premised on the revisionist morality of war is simply *too* fragile. Can we attenuate this fragility, and if so, how? This is what will occupy Part 2 of this book.

In the following chapters, I will argue that the most promising way in which to attenuate this fragility is to propose an alternative conception of the morality of war that denies that what unjust combatants do is *just like* killing the innocent. Nevertheless, the instrumentalist justification I proposed in this chapter will remain necessary, even after having proposed a morality of war different from the revisionist one. Any plausible contestation of the revisionist morality will be limited; it will have to concede that at least some killing done by the unjust side of a war will be very close to murdering the innocent.

The adoption and maintenance of a symmetrical legal permission to target combatants can be justified by the instrumentalist account I proposed here, but only contingently. There is no way to erase the fact that regulating war symmetrically—that regulating war at all—is, indeed, a significant moral compromise.

Now is the time to let go of the revisionist morality—if, as will be most likely, only partially.

[140] John Stuart Mill, *Utilitarianism* (George Sher ed, 2nd edn, Hackett Publishing Company 2001) 63.

PART 2
ATTENUATING INSTRUMENTALISM'S FRAGILITY

The instrumentalist justification of the legal privilege to kill combatants I provided in Chapter III remains quite fragile if unjust combatants are engaged, as revisionists believe, in murdering the innocent. Given the constraints that we face when regulating war, there will always be some work for this instrumentalist justification to do. Yet because my aim is to provide an attractive moral justification for the legal privilege of combatants, we would do well in trying to attenuate the fragility of the instrumentalist justification.

The attempt at attenuating the moral fragility of the instrumentalist rationale is related to a second way in which we can understand our legal rules as morally justified. This type of moral justification demands a greater fit between morality and law than an instrumentalist justification. According to this second type of justification, our legal rules must not only be good at achieving some consequences, they must also track or respect some moral demands. Thus, for example, legal rules on murder must be good (or at least not very bad) at preventing murder *and* they must also track the moral prohibition on murder.

Not all legal rules are like the legal privilege of combatants. Some legal rules do not have a corresponding moral content to track. They regulate behaviour which has no moral status, like the rule that tells us on which side of the road to drive. We thus cannot judge the moral adequacy and value of this kind of legal rule by reference to any moral content—for there isn't one.[1] But even though there is no moral content to track, a moral justification for such a rule can still be provided, because even if the content of the rule is arbitrary (driving on the left or right side of the road) the point of having or obeying the rule may not be morally indifferent (avoiding accidents).[2] In these cases, the content of the

[1] *See* George I Mavrodes, 'Conventions and the Morality of War' (1975) 4 Philosophy & Public Affairs 117, 126.

[2] *See* Jeremy Waldron, *Torture, Terror, and Trade-Offs: Philosophy for the White House* (Oxford University Press 2010) 105; Tony Honoré, 'The Dependence of Morality on Law' (1993) 13 Oxford Journal of Legal Studies 1, 5; Mavrodes (n 1) 126–27. Mavrodes refers to the moral obligation to obey

legal rule is perfectly interchangeable with an alternative one that we would still be morally obliged to follow[3] (eg drive on the left side of the road).

Other legal rules—and this is the case with laws dealing with violence, such as the laws of war—do have some corresponding moral demands to track. I will assume that morality has a role in shaping law, and that there are some moral demands that can be ascertained, at least partially, independently of law—such as the prohibition on killing the innocent—but I do not wish for the justification I will provide to depend on accepting a controversial relationship between law and morality or a full account regarding the details of that relationship.

We might posit that when law has a corresponding (non-consequentialist) moral content to track, its justification cannot entirely depend on its ability to coordinate behaviour and achieve a good result. We might think that, in those cases, our laws must track or stick close to that content. As a result, this second type of legal rule—those that have some moral demands to track—will be morally justified not only due to their ability to achieve some consequences or coordinate certain behaviour; they must also track or reflect that moral content.

However, generally accepting this position about law and morality is not necessary for the purposes of my argument (and I, myself, do not embrace this position and accept some instrumentalist views about law in other areas). It is enough to accept the following, more limited claim. If we accept that there are stringent moral rights, such as rights against harm and the arbitrary deprivation of life, and a very stringent moral prohibition against killing the innocent, we should conclude that there are very weighty reasons to adopt laws that reflect and respect those moral rights and prohibitions. The adoption of those laws might not be possible—as I argued in Chapter III—but the content of our laws should be limited or constrained by certain, very stringent, moral demands.

Adherence to those moral demands is, however, not enough. As I pointed out in Chapter III, legal rules are a special sort of rule, and they must be able to function as such. As Jeremy Waldron points out, we cannot just transfer the content of morality to a positive rule, mainly because moral ideas do not present themselves in law-like form.[4] Moral rules, even the most basic ones (you shall not kill), need to be made into law, and law-making is mostly a

such rules as 'convention dependent-obligations', for their moral force comes from the convention's ability to coordinate behaviour.

[3] *See* Mavrodes (n 1) 126. Mavrodes in fact suggests that the prohibition on targeting civilians is a convention-dependent obligation.

[4] Waldron (n 2) 95.

technical endeavour.[5] As a result, legal rules that track moral content closely enough for them to be justified will not necessarily have the *same* content than the moral rule being tracked, precisely because the need to turn that moral rule into a legal one might prevent that perfect mirroring.

Legal rules have particular standards to meet, and any workable legal rule, even one that closely tracks a moral prohibition, will stray away from it in order to be precisely what is meant to be: a workable legal rule. First, even the integration of an uncontroversial moral rule into the law can be a complex endeavour. Many of the definitions and legal implications that the legislator and judges will need to carry out will extend beyond the point where we can talk of a moral content to track.[6] Second, whatever legal rule we come up with must be able to work as a *legal* rule, that is, as part of a complex and coherent legal system:[7] details have to be settled; rules of evidence, presumptions, and burdens of proof need to be laid down; bright lines must be drawn; operationalized criteria need to be established; and so on.[8] Finally, law-making in pluralist settings (like the international community) will also mean that the content of the legal rules we create are constrained by what we can agree on, which may not always be consistent with our preferred moral considerations.[9]

Thus, the legal privilege of combatants need not perfectly reflect its corresponding moral content; it needs to be able to work as a legal rule for war. Laws of war must be self-administered in times of horror, danger to oneself and others, exhaustion, and fear, among others.[10] Besides the ordinary difficulties there are in specifying a moral rule into a legal provision, these additional hardships must be considered. Laws of war need to be justified in the sense that we do not ask too much nor too little of soldiers; we need to be aware of the reasonableness or unreasonableness of the burdens that the rules place on combatants and their superiors.[11] Additionally, a legal rule designed to apply

[5] ibid.

[6] John Finnis, *Natural Law and Natural Rights* (2nd edn, Oxford University Press 2011) 284.

[7] *See* ibid 282–83.

[8] *See* Waldron (n 2) 96; Jeremy Waldron, 'Torture, Suicide, and Determinatio' (2010) 55 American Journal of Jurisprudence 1, 3–4.

[9] *See* Waldron (n 2) 93; John Gardner, 'Law and Morality' (Social Science Research Network 2008) SSRN Scholarly Paper ID 2344794 4–5 <http://papers.ssrn.com/abstract=2344794> accessed 21 December 2015.

[10] *See* Waldron (n 2) 94.

[11] *See* ibid. Particularly, we need to be aware that we are demanding soldiers not only to kill and be killed in circumstances that, most likely, will not be chosen by them, but that we are also training them to go to war, and thus, to overcome their natural resistance to kill other human beings. That is, we place combatants in situations where they will face both mortal and moral risks. *See* Dave Grossman, *On Killing. The Psychological Cost of Learning to Kill in War and Society* (Open Road Integrated Media 2014); Phil Klay, 'The Citizen-Soldier: Moral Risk and the Modern Military' (*The Brookings Institution* 24 May 2016) <http://csweb.brookings.edu/content/research/essays/2016/the-citizen-soldier.html> accessed 17 August 2022.

during combat cannot take into account every detail that a moral theory would ideally consider,[12] a point that is often also true for 'ordinary' legislation for 'ordinary' times.[13]

All said, legal rules for war must be a sound instrument of their own moral purpose. As I briefly pointed out in Chapter III, they mustn't be directly or indirectly self-defeating. A legal rule that prohibits murder while also providing perverse incentives that encourage people to commit murder fails what I called 'the weak instrumentalist requirement': it is indirectly self-defeating.[14] And so does a legal rule whose obedience results in more murders: it is directly self-defeating.

Now, we already know that the laws of war are not actually tracking what revisionists take their corresponding moral content to be. On the contrary, our laws of war diverge significantly from it. They permit, under revisionism, unjust combatants to murder the innocent. Thus, as it should be clear by now, attenuating the fragility of the instrumentalist justification can only be done to the extent that we are able to challenge the revisionist morality. In particular, we ought to deny that what unjust combatants do is *just like* murdering the innocent. This road still leaves open different ways in which the analogy between murder and what unjust combatants do can be deconstructed. In its strongest form, that deconstruction would lead us to deny all similarities between murder and unjust combatants' killings. In a weaker version, it might point out simply that unjust combatants' killings of just combatants are *somewhat* different from murdering the innocent, and that just combatants might not be *simply* morally innocent.

Justifying the symmetry thesis does not require the strongest possible version of the dissimilarity between murder and what unjust combatants do. We do not have to be committed to the claims that all unjust combatants are morally permitted to kill just ones, that just combatants are not morally innocent, or that killing just combatants is *never* like murder. Weaker claims might be enough, such as arguing that *sometimes*—but importantly, not *rarely*—unjust combatants are morally permitted to kill just ones, that *sometimes*, though not rarely, just combatants are not morally innocent, and/or that *sometimes*, killing

[12] Waldron (n 2) 94.

[13] *See* the discussion between Henry Shue and Jeff McMahan. *See* Henry Shue, 'Do We Need a "Morality of War"?' in David Rodin and Henry Shue (eds), *Just and Unjust Warriors: The Moral and Legal Status of Soldiers* (Oxford University Press 2008); Jeff McMahan, 'The Morality of War and The Law of War' in David Rodin and Henry Shue (eds), *Just and Unjust Warriors: The Moral and Legal Status of Soldiers* (Oxford University Press 2008); Jeff McMahan, *Killing in War* (Oxford University Press 2009).

[14] John Gardner, 'What Is Tort Law for? Part 1. The Place of Corrective Justice' (2011) 30 Law and Philosophy 1, 20.

just combatants is not simply like murder. If these weaker claims are true, the fragility of the instrumentalist justification can be attenuated and the adoption and maintenance of the symmetry thesis can, perhaps, be justified—if only contingently. We thus don't have to embrace the implausible position that just wars are entirely equivalent to unjust ones, and, more importantly for my purposes, that what just combatants do is entirely equivalent to what unjust combatants do. It is not. Yet it may not be entirely dissimilar. In other words, what I will say in the next chapters will amount, at best, to what Barry and Christie call a 'noncategorical' defence of the moral equality of soldiers,[15] which in any event, may be able to provide a good enough defence of their legal equality.

Part 2 of this book will thus focus on attenuating the fragility of the instrumentalist justification by deconstructing the moral analogy between what unjust combatants do and murdering the innocent. In order to do this, I will have to look closely at the revisionist morality and what I take to be its shortcomings. This will occupy Chapters V and VI. Thereafter, in Chapters VII and VIII, I will argue for a different morality of war. This morality of war can attenuate the fragility of the instrumentalist justification, but to a limited extent. It also fits better with the perpetual peace ideal I explained in Chapter III.

The hope is that a different morality of war, along with the instrumentalist justification I proposed in Chapter III, will offer a better justification of the legal privilege to kill than a purely instrumentalist one, provided significant changes are made to our laws of war, and relatedly to international law in general. But such a justification, even if better, is contingent on the fact that we cannot stop war from happening. And it is also contingent on the impossibility of agreeing to better laws. It is, as a result, still a moral compromise—and a significant one.

[15] *See* Christian Barry and Lars Christie, 'The Moral Equality of Combatants' in Seth Lazar and Helen Frowe (eds), *The Oxford Handbook of Ethics of War* (Oxford University Press 2015).

IV

Non-Instrumentalism and the Revisionist Morality

I. Introduction

Before attempting to attenuate the fragility of the instrumentalist justification by contesting the revisionist morality of war and proposing an alternative one, there are three non-instrumentalist justifications of our laws of war that we ought to consider: arguments from role morality, arguments from democratic authority, and Ripstein's Kantian account of war.

Besides their non-instrumentalist character, there is little in common between Ripstein's view and arguments from role morality and democratic authority. Although none of them relies on a consequences-based justification of the laws of war, Ripstein develops a distinctive and complete view about the law and the morality of war by relying on Kant's ideas. His view offers a different approach, and provides a different answer, from the just war tradition—where revisionists are located—and what Ripstein calls the regular war tradition, which views war as a mechanism for dispute resolution.[1] Ripstein, like Haque, believes that the laws of war do not confer a permission to wage an unjust war. They impose distinctive prohibitions, which are specific to war.[2] And Ripstein, like revisionists, believes that killing in pursuit of an aggressive war is a grievous moral wrong.[3] Thus, Ripstein does not aim to justify what I have characterized as a legal privilege to kill the innocent—he does not believe that such a privilege exists. His aim is narrower: to defend combatants' immunity from prosecution and the symmetrical application of *jus in bello* prohibitions on non-instrumentalist grounds. I agree that there are non-instrumentalist reasons why the punishment of some unjust combatants is not warranted, and, more strongly, I think that sometimes unjust combatants act permissibly, as

[1] Arthur Ripstein, *Kant and the Law of War* (Oxford University Press 2021) 8.
[2] ibid 107.
[3] ibid 154.

The Morality of the Laws of War. Marcela Prieto Rudolphy, Oxford University Press. © Marcela Prieto Rudolphy 2023.
DOI: 10.1093/oso/9780192855473.003.0004

I will argue in Chapter VIII. Therefore, I will briefly refer to Ripstein's defence of combatants' immunity from prosecution in that later chapter.

By contrast, arguments from role morality and democratic authority can complement other moral theories of killing in war, such as revisionism. They occupy, in a sense, the same theoretical space. These views argue that there are agent-relative reasons—that emanate either from roles or from the directives of democratic authority—that can justify the symmetry thesis, understood as a permission to participate in war and kill enemy combatants.

In section II, I will discuss arguments from the moral division of labour and role morality. I will argue that this justification is of limited application and cannot justify, on its own, the legal regime if revisionists are right about the morality of war—if they are right that killing just combatants is just like killing the innocent. In other words, these justifications must deny that killing just combatants is just like killing the innocent (which some of them in fact do). In section III, I will examine the argument from democratic authority and argue that this justification, coupled with the moral division of labour, fares better, but faces similar problems to those faced by the view examined in section II. Section IV concludes.

Even if these non-instrumentalist justifications are of limited application and do not work when premised on the revisionist morality, they are nonetheless important. If the alternative morality of war I will propose in the following chapters is successful in deconstructing the analogy between what unjust combatants do and murdering the innocent, these non-instrumentalist justifications will be able to take their place in partially accounting for the symmetrical legal privilege of combatants.

II. The Moral Division of Labour and Role Morality

Among agent-relative reasons, a special kind is related to role morality. Role morality-related reasons arise out of the special roles we occupy in society—in this case, most obviously, the role of soldiers:

> We must distinguish the war from those who took part in it. The war was criminal and has been condemned as such, but the boys must be defended and protected.[4]

[4] Svetlana A Aleksievič, *Zinky Boys: Soviet Voices from the Afghanistan War* (Julia Whitby and Robin Whitby trs, Norton 1992) 193.

As we can see, arguments from role morality try to explain why and within which limits occupying certain roles makes some morally permissible actions morally required, and why and within which limits it makes some morally forbidden actions morally permissible.[5] They aim to separate, in some sense, the wrongness of the actions—'the war was criminal'—from those who are called upon, on the basis of their role, to undertake that action—'the boys must be defended and protected'.

Role morality, then, tries to come to terms with the 'slippery moral surface', as Nagel calls it,[6] of agents who occupy public positions. What do we make of someone like the executioner of Paris, who put to death hundreds of people during the French Monarchy and during the years following the French Revolution? Is he a murderer? Is he an executioner? Is he both?[7]

Arguments from role morality—defended separately by Benbaji and Estlund in the context of war—provide a non-instrumentalist alternative to instrumentalist justifications of our laws of war. Benbaji's and Estlund's arguments aim at defending the legal and the moral equality of combatants by emphasizing the importance of the roles that individuals occupy within social and political structures in shaping our moral obligations, as well as the division of labour between those different roles.

The idea of role morality suggests that individuals may adopt different moral principles depending on the roles they undertake within institutions.[8] If the institution is justified, then people who occupy roles within it have to behave according to the obligations those roles impose, even if some of those obligations conflict with 'ordinary' moral obligations.[9] Actors occupying certain roles are thus not to make all-things-considered evaluations about the goodness or rightness of their actions, but rather, they ought to act on restricted reasons for action, only taking into account a limited or partial set of values, interests, or facts.[10] Moral prescriptions are relative to the role the agent occupies: the agent faces particular moral reasons for action that others, who do not share his role, do not face, and he ought to ignore moral reasons for actions that others, who do not share his role, ought not to ignore.[11]

[5] *See* Arthur Isak Applbaum, *Ethics for Adversaries: The Morality of Roles in Public and Professional Life* (Princeton University Press 2001).
[6] Thomas Nagel, 'Ruthlessness in Public Life' in Thomas Nagel, *Mortal Questions* (Cambridge University Press 2012) 75.
[7] *See* Applbaum (n 5).
[8] Kevin Gibson, 'Contrasting Role Morality and Professional Morality: Implications for Practice' (2003) 20 Journal of Applied Philosophy 17, 17.
[9] ibid 23–24.
[10] Applbaum (n 5) 5–6.
[11] ibid 63.

Role morality is related to the idea of the moral division of labour between political authorities and soldiers (and citizens in general).[12] The moral division of labour refers to the idea of leaving most of the job of securing justice to institutions and authorities, as a result of which it takes 'the business of securing justice off people's plates in their day-to-day lives', for people's constantly thinking about justice and how to adjust their behaviour accordingly seems morally unappealing and potentially self-defeating.[13] People would 'lead freer and better lives' if they could devote most of their concerns to their own affairs while institutions do the job of securing justice.[14] Or, in different terms, people would be left free to pursue agent-relative values in their personal lives while institutions realize the bulk of agent-neutral values.[15]

In the case of soldiers, the moral division of labour operates in a slightly different manner. The focus should not be on the tension between the soldier's personal projects and the demands of justice but on the tension between the soldier's professional role and the demands of justice. Indeed, the point of having certain institutions concerned with the decision to wage war is not that soldiers can 'lead freer and better lives'. The point is simply for them to be free from making a particular decision (to wage or not to wage war) and to concern themselves with the responsibilities attached to their role (following the *jus in bello*, among others). This point, nonetheless—as well as the point of leading freer and better lives—is subsidiary to the main point of the scheme, which is to secure justice, not to secure that people live better and freer lives or that they may disregard certain reasons for actions. On the contrary, *because* the scheme is able to secure justice, agents may be able to live better and freer lives or disregard certain reasons.

In both cases—that is, generally and in the case of soldiers—a social structure that divides the moral labour serves to moderate the requirements of morality.[16] It allows individuals who occupy certain roles within this structure to disregard considerations to which others ought to respond to.[17] Role morality—that certain roles may exclude certain reasons from consideration—is attractive only to the extent that there are other institutions in the background that are taking care of those excluded reasons. That is, it is only attractive to the

[12] *See* Yitzhak Benbaji, 'The War Convention and the Moral Division of Labour' (2009) 59 The Philosophical Quarterly 593.

[13] Liam Murphy, 'Institutions and the Demands of Justice' (1998) 27 Philosophy & Public Affairs 257–58.

[14] ibid 258.

[15] *See* ibid; Thomas Nagel, *Equality and Partiality* (Oxford University Press 1995) 83–86.

[16] Benbaji, 'The War Convention and the Moral Division of Labour' (n 12) 616–17.

[17] ibid.

extent that the moral division of labour is fair and able to secure justice. If it is not, then we may not 'with good conscience' live our personal and professional lives according to those minimal rules.[18] This is so because the point of the scheme—the point of role morality and the moral division of labour—is not to moderate the requirements of morality. It *does* moderate them, but only because their moderation serves the aim of securing justice. Justice comes first; moderation and disregarding certain reasons for action come second.

Both Estlund[19] and Benbaji[20] (and more recently, Benbaji and Statman)[21] have employed these arguments—although Estlund, as we will see, also relies on arguments from democratic authority—to defend the symmetry thesis. They begin with the example of a jailer, who knows that a man is innocent. Is he morally permitted to do his job and put the innocent man in jail? Estlund and Benbaji both think that the jailer acts permissibly if he does so. The jailer is not required, under the circumstances, to let the man go free.[22] Both Benbaji and Estlund extend their arguments from the case of the jailer to that of the executioner,[23] and finally to that of a soldier fighting in an unjust war.[24]

Estlund adds several qualifications to his account, two of which are particularly relevant. First, the operative element which permits and obliges the jailer to carry out the punishment is the nature of the process that generates the command, which in this case is a responsible trial.[25] Second, regardless of the nature of the process that generated the command, there are limits to what the jailer should do. He is not permitted to torture, for example, because such a

[18] Nagel (n 15) 83–84.

[19] *See* David Estlund, 'On Following Orders in an Unjust War' (2007) 15 Journal of Political Philosophy 213.

[20] *See* Benbaji, 'The War Convention and the Moral Division of Labour' (n 12); Yitzhak Benbaji, 'A Defense of the Traditional War Convention' (2008) 118 Ethics 464; Yitzhak Benbaji, 'The Responsibility of Soldiers and the Ethics of Killing in War' (2007) 57 The Philosophical Quarterly 558.

[21] Yitzhak Benbaji and Daniel Statman, *War by Agreement: A Contractarian Ethics of War* (Oxford University Press 2019). Benbaji and Statman's argument is also a contractarian one, though sometimes it leans into some form of rule-consequentialism: 'In contrast, according to contractarianism, in many social circumstances, morality should focus on the rules to be followed rather than directly on the morality of the acts.' Since Benbaji and Statman's view relies on some empirical assumptions about the ability of the rules to protect peace and justice, it seems the view has at least this element in common with rule-consequentialism. *See* ibid 6–7.

[22] In Estlund's account, he is also obligated not to do so. Estlund (n 19) 216.

[23] ibid 219.

[24] ibid.

[25] ibid 217. 'Many people believe that the legal infliction of the punishment of death is never morally justified. I mean to argue only that if capital punishment can be justified for the perpetrators of certain crimes, then under the right conditions the executioner can be permitted to execute knowingly even an innocent person. Moreover, she can have a moral duty to do so. It is crucial, then, to assume for the sake of argument both that capital punishment would have been a just punishment if the convict had been guilty of the crime in question, and that the trial procedure, along with its social and cultural background, is not so irrational or prejudiced that it no longer counts as an adequate legal method for accurately punishing the guilty without punishing the innocent.'

punishment is not even approximately just even if the person were guilty of the alleged crime.[26] But once these conditions are met, that is, once the political and institutional process that produces the commands is duly looking after the question of whether the war is just, Estlund argues that

> [T]he soldier would be wrong to substitute his own private verdict and thwart the state's will. On the other hand, these preconditions are substantial, and the soldier will need to think for himself about whether they are met. The soldier is not exonerated simply because he was following orders. On the other hand, when the state and its procedures are of the right kind the soldier's participation in an unjust war is sanitized precisely because he was following orders.[27]

The soldier in an unjust war is not acting wrongly in obeying orders when the background conditions are right, argues Estlund. If that is so, Estlund continues, then surely he is not acting wrongly in defending himself against the soldiers he is obligated to fight.[28]

Benbaji reaches a similar conclusion using similar examples. The executioner of an innocent person is not morally responsible for killing the innocent, argues Benbaji, for it is not his job to decide on the innocence of the accused. The moral responsibility belongs to the state, and the job of deciding on the man's innocence belongs to the courts. Soldiers are thus not morally responsible for waging an unjust war because it is not their job to decide whether to go to war; it is state officials who are under a duty to attend to the justice of the wars they fight.[29] Roughly, argues Benbaji, morality is moderated by a social cooperative arrangement that divides the moral labour. Individuals are free to attend to their concerns, values, and interests without thinking about the effects of their actions. They are also free to lack information whose obtainment is the duty of other parts of the social structure.[30] People are only obliged to comply with their institutional duties, among which there isn't one that obliges them to judge the justness of the conviction or of the war they are to fight. Tacit acceptance of this convention means that the innocent victims—whether the innocent about to be executed or the innocent just combatant—have no

[26] ibid 216.
[27] ibid 213.
[28] ibid 226.
[29] *See* Benbaji, 'The War Convention and the Moral Division of Labour' (n 12) 612–13. A similar example and position can be found in Benbaji and Statman (n 21) 51–52.
[30] Benbaji, 'The War Convention and the Moral Division of Labour' (n 27) 612–13.

claim against the executioner nor the soldier, because their claims are against the state.[31] The alternative, thinks Benbaji, is exceedingly demanding.[32]

Estlund's and Benbaji's arguments are important but of limited application. They require, first, that there are just institutions doing their job in the background—to which I will return later—and second, Estlund's argument requires that what unjust combatants do is not the moral equivalent of murdering the innocent. His argument cannot work if it is premised on the revisionist morality of war because he does not argue that the division of labour coupled with the demands of different roles make the unjust combatant's act unlike murder, but that because unjust combatants' acts are unlike murder, arguments about the division of labour and role morality have space to do their work.

Estlund explicitly says that his argument works only to the extent that it is sometimes permissible to jail people or apply capital punishment.[33] That is, for wrongs within the practice—accidentally punishing the innocent—to be justified, it requires that the practice is not generally morally bankrupt. If the practice is unjust—because, say, it systematically punishes the innocent—instances of it will not be justified. Under revisionism, an unjust war is, however, precisely like this; it systematically kills the innocent. Thus, the argument would only work to the extent that participating and waging an unjust war is morally permissible, in at least some sense or under certain circumstances. The features of the institution and the institutional reasons on which soldiers act cannot, in his account, 'alter the moral meaning'[34] of the soldiers' actions to the point of making them dissimilar to the murder of innocents.

In response, one might argue that war is sometimes permissible (when it is just), and so, nothing distinguishes instances of mistakenly punishing the innocent from instances of killing the innocent in pursuit of an unjust war. Both practices (punishment and war) are sometimes permissible, and thus, role-holders within the practice will sometimes be justified in doing what is ordinarily forbidden (punishing the innocent; killing the innocent), insofar as the institutions are doing their job in the background. However, it can't be that roles gain or lose their power by simply redescribing the practice those roles belong to. There is no reason to accept 'war' over 'unjust war' as the correct description of the practice. Every practice can be redescribed in a way that makes it sometimes just and sometimes unjust—even torture. It is just a matter

[31] *See* ibid 613.
[32] ibid 611–12.
[33] Estlund (n 19) 219.
[34] Applbaum (n 5) 77.

of describing the latter as non-consensually imposing harm in order to achieve a goal.

Benbaji and Statman's position does aim to redescribe, in at least some sense, the moral meaning of actions such as the murder of the innocent. Both Benbaji[35] and Statman's arguments, and earlier arguments made by Benbaji, rely on a contractarian view about the morality of war, which allows for symmetrical self-defence cases.[36] The moral division of labour, which is a convention that in this case allows for soldiers to disregard considerations of the justness of the war, is fair and (*ex-ante*) beneficial to all, which is why it is universally accepted. Universal acceptance of the convention has normative implications.[37] In the case of soldiers, just combatants waive their claim against unjust combatants by tacitly accepting the division of moral labour conventionalized by the distinction between *jus ad bellum* and *jus in bello*.[38]

Benbaji and Statman believe then that under certain conditions, social rules can alter the distribution of pre-contractual natural rights and duties. Unjust combatants do not wrong just combatants when they kill them and are not responsible for the violation of just combatants' rights.[39] At the same time, Benbaji and Statman require a certain level of decency on the part of the institution or society.[40] They acknowledge that they must assume that a society or organization is not rendered indecent by its application of, say, capital punishment. This caveat poses a problem for the view. It means that we would eventually require an argument as to why a society or organization that allows people

[35] In earlier work, Benbaji defended a similar view, which relied on contractualism and the moral division of labour, by resorting to the idea that the alternative is too morally demanding. In this view, moral requirements are not 'excessively high-minded'; morality does not require moral sainthood, nor does it condemn our natural tendency to promote our self-interest, to pursue our personal projects, or to give priority to the interests of our loved ones. Morality merely restricts our partiality and is sensitive to human cognitive limitations. Benbaji thinks this argument is about what is morally (weakly) permissible, not about what is morally justified. Thus, his earlier arguments rest on a claim about the moderation of morality, which in turn shapes its content, justifies the moral division of labour, and permits unjust combatants to engage in self-defence. At least in this version, it seems that the view is not actually purporting to redescribe actions that would ordinarily count as murdering the innocent and so the justification is not meant to work when premised on the revisionist morality. *See* Benbaji, 'The War Convention and the Moral Division of Labour' (n 12) 611–15.

[36] *See* Benbaji, 'The War Convention and the Moral Division of Labour' (n 12); Benbaji, 'A Defense of the Traditional War Convention' (n 20); Benbaji, 'The Responsibility of Soldiers and the Ethics of Killing in War' (n 20); Benbaji and Statman (n 21).

[37] Benbaji, 'The War Convention and the Moral Division of Labour' (n 12) 611. In Benbaji and Statman's view, what is required for the distribution of natural rights is non-coercive habitual obedience to the rule. Benbaji and Statman (n 21) 49.

[38] *See* Benbaji, 'The War Convention and the Moral Division of Labour' (n 12) 613; Benbaji and Statman (n 21) 5–6.

[39] Benbaji and Statman do not rule out that unjust combatants may rightfully refuse to fight when they are convinced that overall such an act would be morally unjustified. Benbaji and Statman (n 21) 38–59, 131.

[40] ibid 336.

to murder the innocent on a massive scale (even if there are social rules that are generally followed, fair, and mutually beneficial to all participants) would still count as a decent one. It seems implausible that a society in which the moral division of labour could redistribute natural rights to the point of morally redescribing what counts as murdering the innocent would remain a decent society. Indeed, it is doubtful whether natural rights such as rights not to be harmed or killed are rights whose distribution can be altered at all, even under an attractive contractarian account. We might plausibly say that an altered distribution of those rights would, precisely, make the view lose whatever initial appeal it had.

Benbaji and Statman are aware that the distribution of natural rights that involve killing is difficult to justify. They argue that the responsibility for killing the innocent—just combatants—is not simply eliminated by the convention; it is transferred to another role-holder within the social structure.[41] Combatants lose (waive) their right not to be killed by enemy combatants, but they don't lose their rights against the politicians who send the enemy combatants to the battlefield; the rights of just combatants who are killed are violated by the aggressive state.[42] This leaves open the question of what measures, if any, should be taken against the role-holders that bear responsibility for the unjust war.[43]

But the plausibility of a contractarian view on account of which natural rights are re-distributed does not increase because of this. It is difficult, if not impossible, to justify a social convention that does not demand at least some form of accountability for role-holders who choose to employ others—in this case, soldiers—to murder the innocent on a vast scale. The question of accountability, if the view is to remain attractive, cannot be left open; it must be answered in a certain way.

Now, even if unjust combatants' killings are disanalogous to the murder of innocents—and I will argue that sometimes they are—Estlund's and Benbaji's first requirement, that there are just institutions in the background doing their job, makes their argument limited in its application and quite demanding. The requirement entails that arguments from role morality are at its most compelling when applied to armies within legitimate and democratic states. It starts to unravel when we consider non-state actors (particularly, terrorist organizations or warlords) and highly authoritative and oppressive states, because in these cases the moral division of labour will be non-existent or corrupt.[44] In

[41] ibid 64.
[42] ibid 64, 130.
[43] ibid 65.
[44] This does not necessarily mean that these soldiers are morally responsible for the justness of the war. They may be even less responsible than soldiers who are citizens of democratic states, for they have

other words, the moral division of labour is attractive only to the extent that the demands of justice are being satisfied by background institutions.

However, even in democratic states, it is unclear to what extent these background conditions are satisfied. For the question doesn't seem to be whether a society is *overall* just but whether its institutions are doing the job of properly ascertaining whether a war is just or not. At the moment, it is far from clear that they are.

First, states are hardly in a good position to make such a judgment, since their own interests are at stake. Second, the international legal regime itself is also far from satisfying these background conditions. In addition to the problems with the composition and functioning of the United Nations Security Council and its veto system, international law, as I said in Chapter II, does not concern itself too much with whether a war is just or unjust. There is significant lack of overlap between what makes war just and what makes it legal; a legal war can be an unjust one in the current system. This is not to deny that these are the institutions that currently have the responsibility to ascertain whether a war is just or that they should undertake this responsibility as best as possible. But the current institutional arrangements are certainly a problem for arguments based on role morality and the division of moral labour.

Of course, if arguments about role morality and the moral division of labour are to have any purchase at all in non-ideal conditions (which are the conditions of all present states in the world), the requirement of background institutions satisfying the demands of justice cannot be too demanding. Otherwise, the arguments lose all their power. Thus, the requirement of 'background institutions securing justice' has to be weakened somehow. It must leave room for the existence of second-best institutions in looking after, not even securing, the demands of justice—that justice is being attended to, whether substantively or procedurally. It must, in other words, be a weakened requirement that allows for 'good enough' institutions to be sufficient for the arguments of role morality and the division of labour to take off. After all, the institutions of most democratic societies are somewhat corrupt and prone to failing miserably in securing justice.

A response to the above may simply be, well, so much the worse for role morality and the moral division of labour. They are theories for an ideal world, and since that is not the world we live in, there is, in fact, no place for them at all in human affairs. This would mean that individuals, not only combatants, but *all*

no influence on the conduct of politics and they may be under extreme duress. The point is that the argument from the moral division of labour does not work here.

of us, cannot, in good conscience, leave the demands of justice to institutions. We must attend to all reasons; we cannot exclude some of them and be content to attend to those that apply most readily to our professional and personal lives. This position, however, seems at complete odds with the way in which some of us think we can live morally decent lives, and it also condemns us to living an utterly immoral life. A position that demands us to consider first-order reasons for *everything* we do seems implausible. At least *some* reliance on institutions doing their job might be justified and not simply misguided or self-serving optimism. This is not to say that there is no room for the improvement of our institutions, and our reliance on them may well be too much and at a too heavy price. It is just to point out that role morality has a place in our moral lives—limited, for sure—that is not denied by all the many failures of our institutions in securing justice.

But even a weakened requirement of institutions doing their job in the background poses difficulties for the view in the context of war. For it is not only that the current background conditions are not ideal. It is also that the distribution of roles and the division of moral labour in the international system are not even designed to satisfy the demands of justice. In order for these arguments to gain purchase, we would need to significantly reform our international legal regime in a way that ensured that legal wars were also just ones. This would require enacting new definitions, introducing new requirements for the use of force, and so on.

There is, additionally, one final problem. Even if those background conditions were satisfied, the answer to the question of whether the soldier can serve in a war he reasonably believes to be unjust simply because of the role he occupies is not obviously a positive one. There is, after all, a limit to the power of roles and the moral division of labour. I see no definitive reason why that limit should be in assessing whether a war is just or unjust, or even, whether the war is legal or illegal. After all, we do demand that soldiers refuse to comply with orders that amount to war crimes, crimes against humanity, and genocide.

In other words, the problem with justifications based on role morality and the moral division of labour is that they must distinguish between killings of the innocent committed in unjust wars, which can be justified, and killings of the innocent that are the result of what the law defines as international crimes, which cannot be justified. But if both are instances of killing the innocent, it is hard to see how this could be done in a principled manner. In the end, a point will always come where there will be no substitute to refusal or resistance,[45]

[45] Nagel (n 6) 90.

and it is not clear why fighting a war we believe to be unjust is not where the line should be drawn. As Nagel remarks, the ideas usually associated to role morality form a 'heady and sometimes corrupting brew'.[46]

The problem is that the conflict between what the role demands of soldiers—obedience and service in an unjust war—and 'ordinary' moral demands is not best understood as a conflict between the personal moral judgments of combatants and the demands of the role but as a conflict between what Applbaum calls the demands of the role and political morality.[47] Yet political morality is precisely where the role and its demands find their substantive justification, legitimate authority, and force of obligation.[48] A civil servant, or, in this case, a soldier, cannot be neutral with respect to *political* morality. She must engage in its interpretation and fulfilment.[49] An unjust war—at least, an egregiously unjust one—might go precisely against the political morality of the system to which the soldiers belong. So how can the soldier disregard those considerations?

This problem, as David Luban notes, is not really an objection against the division of moral labour, but against the power of roles.[50] The division of moral labour is retained, in the sense that we can still agree that there are reasons for assigning different duties to different roles and institutions.[51] What changes is the claim that the division of moral labour holds over the soldier's moral judgment: it might not be as absolute as we thought.[52] Some clear-cut cases—most likely to be egregiously unjust wars—may demand resistance to comply; in others, the soldier might be justified in relying on the role of background institutions.

Despite these limitations, arguments from role morality and the division of labour can become significantly stronger if they also rely on arguments about legitimate political authority. They are not then merely arguments about background institutions securing justice but also about the legitimate political authority of those institutions in both securing and deciding how to secure justice, as Estlund himself suggests.[53] If this argument works—and I think it does, although barely when it is premised on the revisionist morality—the soldier might plausibly judge that he does not have the moral authority to act

[46] ibid 76.
[47] Applbaum (n 5) 68.
[48] ibid.
[49] ibid 69.
[50] *See* David Luban, *Lawyers and Justice: An Ethical Study* (Princeton University Press 1988) 144–45.
[51] ibid.
[52] ibid.
[53] *See* Estlund (n 19); David Estlund, *Democratic Authority: A Philosophical Framework* (Princeton University Press 2009).

on his own judgments.[54] The judgment has already been made by democratic institutions.

III. Democratic Authority

Both Estlund[55] and Renzo[56] have advanced arguments from democratic and political authority in the context of war. The argument is obviously limited in its application to soldiers that belong to democratic and legitimate states.

The basic idea behind the argument of democratic authority is that we are obliged to comply with the directives of democratic institutions—even if we disagree with them—precisely because of their democratic provenance.[57] It is thus not an argument about the legitimacy of democratic directives but about their authority. These two can come apart. Authority refers to the moral power of an agent to require or forbid actions by others.[58] Meanwhile, legitimacy refers to the moral permissibility of the state's issuing and enforcing its commands.[59]

Democratic authority then focuses on the subjects' duty to obey the commands of democratic institutions. This duty of obedience is content independent—within certain limits—and pre-emptive.[60] It is content independent because the duty to obey does not depend on the contents of the directive—again, within limits—but merely on its democratic provenance.[61] And it is pre-emptive because other considerations in favour of a different course of action must be put aside by the agent in her decision-making.[62]

There are different answers as to why commands or directives issued by democratic institutions may exercise authority over their subjects. That is, there are different ways in which we can explain why and how the democratic nature of certain directives is important in generating a duty of obedience. There are instrumentalist conceptions of democratic authority (which usually make no special mention of democracy), consent theories of democratic

[54] Applbaum (n 5) 69.
[55] *See* Estlund (n 19); Estlund (n 53).
[56] Massimo Renzo, 'Political Authority and Unjust Wars' (2019) 99 Philosophy and Phenomenological Research 336.
[57] *See* Tom Christiano, 'Democracy' (*Stanford Encyclopedia of Philosophy*, 27 July 2006) <https://plato.stanford.edu/archives/fall2018/entries/democracy/> accessed 17 August 2022.
[58] Estlund (n 53) 2.
[59] ibid.
[60] *See* Tom Christiano, 'Authority' (*Stanford Encyclopedia of Philosophy*, 2 July 2004) <https://plato.stanford.edu/archives/spr2013/entries/authority/> accessed 17 August 2022.
[61] ibid.
[62] ibid.

authority,[63] equality-based or egalitarian ones,[64] and theories supported on epistemic proceduralism.[65] Related to this discussion, there is the more general discussion about political obligation—whether there is a general duty to obey the law as such and why.[66] I will not explore them here, nor will I take a stance on which one is better. It is enough to assume, for my purposes, that democratic decisions *can*—morally speaking—generate duties of obedience in their subjects. We don't need to decide on a particular theory, but we do need to grapple with their limits.

At first glance, there might be a natural resistance against arguments from democratic authority. Surely the morally right action trumps whatever we may say in favour of democratic authority. If we know what the morally right action is, why would we be obligated to follow the democratically enacted directive? Presumably, we would be wrong in doing so. And indeed, in some cases this will be true. The argument about democratic authority has some important limitations; it is not able to sanitize any morally wrong action. Theorists have argued that the same principle that gives democratic authority its power also grounds its limits: the principle of equality on which the argument for democratic authority relies also entails a set of liberal rights.[67] And commands or directives that undermine democratic institutions are also outside those limits.[68]

In thinking about those limits, we once again run into the problem posed by the combination between a duty or permission to participate in an unjust war—here, given by democratic authority—and the revisionist morality. If killing just combatants is just like killing the innocent, the argument from democratic authority seems to lose most of its initial plausibility. Yet, coupled with the argument from role morality and the moral division of labour, I think it does have some plausibility. It may work in cases of wars that are not egregiously or obviously unjust and in wars that are unjust on both sides (and it will also work if the arguments I will provide against the revisionist morality are successful).

We can start thinking about the power of democratic authority in shaping our moral duties by thinking about the moral position of the executioner in a society that disagrees on the question of capital punishment, yet permits it. Is

[63] Christiano, 'Democracy' (n 57). Renzo, however, takes a different approach and believes that the command gives a presumptive, not an exclusionary reason. Renzo (n 56).

[64] *See* Daniel Viehoff, 'Democratic Equality and Political Authority' (2014) 42 Philosophy & Public Affairs 337.

[65] Estlund (n 53).

[66] *See* Richard Dagger and David Lefkowitz, 'Political Obligation' (*Stanford Encyclopedia of Philosophy*, 2014) <https://plato.stanford.edu/entries/political-obligation/> accessed 10 August 2022.

[67] Christiano, 'Democracy' (n 57).

[68] Applbaum (n 5) 70.

the executioner a murderer? And we can think about the judge called to apply the death penalty in that very same society. Ought the executioner and the judge be guided by their moral views on capital punishment, rather than by laws—very flawed laws—that regulate their conduct as both executioners and judges?

Of course, we are squeamish about the executioner himself. Who, we might think, would even choose to become an executioner? Yet this is primarily a question about the executioner's moral character, which is not our main concern. We do not aim to judge whether the executioner has some murderous inclinations that explain how he came to be one in the first place. Rather, the question is whether—regardless of the existence of those inclinations about which we feel squeamish—he is to be rightfully considered a murderer by those of us who oppose capital punishment. The same applies to the judge—whose decision to become a judge is harder to put into question—whom we can imagine judiciously applying the laws of capital punishment when he is required to do so, even though he may abhor the death penalty and actively protest against it in his spare time. What do we make of him? Is he fundamentally morally misguided? And what do we make of the rogue judge, who guides herself by her, to her mind, morally correct views? She does not apply capital punishment, not even when required by the law.

These examples are difficult to judge, mainly because some of us so deeply sympathize with the rogue judge's views. A different example might put things into perspective. Suppose there is a doctor in a small-town who practices abortions. She sees it as her duty to provide this medical service, which is legal in this imaginary country with public healthcare and a functioning democracy. A second doctor, in a different town, refuses to practice abortions, in spite of his being the only obstetrician and gynaecologist in town. Plain and simple, he thinks abortion is murder. And, finally, a third doctor living in a coastal town nearby, who also thinks abortion is murder, does, in fact practice abortions. As the only obstetrician and gynaecologist in town, and as part of a public healthcare system, she thinks it is her duty to provide this medical service, even if she thinks it amounts to murder. Now, what do we make of these doctors?

What is unsettling about these examples is that there seems to be something admirable about the doctor who, while thinking that abortion is murder, practices abortions because of the democratic authority of the laws, even if—and perhaps, precisely *because*—she thinks that abortion is murder. And there might be something admirable about the judge who applies capital punishment when the laws say that she should. This is not just a matter of their moral character: we might also be able to judge them as doing what is right; what democratic institutions and their roles demand of them. They are free, we might say, to protest against those decisions outside of those roles.

Of course, we might argue that these doctors have the wrong moral views about abortion, while the judge and the executioner have the right ones about capital punishment. There is nothing admirable about the doctor who practices abortion while thinking abortion is murder that is about following the directives of democratic authority. It is simply good that she thinks herself bounded to them, but only because they lead her to do the right thing. By contrast, the executioner and the judge are simply in the wrong: there is nothing to say in favour of their following the directives of democratic authority. Nothing at all—it is misguided, harmful, and perpetuates evil.

Capital punishment is a very tricky example, very much at the margins—most likely outside the margins—of the reach of a plausible argument in favour of democratic authority. We would certainly find nothing to admire in a judge who dutifully sent people off to concentration camps because the law says he should. There is a place for activism, even within roles and even if limited to the most egregious wrongs.[69] Still, at least historically, individuals who have found themselves in need of deciding between the demands of their role and what they believe to be a grievous moral wrong have struggled. Perhaps the most well-known and carefully examined examples come from Robert Cover's *Justice Accused*, where he describes the 'squirming' and 'untold personal agony' undergone by judges who opposed slavery when they had to decide on cases concerning slavery.[70] And even in spite of that agony and struggle, the judiciary, 'almost uniformly, marched to the music, steeled themselves, and hung Billy Budd'.[71] Some of those judgments that 'marched to the music' acknowledged that although slavery was decidedly against natural law, it was left to the people to decide, not to judges.[72] A disregard of the law by the judiciary could potentially 'undermine and overturn the social compact'.[73] Indeed, sometimes these judges thought that there were strong reasons for sticking closely to the legal rules, even in spite of these laws' clear divergence from what was morally required.

[69] To be clear, I think that abortion bans are one of them.

[70] *See* Robert M Cover, *Justice Accused: Antislavery and the Judicial Process* (Yale University Press 1975).

[71] ibid 5–6.

[72] ibid 119–21.

[73] ibid. Cover goes on to try to articulate what underlies the judges' restraint (a tension between moral values served by anti-slavery and those served by fidelity to the formal system) and notes that even antislavery forces disagreed about what judges ought to do in slavery cases—whether they should apply natural law or positive law.

There are, nonetheless, many cases where the plausibility of democratic authority is stronger and which are outside of the landscape painted by morally egregious wrongs. The tax collector who thinks that billionaires should pay a 10 per cent wealth tax might be right. Yet we would likely frown upon his going around using his office to actually *collect* this tax from the billionaires under his radar. The water company agent who thinks that the law limiting water usage to stop climate change is too weak and that *more* needs to be done, cannot just go around limiting people's water usage even further. The police officer, who is convinced that people shouldn't drive unless they are impeded from using public transportation, cannot just go around fining those who drive even if not strictly necessary. And while someone who assigns food or food stamps in state-funded programmes might think that being vegan is morally required, she cannot just refuse resources to those who eat meat. And finally, think about our rogue judge again. It turns out, she has a full range of moral views. And in order to be consistent, she has now decided that she will no longer apply the law: she is required, she claims, to apply what morality demands. No less can be expected of her.

These people—their actions seem morally dubious now. A while ago, I said, quoting Nagel, that the ideas associated to role morality form a heady and sometimes corrupting brew. But the brew formed by the absence of role morality, and in this case, by the absence of democratic authority, is also heady. It is the brew of zealotry.

Thus, even with its limitations, the argument from democratic authority, coupled with arguments from role morality and the moral division of labour, has some power. It is not enough to support a duty to obey the most egregiously unjust directives,[74] but its power and scope are larger than what may seem at first sight. In the case of unjust combatants, the authority of the directive might be powerful enough to judge their actions as, at the very least, permitted. Unless one is commanded to fight in a war that is egregiously unjust—like a genocidal war or a war that is a naked act of aggression—the judgment of whether a war is unjust is complex. Unjust wars are unjust in different ways: some lack a just cause; some fail the proportionality requirement; sometimes, there is a just cause but the just cause might not have been the reason why the country has

[74] Renzo, for example, argues that members of a legitimate state have a *pro-tanto* presumptive duty to obey the command to fight in a war, unless they are justified in believing that the authority has made a mistake in ordering them to fight and they are sufficiently confident of that belief. In that case, they would have the duty to disobey. Renzo (n 56). And Burri also thinks that arguments like the ones defended by Benbaji and Statman are limited to non-culpable breaches of *ad bellum* rules. *See* Susanne Burri, 'If You Care About a Rule, Why Weaken Its Enforcement Dimension? On a Tension in the War Convention' [2022] Law and Philosophy 1.

decided to go to war, and so on. In these cases, the argument from democratic authority will be at its most powerful. It is only in the most egregious cases of wrongfulness that it will fail to do its job. Thus, even if limited to democratic societies, this argument provides a reason for combatants to obey the directive to participate in an unjust war.

However, the objections I posed in section I about the inattention of the international legal system to the justness of war and the partiality of states also have force here. Even legitimate and democratic states are not in the best position to ascertain whether they are fighting just wars. Thus, we should not overestimate the force of these non-instrumentalist justifications in generating duties or permissions to participate in unjust wars. They are most likely to be of limited application. And they will fail in cases of egregiously unjust wars.

IV. Conclusion

Within certain limits, and when not premised on the revisionist morality, the arguments above—the division of moral labour, role morality, and democratic authority—can explain why unjust combatants may be sometimes permitted to participate in an unjust war.

These arguments can also gain strength due to the epistemic limitations that individuals might face in judging whether a war has a just cause. Most of the time, decisions to go to war will be taken in light of information that is not publicly available. In some of these cases, it might be reasonable to trust authorities, and agents will thus be evidence-relative justified in believing that their war has a just cause. However, this is an argument about evidence-relative justification and deference to experts.[75] It is not an argument about the special permissions granted by institutions to certain individuals so that they can do what is morally wrong, nor is it an argument about the moral authority of certain institutions. After all, if the opportunity of judging the justness of the war is available to someone, she should make use of it.

Yet if revisionists are right and what unjust combatants do is simply like murdering the innocent, the purchase of non-instrumentalist justifications such as the above is quite limited. Thus, if we are to attenuate the fragility of the instrumentalist rationale, we must move on and examine the revisionist morality itself. This will be the task in the following chapters.

[75] *See* Applbaum (n 5) 237–38.

V

Just Wars, Just Combatants, Just Killings?

I. Introduction

This chapter is the first step in dismantling the revisionist morality of war. The aim is to attenuate the fragility of the instrumentalist justification of the legal privilege of combatants. And this is where we start: with the revisionist morality and its shortcomings.

That just combatants have a moral justification for killing unjust combatants is the view held by most of the literature on the morality of war.[1] Indeed, revisionists think that unjust combatants are liable to be killed in self-defence: they forfeit their rights against harm,[2] and because they forfeit their rights against harm, they are not allowed to defend themselves and are not wronged when killed. The revisionist morality of war, in its most prominent versions, has some distinctive features: it relies on certain methodological commitments (reductivism and individualism) and posits a particular theory of self-defence (based on the forfeiture of rights). It aims, precisely, to provide a complete account of the morality of killing in war—yet it ends up being quite an incomplete account of the morality of war. And in order to escape its limitations, it must let go of its methodological commitments or risk becoming an implausible theory about killing in war.

If the arguments in this and the next chapter are successful, the revisionist morality of war leaves a vast space of killing in war unaccounted for, morally speaking. It is within this vast space that we can deconstruct the analogy between what unjust combatants do and the murder of the innocent by proposing

[1] As noted before, Tadros says that 'it is uncontroversial that combatants on the unjust side are liable to be harmed'. Victor Tadros, 'Orwell's Battle with Brittain: Vicarious Liability for Unjust Aggression' (2014) 42 Philosophy & Public Affairs 42, 55.

[2] Among the ones who hold a forfeiture-based view of self-defence in war, *see* Jeff McMahan, *Killing in War* (Oxford University Press 2009); Cécile Fabre, *Cosmopolitan War* (Oxford University Press 2012); Adil Ahmad Haque, *Law and Morality at War* (Oxford University Press 2017); Helen Frowe, *Defensive Killing* (Oxford University Press 2014). On forfeiture *see* Judith Jarvis Thomson, 'Self-Defense and Rights' (The Lindley Lecture, University of Kansas, 1976) <https://kuscholarworks.ku.edu/bitstream/handle/1808/12392/Self-Defense%20and%20Rights-1976.pdf;sequence=> accessed 18 August 2022; Judith Jarvis Thomson, 'Self-Defense' (1991) 10 Philosophy & Public Affairs 283; Judith Jarvis Thomson, *The Realm of Rights* (Harvard University Press 1990) 361–73.

The Morality of the Laws of War. Marcela Prieto Rudolphy, Oxford University Press. © Marcela Prieto Rudolphy 2023.
DOI: 10.1093/oso/9780192855473.003.0005

an alternative morality of war. I leave somewhat open how much purchase the revisionist morality has after I have argued against it. The revisionist morality might be entirely mistaken but I do not go that far in the following pages. However, it does not matter whether the revisionist morality is, in fact, completely mistaken. It is enough, for my purposes, to partially—if significantly—displace it.

Although I will raise some doubts about revisionism's methodological commitments, my main disagreement with revisionists is substantive. It is about the plausibility of a forfeiture or liability-based view of self-defence in accounting for most killing in war; it is about the claim that most unjust combatants are liable to substantial and even lethal harm. The most prominent version of this view is McMahan's, and I will engage mostly with his arguments. However, some of what I say is also applicable to other similar forfeiture-based views—such as Frowe's, Fabre's, and Tadros'—and to views that are not entirely revisionist, because they do not challenge the prohibition on killing civilians, but are, however, based on forfeiture—such as Haque's. When necessary, I will address some of their arguments directly.

The argument against the plausibility of this view has two different, but related, parts. In this chapter, I will argue that a self-defence view committed to reductivist individualism fails to account for most killing in war. In order to do so—that is, in order to expand its justificatory reach—it would have to let go of its methodological commitments (which is something that it sometimes inadvertently does). In Chapter VI, I will argue that a plausible forfeiture-based view cannot make many—and even most—unjust combatants liable to significant or lethal harm. Because the two arguments are related, I sometimes anticipate the problems with forfeiture in this chapter, and leave some arguments about the consistency of the view in holding on to reductivism and individualism for Chapter VI.

This chapter begins with a brief description of revisionism, understood both as a substantive and methodological position (section II). It focuses on the views of Jeff McMahan, Helen Frowe, and Cécile Fabre. These authors hold slightly different views of the morality of killing in war,[3] but they find common ground in their methodological commitments—they are individualists and

[3] I do not include Adil Haque as a revisionist, even though he is committed to individualism and reductivism, because his conclusions can be seen as occupying a middle ground between revisionist and traditionalist views. Haque thinks that most civilians from the unjust side are not liable and that not all unjust combatants are liable. However, because he holds a forfeiture account of self-defence and is committed to individualism and reductivism, I occasionally discuss his position regarding the morality of killing in war. *See* Haque (n 2).

reductivists—and hold a liability-based view of self-defence.[4] They also think that most unjust combatants and some civilians from the unjust side are liable to harm. Section III focuses on revisionism's limitations as a complete account of the morality of killing in war. Section IV concludes.

II. Revisionism

The debate between revisionists and traditionalists has grown increasingly complex and, as a result, the debate about how to categorize these views properly has done so too.[5] Traditionalism, associated to Walzer's view,[6] defends the legal equality of combatants and the immunity of civilians from attack as consistent with the morality of war. Walzer's account of this morality is somewhat collectivist and allows for institutional arrangements to redistribute certain rights. Revisionism, made prominent by Jeff McMahan,[7] challenges all these claims. McMahan argues that both the symmetry thesis and the immunity of civilians are false as a matter of morality, rejects the power of institutional arrangements to redistribute rights against harm, and endorses individualism (as opposed to Walzer's more sympathetic view toward certain forms of collectivism) and reductivism (as opposed to Walzer's position that the morality of war is *sui generis*) as methodological commitments.

McMahan's revisionism—and by extension, revisionism in general—can be understood as having a substantive and a methodological strand. The substantive strand consists of the denial of the symmetry thesis and of civilian immunity from attack; the methodological strand consists of reductivism and individualism. I will refer to these as substantive and methodological revisionism, respectively.

The relationship between substantive and methodological revisionism is only contingent. Most substantive revisionists are also methodological revisionists: they are committed to individualism and reductivism.[8] They have thus

[4] Tadros and Frowe, however, also hold a different view of liability, which is not related to forfeiture but to the existence of a morally enforceable duty. And Tadros has recently distanced himself from individualism. *See* Victor Tadros, 'Duty and Liability' (2012) 24 Utilitas 259; Frowe, *Defensive Killing* (n 2); Victor Tadros, *To Do, to Die, to Reason Why: Individual Ethics in War* (Oxford University Press 2020).

[5] *See* illuminating distinctions by Seth Lazar, 'Method in the Morality of War' in Seth Lazar and Helen Frowe (eds), *The Oxford Handbook of Ethics of War* (Oxford University Press 2015); Tadros, *To Do, to Die, to Reason Why* (n 4).

[6] Michael Walzer, *Just and Unjust Wars: A Moral Argument with Historical Illustrations* (5th edn, Basic Books 2006).

[7] McMahan, *Killing in War* (n 2).

[8] Frowe, *Defensive Killing* (n 2); McMahan, *Killing in War* (n 2); Fabre (n 2).

held on to self-defence as the justification for killing in war under the understanding that war is composed of a series of individual acts (they are individualists), each of which can be justified by self-defence (they are reductivists).[9]

There are, however, notable exceptions to this generalization—different combinations between the claims of substantive and methodological revisionism. Some authors have tried to defend some features of the traditionalist view—most usually, the principle of civilian immunity—while embracing individualism and reductivism.[10] In recent work, Tadros has distanced himself from individualism,[11] and Walen has embraced substantive and methodological revisionism while developing a new model of thinking about rights.[12]

A. Methodological Revisionism

Methodological revisionism consists of individualism and reductivism. Both individualism and reductivism are hard to parse as methodological commitments. Individualism, as Seth Lazar explains, can be evaluative or descriptive.[13] Evaluative individualists think that the moral significance of a collective is reducible to its contribution to the well-being of the individuals who compose it.[14] Descriptive individualists think that any act that may appear to be collective is reducible to individual acts.[15] As Lazar points out, the debate between individualists and collectivists is unlikely to be solved just by thinking about war.[16] Tadros finds the contrast between individualists and collectivists unhelpful.[17] He thinks that there are different questions to ask about groups and their relation to individuals in the ethics of war, such as whether groups can have permissions and duties, whether liability to be harmed can be grounded in facts about groups, and so on.[18]

The difficulty for a purely individualist view about war is to move away entirely from some form, even if weak, of collectivism. As Lazar notes, only at the collective level can we speak of a just or unjust

[9] *See* Lazar, 'Method in the Morality of War' (n 5). Among those who hold this view see, McMahan, *Killing in War* (n 3); Haque (n 3); Frowe, *Defensive Killing* (n 2); Fabre (n 2).

[10] Seth Lazar, *Sparing Civilians* (Oxford University Press 2015); Haque (n 2).

[11] Tadros, *To Do, to Die, to Reason Why* (n 4).

[12] Alec Walen, *The Mechanics of Claims and Permissible Killing in War* (Oxford University Press 2019).

[13] Seth Lazar, 'War' (*The Stanford Encyclopedia of Philosophy*, 3 May 2016) <https://plato.stanford.edu/archives/spr2017/entries/war/> accessed 18 August 2022.

[14] ibid.

[15] ibid.

[16] ibid.

[17] Tadros, *To Do, to Die, to Reason Why* (n 4) 9.

[18] ibid.

war.[19] Without an appeal to this level, the question 'was the war just?' has no meaningful form.[20] If wars are reduced to a set of individual actions—as some individualists would have it—we can only say whether each action is or is not justified. Thus, to say that the war is just, we must either say that all the actions that compose it are justified, individually considered (which seems exceedingly demanding), or that the majority of individual actions that compose it are justified (which seems arbitrary).[21]

Frowe, who is committed to individualism, has conceded that the individualist approach entails that a war can be properly described as justified only if each composite action is justified.[22] She thinks, however, that this is not an objection against individualism. What the approach truly holds is that the overall description of a war as justified is not terribly interesting.[23] There can be no meaningful single verdict on those actions taken together.[24] What has moral interest, she seems to suggest, are the actions of each individual: of the individual leader who decides to go to war, of the individual soldier who has to decide whether to fight, and so on.[25] There is simply no point in attempting to pronounce a war just as a whole.[26] This version of individualism would also put pressure on the distinction between *jus in bello* and *jus ad bellum*. We could no longer ask whether the resort to force is just without resorting to some level of collectivism.

Reductivism is also ambiguous, and consequently the debate between reductivism and exceptionalism (the idea that the morality of war is *sui generis*) may be partly overblown, as Lazar suggests.[27] In general terms, reductivism claims that there is only one morality: war is not a special moral sphere with special moral rules but only a particular aspect of ordinary life where we apply our ordinary moral rules.[28] The conditions of war change nothing at all; they simply make it more difficult to ascertain relevant facts.[29] Consequently, the rules governing killing in war are simply the rules governing killing between

[19] Individualists do not think there is a morally significant collective dimension in war. They think war can be wholly understood from a moral perspective by understanding each individual act. For the distinction and explanation *see* Lazar, 'Method in the Morality of War' (n 5); Lazar, 'War' (n 13).

[20] *See* Lazar, 'Method in the Morality of War' (n 5) 14.

[21] ibid 16.

[22] Helen Frowe, 'The Just War Framework' in Seth Lazar and Helen Frowe (eds), *The Oxford Handbook of Ethics of War* (Oxford University Press 2015) 55.

[23] ibid.

[24] ibid.

[25] ibid.

[26] ibid.

[27] Lazar, 'War' (n 13).

[28] Frowe, *Defensive Killing* (n 2) 124.

[29] ibid.

individuals.[30] Most obviously, revisionists think, these are the rules of self-defence and defence of others.[31]

An alternative version of reductivism holds that killing in war must be justified by the same properties that justify killing outside of war, that is, that justify killing in self-defence.[32] Understood in this way, reductivism is not just a methodological commitment.[33] It is also a substantive position about the properties that matter in justifying killing. Methodological reductivism—discerning the ethics of war by discerning the ethics of killing in self-defence—can only take off once we have accepted a substantive position about killing in self-defence whereby properties that are often present when killing in war—but absent when killing in self-defence—do not have any moral significance.

This substantive position is adopted by most revisionists,[34] but it need not be. If we understand reductivism as committed to the idea that the properties that justify killing in war must be the same as the properties that justify killing in self-defence,[35] there is room for arguing that certain properties are more often present in war (eg coercion, hierarchical organization, high number of individuals involved), thus resulting in a modified version of the morality of self-defence. This version of reductivism would embrace some differences between the two moralities as a contingent result of how often certain properties, identified as important in justifying killing, are present in different contexts.

The substantive reductivist position—which claims that properties often present in war but absent in self-defence have no moral significance—is difficult to defend. Ordinary moral rules vary significantly among themselves. There is the morality of promising, the morality of euthanasia, the morality of friendship, the morality of romantic partnership, and so on. All of these 'spheres of morality' have some features in common, but they also have important differences.

Perhaps the reductivist argument goes like this: all spheres of life where we intentionally harm individuals without their consent must be governed by the same moral rules. Much rests on the meaning of 'intentionally' and 'harm', since both have the potential of swallowing up other areas of morality where we might think that different rules apply (is lying to someone imposing a

[30] ibid.

[31] ibid.

[32] Lazar, 'War' (n 13). This seems to be Tadros' and Fabre's positions. *See* Tadros, *To Do, to Die, to Reason Why* (n 4) 12–13, 27; Fabre (n 2) 7.

[33] The debate between individualism and collective is also substantive, for similar reasons.

[34] *See*, eg, Helen Frowe, 'The Moral Irrelevance of Moral Coercion' (2021) 178 Philosophical Studies 3465; and Tadros, *To Do, to Die, to Reason Why* (n 4).

[35] Lazar, 'War' (n 13). This seems to be Tadros' and Fabre's positions. *See* Tadros, *To Do, to Die, to Reason Why* (n 4) 12–13, 27; Fabre (n 2) 7.

non-consensual harm? Not fulfilling a promise?) But even if we find a coherent way of setting up these limits, the fact that two spheres of human life can be described similarly—as the intentional imposition of non-consensual harm—does not really provide a conclusive argument that both spheres are subject to the same moral rules. Friendship, parenting, and romantic partnership can all be described as loving relationships that have intrinsic value, yet the moral rules for friendship, romantic partnership, and parenting are different. They are different not because that description—that they are loving relationships with intrinsic value—is wrong, but because that description does not capture the differences between those relationships. To put it in terms of moral properties, friendship-based duties are justified and shaped by moral properties that also justify parental or romantic partnership-based duties. But it is likely that those duties also respond to different moral properties, some of them particular to friendship, parenting, and so on. Thus, a complete argument for reductivism must show that the differences between killing in war and killing in self-defence do not have any bearing on the moral rules that regulate them; that there are no features particular to war that can act as moral properties in justifying killing—that they have no moral significance at all. This seems very unlikely.[36]

War has certain characteristics that make it peculiar. Michael Walzer has set forth some of them. First, he argues that the circumstances of war and of the battlefield are highly coercive.[37] Second, war is an intensely collective and collectivizing experience. People don't go to war only to defend themselves. They are members of a collective to which they often attach great value, and they are engaged in a project which is not entirely their own.[38] Strategic decisions on the battlefield are made not with great attention to individuals but to what they can accomplish together.[39] Third, war is a world of radical and pervasive uncertainty.[40]

There is certainly room for expanding this list, as Walzer himself acknowledges.[41] I would only add the observation that all the features Walzer mentions are precisely what make war *war* and not something else entirely. It is difficult to imagine how the circumstances of the battlefield could ever be non-coercive: it is in the nature of combat that some decisions have to be made in

[36] Lazar agrees that reductivism may be incomplete. *See* Seth Lazar, 'Just War Theory: Revisionists versus Traditionalists' (2017) 20 Annual Review of Political Science 37, 39.
[37] Walzer (n 6) 342–42.
[38] *See* Moshe Halbertal, *On Sacrifice* (Princeton University Press 2012) 87.
[39] Walzer (n 6) 341–42.
[40] ibid 344.
[41] ibid 339.

very little time, sometimes while being shot at. It seems impossible for armies and armed groups not to be somewhat coercive (though not necessarily as coercive as they are now, as we will see later): their hierarchical organization and division of labour is explained by the demands of efficiency in waging a war. The collective and collectivizing aspects of war also seem inescapable. Though we can imagine that a lonely individual could, with many weapons, conduct large-scale attacks on their own, it is uncertain whether we would call this war. War usually requires many people doing something together, sometimes to the detriment of their individualities. Finally, uncertainty is another feature of war that seems unlikely to disappear.

Perhaps these features of war are reasons *not to fight wars*. However, they are hardly features that we can cast aside when thinking about the morality of war, which is what reductivism tends to do. If what I say in the following pages is correct, these features are quite important: it is inevitable that when many individuals do something together within hierarchically organized groups, individual contributions will be smaller, and knowledge and responsibility will be diluted. These features of war, I will argue in Chapter VI, make the liability of most unjust combatants implausible. They are, then, properties of moral significance.

Reductivism can also be understood as denying that the just war or the just cause themselves can act as *sui generis* moral justifications, different from self-defence, for imposing harm on other individuals. Some versions of individualism will also deny this. Since a justification premised on the just war or the just cause seems collective or, at the very least, requires aggregation of some kind, an individualist view would reduce it to its composite parts: several instantiations of killing in self-defence. That is, a just cause is nothing but the aggregation of several instantiations of killing in self-defence. And a reductivist view cannot claim that the just war or the just cause provide a distinctive moral justification for harming individuals without ceasing to be reductivist.

B. Substantive Revisionism: Liability and Self-Defence

Substantive revisionism advances two main arguments. First, the symmetry thesis is morally false: unjust combatants are liable to harm, whereas just combatants are not.[42] This argument has become predominant in the literature,

[42] McMahan, *Killing in War* (n 2); Fabre (n 2); Frowe, *Defensive Killing* (n 2); Tadros, *To Do, to Die, to Reason Why* (n 4); David Rodin, *War and Self-Defense* (Clarendon Press 2002).

even amongst scholars who reject other parts of the revisionist view.[43] Second, the principle of civilian immunity is also false: some civilians from the unjust side of the war are liable to harm.[44]

Most revisionists hold a theory of self-defence that is liability-based: unjust combatants can be killed in self-defence because they are morally liable to be killed. But what does being liable mean?[45]

1. Liability to harm

Individuals can be liable to different things,[46] but it is liability to lethal and sub-lethal harm that is relevant to the ethics of killing in war. The most important feature of liability is that if someone is liable to a certain amount of harm, they are not wronged when that harm is non-consensually imposed on them.[47] As a result of the no-wronging nature of this kind of harm, those who are liable have no justified complaints against the agent who has harmed them; their rights are neither violated nor infringed.[48] Liability is relevant because most think that if someone is liable to defensive harm, they are not entitled to engage in counter-defence.[49] Liability may also have implications regarding whether third parties may intervene in a conflict and on which side, the need to apologize or engage in other compensatory acts, and appropriate reactive attitudes.[50]

In the ethics of war, liability is usually employed to mean forfeiture.[51] Thus, when revisionists say that unjust combatants are *liable* to be killed or harmed by just combatants, they are simply saying that unjust combatants have *forfeited* their rights against harm or death.[52] Because they have forfeited those

[43] See, eg, Haque (n 2); Lazar, *Sparing Civilians* (n 10).

[44] McMahan, *Killing in War* (n 2); Fabre (n 2); Frowe, *Defensive Killing* (n 2); Tadros, *To Do, to Die, to Reason Why* (n 4).

[45] As Tadros notes, philosophical work on liability is still in its infancy. A significant part of what follows will reflect that. *See* Tadros, 'Orwell's Battle with Brittain' (n 1) 47.

[46] ibid 48.

[47] Ibid.

[48] Lazar, *Sparing Civilians* (n 10).

[49] Jeff McMahan, 'Self-Defense against Justified Threateners' in Gerald Lang and Helen Frowe (eds), *How We Fight: Ethics in War* (Oxford University Press 2014) 117; Frowe, *Defensive Killing* (n 2) 1; David R Mapel, 'Moral Liability to Defensive Killing and Symmetrical Self-Defense' (2010) 18 Journal of Political Philosophy 198, 200.

[50] *See* Jonathan Quong, 'Liability to Defensive Harm' (2012) 40 Philosophy & Public Affairs 45, 45–46.

[51] McMahan, *Killing in War* (n 2) 10. Ferzan makes the same point: 'Although the language on the surface is "liability," the moral principle at work is one of rights forfeiture.' Kimberly Kessler Ferzan, 'Forfeiture and Self-Defense' in Christian Coons and Michael Weber (eds), *The Ethics of Self-Defense* (Oxford University Press 2016) 233. But Tadros, for example, thinks that there are (at least) two types of moral liability, one based on forfeiture and the other on the liable person's having an enforceable duty to serve certain ends. *See* Victor Tadros, *The Ends of Harm: The Moral Foundations of Criminal Law* (Oxford University Press 2011).

[52] *See* Uwe Steinhoff, 'The Liability of Justified Attackers' (2016) 19 Ethical Theory and Moral Practice 1015, 1016; Jeff McMahan, 'The Basis of Moral Liability to Defensive Killing' (2005) 15 Philosophical Issues 386, 386; David Rodin, 'Justifying Harm' (2011) 122 Ethics 74, 75; Frowe, *Defensive Killing* (n 2) 4;

rights, they are not wronged when harmed or killed.[53] Liability is thus a convenient way to say that someone has forfeited their rights. As McMahan puts it:

> to attack someone who is liable to be attacked is neither to violate nor to infringe that person's right, for the person's being liable to attack *just is* having *forfeited* his right not to be attacked, in the circumstances.[54]

Because liability is just a synonym for forfeiture, I will use these two terms interchangeably. However, the usage of liability to mean forfeiture raises some difficulties, as it tends to muddle what forfeiture can explain clearly under the language of rights. If we all have a right not to be killed, how could it be that someone's killing us in self-defence doesn't wrong us? The answer is precisely that we may have lost that right: it may have been forfeited. Thus, we come to lack a claim-right we originally had, and its Hohfeldian correlative, a duty, may be absent as well. In its place, there may be a privilege to kill us in self-defence, which is precisely the correlative of a no-right.[55] Using 'liability' here to mean 'forfeiture' introduces unnecessary confusion, for moral liability cannot be understood as equivalent to legal liability under Hohfeld's scheme.[56] In Hohfeldian terms, liability is the correlative of a power, and the opposite of an immunity. B will have a liability when B has volitional control over some super-added fact or group of facts that may change a given legal relation—that is, when B has a legal power.[57] Thus, having a liability is to be subject to having one's first-order normative relations altered by another party.[58]

If we examine the relation between the victim and her aggressor in respect to her killing him in self-defence,[59] given that the aggressor is morally liable (has a liability to be killed), the victim should have a correlative power. But

Seth Lazar, 'The Responsibility Dilemma for Killing in War: A Review Essay' (2010) 38 Philosophy & Public Affairs 180, 183. Tadros, 'Orwell's Battle with Brittain' (n 1) 50; Tadros, 'Duty and Liability' (n 4) 260–61; McMahan, *Killing in War* (n 2).

[53] Tadros, 'Orwell's Battle with Brittain' (n 1) 50; Tadros, 'Duty and Liability' (n 4) 260–61; Frowe, *Defensive Killing* (n 2) 3–4; McMahan, *Killing in War* (n 2).

[54] McMahan, *Killing in War* (n 2) 10.

[55] *See* Wesley Newcomb Hohfeld, 'Some Fundamental Legal Conceptions as Applied in Judicial Reasoning' (1913) 23 Yale Law Journal 16; Wesley Newcomb Hohfeld, 'Fundamental Legal Conceptions as Applied in Judicial Reasoning' (1917) 26 Yale Law Journal 710.

[56] Ferzan briefly notes this issue too. *See* Ferzan (n 51) 253.

[57] Hohfeld, 'Some Fundamental Legal Conceptions as Applied in Judicial Reasoning' (n 55) 44.

[58] *See* Rodin, *War and Self-Defense* (n 42) 20–21.

[59] *See* John Finnis, 'Some Professorial Fallacies about Rights' (1971) 4 Adelaide Law Review 377, 377; Andrew KW Halpin, 'Hohfeld's Conceptions: From Eight to Two' (1985) 44 The Cambridge Law Journal 435, 439.

this doesn't work. If liability is the correlative of a power, it would imply that the other party in the relation—that is, the victim—has volitional control over facts that may change some first-order normative relation:[60] the aggressor's right not to be killed. Yet she does not. It is usually the *liable* individual, that is, the aggressor, who has volitional control over who forfeits a right, not the victim. The victim does nothing and has no power to change the rights and duties of the aggressor.

Neither can moral liability be understood as the opposite of a moral immunity. An immunity means that a party is not subject to having her first-order relations altered by a second party: in this case, that someone's right to life is not subject to modification by a second party. However, we don't generally think other people have powers to alter rights such as other people's right to life. Indeed, to be 'non-liable' simply means that one has not forfeited one's rights because one has done nothing to trigger forfeiture. This, of course, is just the specification of a general right we all have: a right not to be killed, period. To be non-liable is simply to have your claim-right to life (meaning, among other things, a right not to be killed).

Precisely because the liable individual is the one with volitional control over the forfeiture of her own rights, Ferzan argues that liability (ie forfeiture) should be understood as a negative normative Hohfeldian power.[61] If she is right, and moral liability is a normative power, the language of liability is quite misleading because liability, as it turns out, is not a liability but its precise opposite: a power.

Of course, we could just say that Hohfeld's scheme shouldn't apply to moral concepts and that, as a result, these worries are irrelevant. Although I agree that the application of Hohfeld's scheme to moral concepts is not always possible nor straightforward, the forfeiture of rights is one of the cases which can be easily accommodated by Hohfeld's framework. It is thus somewhat unfortunate that the usage of 'liability' to mean 'forfeiture' causes unnecessary confusion.

In spite of this slight disadvantage, I will not drop the terms 'liability', 'liable', and 'non-liable' but will use them interchangeably with the term 'forfeiture'. This choice is just a matter of employing the terms that have become pervasive in the discussion about the ethics of war.

[60] Rodin, *War and Self-Defense* (n 42) 14–48.

[61] Ferzan (n 51). The requirement of some volitional control implies, of course, that people who pose involuntary threats do not forfeit their rights, which is Ferzan's position.

2. The moral responsibility account of liability

Like every self-defence theory based on forfeiture, the revisionist morality of war has two obstacles to overcome. First, it needs to defend forfeiture itself. Second, it needs to provide a convincing reason that may ground (or trigger) forfeiture; that is, it needs to explain what someone must do to lose their rights. Usually, most of the focus is on the latter, and we are left without an account of why we should endorse forfeiture in the first place. Perhaps the use of 'liability' explains why the challenges against forfeiture itself have been partially neglected, or at least why the challenges against forfeiture are not dealt with independently.[62] In any case, the debate about the ethics of war does not focus on why forfeiture is the proper moral explanation: most of the discussion is centred around what grounds or triggers forfeiture.

Although liability theories of self-defence have become dominant, and they are practically the consensus in the ethics of killing in war, forfeiture-based accounts of killing in self-defence were once controversial.[63] Many objections have been posed against forfeiture accounts, mostly related to the unattractiveness of a theory whereby people may lose basic rights against harm. As David Rodin explains, liability justifications involve moral considerations that affect the underlying grounds for the possession of a particular right (in this case, of rights against being killed or harmed).[64] But what seems to ground the possession of these basic rights has nothing to do with whether agents behave according to morality; we have basic rights, such as the right to life, because we are human beings (or, at the very least, sentient creatures).[65] One might worry that the forfeiture of our most basic rights puts into question the very moral status of those who have forfeited their rights.[66]

Although I'm sympathetic to these concerns about forfeiture, I will not attack forfeiture as a general explanation of impositions of non-consensual harms nor will I argue that the revisionist theory of self-defence must be

[62] Though as an exception in recent years, *see* Gerald Lang, 'Why Not Forfeiture?' in Helen Frowe and Gerald Lang (eds), *How We Fight* (Oxford University Press 2014); Massimo Renzo, 'Rights Forfeiture and Liability to Harm' (2017) 25 Journal of Political Philosophy 324; Michelle Madden Dempsey, 'Against Liability' in Christian Coons and Michael Weber (eds), *The Ethics of Self-Defense* (Oxford University Press 2016); Ferzan (n 51).

[63] Lang (n 62); Whitley Kaufman, 'Is There a "Right" to Self-Defense?' (2004) 23 Criminal Justice Ethics 20.

[64] David Rodin, 'The Lesser Evil Obligation' in Saba Bazargan and Samuel C Rickless (eds), *The Ethics of War* (Oxford University Press 2017) 31.

[65] *See* Kaufman, 'Is There a "Right" to Self-Defense?' (n 63); Whitley Kaufman, 'Self-Defense, Innocent Aggressors, and The Duty of Martyrdom' (2010) 91 Pacific Philosophical Quarterly 78; Thomas Scanlon, *What We Owe to Each Other* (Belknap Press of Harvard University Press 1998).

[66] See, eg, these remarks by Locke, where he compares those who have forfeited their rights with animals. John Locke, *Two Treatises of Government: And a Letter Concerning Toleration* (Ian Shapiro ed, Yale University Press 2003) 2.10.

completely discarded. However, some of the issues I will point at in this and the next chapters are inevitably entangled with what is problematic about forfeiture. The problems of forfeiture are only aggravated when thinking about the moral status of unjust combatants. The one that will come to the surface more often is how disproportionate the forfeiture of basic rights against substantial harm is in relation to what unjust combatants do: it is disproportionate to their degree of moral responsibility, to the amount of harm they threaten, to their contributions to the unjust war, and so on.

Forfeiture theories of self-defence, as I said above, need to defend not only forfeiture: they also need to provide an account of what a person must do in order to forfeit their rights. There are different answers to this question,[67] but several revisionists have endorsed the 'responsibility account' of liability, whereby people forfeit their rights by being morally responsible for an objectively wrongful threat of harm.[68] This doesn't require those morally responsible to be posing the threat themselves.[69] If revisionists are right, and unjust combatants are liable, then it follows that they are not morally permitted to fight back, and that they are not wronged when they are killed.[70]

III. Liability, Self-Defence, and their Limitations

An account of the morality of war—such as the revisionist view—that aims to be a complete account of the morality of killing in war by only relying on

[67] Quong distinguishes between the moral responsibility account; the evidence-relative account, whereby a person can only be liable when they are responsible for a threat of harm to the innocent as a result of an action that is morally impermissible on the basis of the available evidence; the culpability account; Tadros' account, which is not forfeiture-based; and Quong's own moral status account, whereby liability is determined by whether one behaves as if other people are liable to the harms one might impose. Quong, 'Liability to Defensive Harm' (n 50) 46–48.

[68] McMahan, *Killing in War* (n 2); Frowe, *Defensive Killing* (n 2); Haque (n 2); Fabre (n 2). Note that Frowe also relies, in some cases, on Tadros' account of liability.

[69] This is McMahan's formulation of the principle. *See* McMahan, *Killing in War* (n 2) 7–15. Frowe speaks of indirect causes, that is, someone whose actions, movements, or presence are causally upstream of the direct threat to the victim. Helen Frowe, 'Self-Defence and the Principle of Non-Combatant Immunity' (2011) 8 Journal of Moral Philosophy 530, 535–40; Frowe, *Defensive Killing* (n 2) 73–78. Finally, Haque talks about killing 'indirectly'. Indirect killers kill someone by inducing those who they effectively control to directly kill someone else. Haque (n 2) 64–65.

[70] But, as I have said before, there are other ways to ground forfeiture as well. Holding a forfeiture account of liability we also find: Saba Bazargan, 'Defensive Liability Without Culpability' in Christian Coons and Michael Weber (eds), *The Ethics of Self-Defense* (Oxford University Press 2016); Frowe, *Defensive Killing* (n 2); Haque (n 2); Gerald Lang, 'What Follows from Defensive Non-Liability?' (2017) 117 *Proceedings of the Aristotelian Society*; Lazar, *Sparing Civilians* (n 10); Michael Otsuka, 'The Moral-Responsibility Account of Liability to Defensive Killing' in Christian Coons and Michael Weber (eds), *The Ethics of Self-Defense* (Oxford University Press 2016); Rodin, *War and Self-Defense* (n 42); Quong, 'Liability to Defensive Harm' (n 50); Christian Coons and Michael Weber, 'The Ethics of Self-Defense' in Christian Coons and Michael Weber (eds), *The Ethics of Self-Defense* (Oxford University Press 2016) 1.

self-defence faces a daunting task. This is because fighting a war involves targeting enemy combatants not only when they are imminently threatening other combatants with grievous harm but also when they are sleeping, retreating, advancing, or simply staying in place; when they are unlikely to contribute to the war in any way; when they do not have a combat function; and even when it is uncertain that they may pose threats in the future. This is, in fact, consistent with the legal rules of targeting during armed conflict.[71] The extension of cases to which an account of the morality of killing in war based on self-defence ought to apply is thus quite significant.

Of course, such a moral view could reduce its scope of application and argue for the impermissibility of some actions—most obviously, of the killings of non-threatening unjust combatants. However, such a view would quickly become an argument for some form of pacifism—it is impossible to fight a war targeting only those who are posing imminent threats to others—or a partial account of the morality of war that needs to be completed with some other moral justification, different from self-defence.

Reductivist individualists are not the only ones who think self-defence is important when thinking about killing in war.[72] They are, however, the only ones to claim that all killing in war can be understood as the aggregation of several individual instantiations of self-defence. Revisionism's extension, then, purports to be quite wide.

Because the theory of self-defence held by revisionists is based on liability, it must put forward a theory of self-defence that makes most unjust combatants liable if it wants to account for killing in war. Otherwise, the theory is cornered into pacifism.[73] Yet, as we will see, that most unjust combatants are liable is a difficult claim to support without abandoning revisionists' methodological commitments.

[71] Yoram Dinstein, *The Conduct of Hostilities under the Law of International Armed Conflict* (Cambridge University Press 2004) 42. Members of the armed forces are permissible targets, except for prisoners of war, *hors de combat*, medical and religious personnel; and members of the armed forces and military units assigned to civil defence organizations. *See Geneva Convention for the Amelioration of the Condition of the Wounded, Sick and Shipwrecked Members of Armed Forces at Sea (Second Geneva Convention)* (hereinafter, GCII) 12 August 1949, 75 UNTS 85, art 12; *Protocol Additional to the Geneva Conventions of 12 August 1949, and Relating to the Protection of Victims of International Armed Conflicts (Protocol I)* (hereinafter, API) 8 June 1977, 1125 UNTS 3, art 41. According to article 41, a person is *hors de combat* if 'a) he is in the power of an adverse Party; he clearly expresses an intention to surrender; or c) he has been rendered unconscious or is otherwise incapacitated by wounds or sickness, and therefore is incapable of defending himself, provided that in any of these cases he abstains from any hostile act and does not attempt to escape'. API I, art 42. *Geneva Convention for the Amelioration of the Condition of the Wounded, Sick and Shipwrecked Members of Armed Forces at Sea (Second Geneva Convention)* (hereinafter, GCII) 12 August 1949, 75 UNTS 85, art 12. For more details, see Chapter I.

[72] Walzer, for example, uses the domestic analogy. *See* Walzer (n 6).

[73] On this, *see* Lazar, 'The Responsibility Dilemma for Killing in War' (n 52).

Given that the extension or scope of the revisionist theory of self-defence is so wide, I will approach it in steps. First, I will assess it as applied to the killing of aggressors (ie individuals who are currently attacking or imminently threatening to attack other individuals). Then I will address its application to the cases of enemy combatants who are not currently engaged in aggressive behaviour towards other individuals. Finally, I will move outside the two-person case of self-defence and directly discuss the problem that the number of aggressors poses for the theory.

A. Killing Aggressors: Liability, Combatants, and the Villainous Aggressor

That we are permitted to defend ourselves against those who are culpably attacking us is one of our most widely shared moral intuitions. Except for absolute pacifism, all moral theories permit at least some form of defensive action against people who are presently engaged in aggressive behaviour toward others. Practically all legal systems mirror this intuition by including self-defence as a defence against murder. Yet though we agree that we may kill in self-defence, there is no agreement as to why we may do so.[74]

Disagreement is also widespread in relation to the scope of the permission. Although everyone agrees it covers the villainous aggressor (someone who is attacking an innocent and is culpable for his behaviour) but not the innocent bystander (someone who is not attacking me but whose death could divert or stop someone from harming me)[75], we disagree about whether it permits killing innocent aggressors (someone who has the intention to kill me but is not morally responsible for having formed this intention)[76] and innocent threats (someone who threatens my life even though he has no intention to kill me and exercises no agency).[77] On top of this, we also disagree about which people

[74] Positions range from rights-based (Thomson, 'Self-Defense' (n 2); Thomson, 'Self-Defense and Rights' (n 2); Rodin, *War and Self-Defense* (n 42); McMahan, *Killing in War* (n 2)) to necessity or lesser evil-based (Paul H Robinson, 'A Theory of Justification: Societal Harm as a Prerequisite for Criminal Liability' (1975) 23 UCLA Law Review 266), and forced choice between lives (Phillip Montague, 'Self-Defense and Choosing between Lives' (1981) 40 Philosophical Studies 207) On this categorization, *see* Rodin, *War and Self-Defense* (n 42) 51–57.

[75] There are further distinctions: culpable bystander is someone who has committed a wrongful act but is not *now* committing a wrong. *See* Whitley Kaufman, 'Torture and the "Distributive Justice" Theory of Self-Defense: An Assessment' (2008) 22 Ethics & International Affairs 93, 98.

[76] The three standard examples of innocent aggressors are the mentally ill or intoxicated aggressor, the mistaken aggressor, and the juvenile aggressor. *See* Kaufman, 'Self-Defense, Innocent Aggressors, and The Duty of Martyrdom' (n 65).

[77] For the distinction *see*, Thomson, 'Self-Defense' (n 2); Jonathan Quong, 'Killing in Self-Defense' (2009) 119 Ethics 507, 507; Kaufman, 'Self-Defense, Innocent Aggressors, and The Duty of Martyrdom'

count as innocent aggressors, innocent threats, and bystanders.[78] Finally, although there is general agreement that self-defence is permitted only when it is necessary to avert the threat, when the threatened act of harm is imminent, and the response is proportional, there is also disagreement about what these requirements mean,[79] and whether they are external or internal to liability.

As said before, revisionists hold a particular view of self-defence. As all theories of self-defence do, it argues that we may defend ourselves and others against the villainous aggressor. Revisionists claim that the aggressor is liable—he has forfeited his rights—by posing an objectively unjust threat of harm for which he is morally responsible. The villainous aggressor is, indeed, the central case of self-defence, and in Judith Jarvis Thomson's classic formulation it goes something like this:

> [Y]ou are standing in a meadow, innocently minding your own business, and a truck suddenly heads toward you. You try to sidestep the truck, but it turns as you turn. Now you can see the driver: he is a man you know has long hated you. What to do? You cannot outrun the truck. Fortunately, this is not pure nightmare: you just happen to have an antitank gun with you, and can blow up the truck. Of course, if you do this, you will kill the driver . . . .[80]

As posed above, the case of the villainous aggressor seems like an odd fit with most of the killing that happens in war. I readily grant that at least *some* unjust combatants will be, morally speaking, like the villainous aggressor and *some* just combatants will be, morally speaking, like the victim. Some unjust combatants may be attempting to kill just combatants while being fully culpable for their actions, not acting under duress, correctly believing that they are fighting an unjust war and that the enemy's side is fighting a just one, and having no reason to believe that their participation in the unjust war is permissible nor that their lives are in imminent danger. If this sounds a bit unreal then I have partially made my point, because even though some instances of killing in war

(n 66). Burri distinguishes between profiting and non-profiting threats. *See* Susanne Burri, 'The Toss-up between a Profiting, Innocent Threat and His Victim' (2015) 23 Journal of Political Philosophy 146.

[78] *See*, eg, Frowe, *Defensive Killing* (n 2); Michael Otsuka, 'Killing the Innocent in Self-Defense' (1994) 23 Philosophy & Public Affairs 74; Noam J Zohar, 'Innocence and Complex Threats: Upholding the War Ethic and the Condemnation of Terrorism' (2004) 114 Ethics 734.

[79] *See* Rodin, *War and Self-Defense* (n 42) 40–43.

[80] Thomson, 'Self-Defense' (n 2) 283.

will stay close to the case of the villainous aggressor, and thus, will be morally justified,[81] many will likely not.

Hence, the first problem that we find in holding on to self-defence as the primary moral explanation of killing in war is that many (if not most) unjust combatants are significantly different from villainous aggressors, at least in three different ways.

1. Unjust combatants and the threats they pose

The first way is related to the unjustness of the threat that unjust combatants pose. If we want to hold on to individualism, it is essential that unjust combatants have forfeited their rights not to be killed on account of something about them. What, exactly, is this? The argument from liability proceeds like this: unjust combatants forfeit their rights not to be killed by being morally responsible for an objectively unjust and harmful threat, whereas just combatants, by posing a justified threat, do not.[82]

Now, this is a bit tricky, because though revisionists claim to be individualists, in most cases,[83] the only way to understand why the threat posed by the unjust combatant is unjust (and thus, gives way to the forfeiture of rights) is by reference to the unjustness of the war his collective is waging—not in reference to the actions of the combatant himself. The same is true for the just combatant; her threat is justified also in reference to the collective on behalf of which she is fighting. In other words, some appeal to the collective must be made. Though in some cases we can predicate combatants wrong or right moral action without reference to the collective,[84] we can only say a soldier is fighting in a just or unjust war in reference to what their collective is doing. Soldiers stand in front of each other not only as individuals but also as part of a bigger entity in reference to which their actions are to be understood.

Thus, the first difficulty in employing a theory of self-defence to explain why unjust combatants are liable to be defensively harmed is that unjust combatants—unlike the villainous aggressor—are liable not due to the

[81] In these cases, the legal privilege of the just combatant will be justified on self-defence. I don't mean by this that if taken to court, the combatant would have to put forward the defence of self-defence; he would not. He would claim the defence most soldiers have in the corresponding domestic criminal law system for killing other combatants (fulfilling a duty, killing in war) or claim that they are privileged combatants. What I mean is that his killing is morally justified on self-defence grounds, even if legally the proper defence to apply is a different one.

[82] *See* McMahan, *Killing in War* (n 2); Frowe, *Defensive Killing* (n 2); Fabre (n 2).

[83] I say most because McMahan, for example, considers cases where the threat posed by the unjust combatant is justified due to lack of justification for the just combatant's threat (eg the latter is committing a war crime). *See* McMahan, *Killing in War* (n 2).

[84] For example, war crimes.

unjustness of the individual threat they are posing towards other individual combatants (their micro-threat)[85] but due to the unjustness of the war they are fighting. Or, in other words, the reason why the threat they pose is unjust is because their collective is fighting an unjust war and their opponents, a just one. This poses a problem in terms of the methodological commitments of revisionists: it strays away from individualism. This is because the moral status of the unjust combatant's threat seems to depend largely on the moral status of the threat posed by the collective to which he belongs. Thus, at the very least, we need an account of the sort of connection that must exist between the micro-threat of the unjust combatant and the macro-threat of the unjust war; an account that explains why and under which circumstances the unjustness of the war is transferred to the micro-threat of the unjust combatant and under which circumstances the unjust combatant is morally responsible for it.

2. Unjust combatants and moral responsibility

The second way in which unjust combatants are different from villainous aggressors is in terms of their responsibility. Though sometimes combatants may have full moral responsibility for the micro-threats they are posing, there is an important distance—in terms of moral responsibility—between them and what renders their threat unjust, which is not present in the case of the villainous aggressor. Most unjust combatants have little moral responsibility—and certainly no culpability—for the unjustness of the war: they do not decide to go to war, and they may have not decided to participate in it either. In other words, their micro-threat is deemed unjust by its relation to a macro-threat for which most of them are not culpable or morally responsible, or about which they may be reasonably mistaken.[86] This is not the case with the villainous aggressor: we judge the unjustness of his action in reference to a threat he is culpable for and that he chooses to pose. And the aggression is initiated by the same person who will bear the harm of having initiated it—again, the villainous aggressor.

[85] I follow Zohar's distinction between micro-threats and macro-threats. *See* Zohar (n 78) 737.

[86] *See* ibid. The common law doctrine of 'felony murder', whereby someone who is committing a felony will be guilty of murder if someone dies during it, even if the death is accidental and unforeseeable, could provide the beginnings of an argument in favour of a more expansive understanding of moral responsibility which could potentially include unjust combatants. Nevertheless, because the doctrine has been abolished in most common law jurisdictions (and does not exist in civil ones) for its unfairness, I will not deal with it. *See*, eg, Paul H Robinson and Tyler Scot Williams, *Mapping American Criminal Law: Variations across the 50 States* (Praeger 2018) Chapter V; William Bald, 'Rejoining Moral Culpability with Criminal Liability: Reconsideration of the Felony Murder Doctrine for the Current Time' (2017) 44 Journal of Legislation 239.

These differences between the villainous aggressor and combatants suggest that many unjust combatants are not culpable, but they do not preclude the possibility that some of them might be, even if they did not decide to initiate the unjust war. A combatant who willingly participates in what they reasonably and correctly believe is an egregiously unjust war, even when it would not be particularly costly to refuse to fight, might have a level of culpability similar to that of the villainous aggressor. As a result, they might become liable when posing threats to the innocent. But, again, many unjust combatants are not like this.

Now, there are ways in which the scope of the permission to kill in self-defence can be extended beyond the central case of the villainous aggressor. Obviously, because of the differences between villainous aggressors and combatants, this must be done in a theory about killing in war. And it may be done in two ways.[87] Only the second one is attractive to revisionists. But even though it is attractive to revisionists, it is far from being an attractive moral view, as we will come to find.

First, we can say it is permissible to defensively kill even those who are not morally responsible for posing a threat. Indeed, most of the literature on self-defence and most legal systems agree that it is permissible to defend oneself against innocent aggressors (the reasonably mistaken, those who are not responsible for their actions due to mental illness or intoxication,[88] and very young children) and do not rule out force against innocent threats.[89] If this is correct, self-defence can ground the permission to kill unjust combatants who, though presently engaged in aggressive behaviour toward others, would be unlike the villainous aggressor: they might have a subjective justification for posing the threat either because they reasonably believe they are being unjustly attacked or because they reasonably believe their war is just. The same applies to unjust combatants who are mentally ill or intoxicated, young children (depressingly relevant to war), and innocent threats. But extending liability to innocent aggressors makes the theory less appealing. A plausible and morally attractive view of forfeiture ought to ground liability to be killed on something

[87] There is a third way, which I don't consider here, which is how defence of others can extend our range of justified killers. For every unjust combatant that can be permissibly killed in self-defence, we may think that not only is the just combatant presently under threat allowed to kill him, but so are his comrades in arms, on account of defence of others. This certainly extends the number of justified killers, but it doesn't extend the number of those who can be killed.

[88] The unfortunate short term in the literature is 'the psychotic aggressor'. I will not use this term and use instead 'someone who is not morally responsible for their actions due to mental illness or intoxication'.

[89] *See* Kaufman, 'Self-Defense, Innocent Aggressors, and The Duty of Martyrdom' (n 65). Against killing the innocent aggressor *see* Otsuka (n 78); Rodin, *War and Self-Defense* (n 42).

for which the agent has at least *some* degree of moral responsibility. If an individual does not have *any* moral responsibility for the threat they pose, it seems implausible that they may forfeit their rights.[90]

This leads us to the second way in which the scope of self-defence within the responsibility domain can be extended, and it is precisely the route revisionists have taken. If revisionism is to provide a complete account of killing in war that is based on liability to defensive harm, it has to argue that most unjust combatants are liable—that they have forfeited their rights against death or, at the very least, significant harm. Otherwise, the theory becomes an argument for pacifism.[91] And it looks like the most plausible—but, as we will see, not that plausible—way of doing this is to propose a theory where the threshold or degree of moral responsibility required to render someone liable to defensive harm is very low. This expansion, however, comes at a heavy price: it makes civilians liable too.

This problem constitutes what Seth Lazar has called 'the responsibility dilemma'. Reductivist individualists must choose between two unpalatable alternatives.[92] If they endorse a low threshold of responsibility for liability, civilians become liable. If they endorse a high threshold of responsibility for liability, they make too few unjust combatants liable, which makes a just war impossible to fight permissibly. A 'moderate' threshold of responsibility has the same problem: too many combatants will be non-liable; too many civilians, liable.[93]

Of course, holding a high threshold of responsibility does not inevitably lead to pacifism. We could provide an additional argument to explain why and under what circumstances we can intentionally kill the non-liable.[94] One way

[90] Burri, for example, suggests that providing us with significant control in our role as would-be right violators pushes us towards culpability-based views of liability. Susanne Burri, 'Defensive Liability: A Matter of Rights Enforcement, Not Distributive Justice' [2021] Criminal Law and Philosophy 1, 11.

[91] *See* Lazar, 'The Responsibility Dilemma for Killing in War' (n 52).

[92] *See* ibid; Lazar, *Sparing Civilians* (n 10) 8–12; Seth Lazar, 'The Morality and Law of War' in Andrei Marmor (ed), *Routledge Companion to the Philosophy of Law* (Routledge 2012) 364, 365.

[93] Lazar, *Sparing Civilians* (n 10) 10. This is related (but not the same) to the idea that liability is comparative. It is not the same because one could think that liability is comparative but that comparisons can only be made after reaching a high enough threshold of responsibility. Conversely, a non-comparative account of liability could perfectly well settle the threshold of responsibility very low. As Lazar explains, for liability to be comparative means that it arises only when harm is unavoidable and must be distributed. What matters in determining liability is the comparison between the responsibility of each of those threatened by an unavoidable and indivisible harm. Even if one bears some responsibility for such harm, one is not liable to suffer it if someone else is more responsible. The downside of this is that if harm is unavoidable and indivisible, even a small degree of responsibility can render someone liable, as long as it is more than the responsibility of the other person. I will come back to this later. *See* Lazar, *Sparing Civilians* (n 10); Lazar, 'The Responsibility Dilemma for Killing in War' (n 52); McMahan, 'The Basis of Moral Liability to Defensive Killing' (n 52); Jeff McMahan, 'Who Is Morally Liable to Be Killed in War' (2011) 71 Analysis 544.

[94] But Rodin, for example, who holds a high threshold of liability, is sceptical about a liability view of self-defence's ability to justify killing in war. *See* Rodin, 'Justifying Harm' (n 52) 110; Rodin, *War and Self-Defense* (n 42).

of doing this would be to argue that there is something about a just war that makes it permissible. Reductivists, however, cannot take this way out, for they deny that the morality of war is *sui generis*. Some of them have thus endorsed a lesser-evil justification to explain why killing the non-liable may be permissible. I will discuss this solution in the next chapter.

Now, how low is the threshold of responsibility most revisionists argue for? Perhaps an example will be illustrative:

> Conscientious Driver. A person who always keeps her car well maintained and always drives carefully and alertly decides to drive to the cinema. On the way, a freak event that she could not have anticipated occurs that causes her car to veer out of control in the direction of a pedestrian.[95]

According to McMahan, because the driver knows that driving is an activity with a very tiny risk of causing great harm, she is liable to be killed to prevent her from harming the pedestrian, who is an innocent bystander.[96] Now, we may think that the pedestrian is permitted to defensively kill the driver. But liability claims have additional implications: because the driver is liable, she cannot defend herself against the pedestrian's attack,[97] or, at the very least, the pedestrian does not wrong the driver when he kills her. That the driver is liable means that she has forfeited her rights.

Thus, as a general principle, McMahan's account of liability would imply that anyone engaged in an activity, knowing or having reason to know that there is a chance (however small) that it will cause harm to an innocent will become liable to defensive harm once that risk materializes.[98] Individuals who fall outside this threshold of moral responsibility are those who are non-culpably and invincibly ignorant about the fact that they pose any kind of threat or risk of harm to others. For example, A is typing on his computer. Unbeknownst to him, Villain has rigged the computer to detonate a bomb that will kill an innocent (thesis writer).[99]

[95] McMahan, *Killing in War* (n 2) 165.

[96] *See* ibid 165–67.

[97] McMahan is uneasy about this implication, and considers whether cases in which the differences between the responsibility of different agents are too small may give rise to symmetrical cases of self-defence. ibid 178–82.

[98] Also note the dilemma that the moral responsibility view faces regarding the notion of foreseeability. *See* Jonathan Quong, *The Morality of Defensive Force* (Oxford University Press 2020) 29–32.

[99] McMahan, *Killing in War* (n 2) 168–69. The problem with this example, however, is that once we have imagined that there is the possibility of someone rigging our laptops so that it kills people, we become very much like the Conscientious Driver. Burri, however, argues that a principled line between Conscientious Driver and examples such as Thesis Writer can be drawn, but that the responsibility

This account—known as the moral responsibility account—is the predominant account of killing in war, probably because any theory that sets the threshold of responsibility higher would have a difficult time making most unjust combatants liable.[100] Yet the moral responsibility account seems implausibly broad: most daily activities would have the potential of making us liable to be killed.[101] Moreover, even if the pedestrian may defend himself, it seems implausible that the driver cannot do so as well. She is too different from the villainous aggressor: her moral responsibility for the threat of harm is far too small to give way to the forfeiture of rights against significant harm.

The combination of a forfeiture-based view of self-defence with a low threshold of responsibility is, as a result, quite counterintuitive. The source of the tension is the idea that we can lose rights against significant and even lethal harm simply because we have a very small degree of responsibility for the threats we pose. Forfeiting one's most basic rights has significant implications: liable individuals cannot engage in counter-defence, have no claims against their killer, and deserve no compensation or apology. They are not wronged. But merely having a small degree of moral responsibility for the threat one poses seems unlinked, in a relevant sense, to these implications. They seem completely disproportionate to the degree of responsibility the driver has over the course of events.

Let us ignore these concerns about the plausibility of forfeiture in cases such as these (I will return to them in the next chapter). Even if we accepted a low threshold of responsibility to trigger liability, where does this leave us in war?

As I said before, lowering the threshold of responsibility makes a significant number of civilians liable to be killed or harmed. Some bite the bullet: indeed, they say, some (or most) civilians from the unjust side of the war are liable.[102] Others argue that although they are liable, or although they have moral responsibility for the unjust war (and sometimes, more so than combatants), it is almost never necessary nor proportional to kill civilians to win the war.[103]

I take it that making civilians liable is an unpalatable consequence. It also puts into question the intuitiveness of the view. Making civilians liable to

account fails to have an adequate rationale. Susanne Burri, 'Morally Permissible Risk Imposition and Liability to Defensive Harm' (2020) 39 Law and Philosophy 381.

[100] Quong sets a higher threshold in his moral status account but it hasn't been applied to war so far. I'll come back to this in Chapter VI. *See* Quong, *The Morality of Defensive Force* (n 98).

[101] Quong also finds this result counterintuitive. *See* ibid 26–29.

[102] Frowe, 'Self-Defence and the Principle of Non-Combatant Immunity' (n 69) 543; Frowe, *Defensive Killing* (n 2).

[103] *See* McMahan, *Killing in War* (n 2) 225–35. Because necessity and proportionality are internal to liability in McMahan's account, this may render them non-liable.

significant harm runs against a deeply held intuition, one that is often stronger than intuitions we may have about examples such as the conscientious driver. But leaving aside this objection, there is still a third way in which unjust combatants are significantly different from villainous aggressors (and are different from the conscientious driver): some of them are not posing any threats to other individuals. They may be sleeping, resting, travelling from one place to the next, repairing machinery or buildings, etc. Thus, a vast array of killings in war still need to be justified on the basis of self-defence. I turn to these now.

B. Killing Enemy Combatants

A low threshold of responsibility to trigger liability does not make the problem of non-threatening combatants—that is, unjust combatants who are not posing a micro-threat to just combatants—disappear. If revisionism is to account for these, it needs to be expanded even more.[104] Revisionists cannot claim that the threat is the unjust war and that every soldier, no matter what he is doing, is responsible for that threat or contributing to it—such a claim would stray away from individualism.[105] And if we want to stick to reductivism, we cannot say that there is something about war that makes it permissible to kill non-threatening people.

Revisionists have attempted to solve the issue of the non-threatening unjust combatant by weakening the required proximity between the aggressor and the threat. B is liable to defensive harm if she has moral responsibility for a threat, even if she is not currently posing the threat herself.[106] Imagine a sergeant who is resting while her platoon is targeting some just combatants. Here, we might say that the sergeant is responsible for the threat her squad is posing (she probably ordered them to do so) and thus, she can be killed in self-defence.

[104] To my knowledge, only Zohar has pressed this point. *See* Zohar (n 79). But some have examined this self-defence requirement with great attention, particularly regarding people (mostly women) who kill their abusive spouses. In these cases, some argue that the imminence requirement should be relaxed. *See*, eg, Marcia Baron, 'Self-Defense: The Imminence Requirement' in Leslie Green and Brian Leiter (eds), *Oxford Studies in Philosophy of Law*, vol 1 (Oxford University Press 2011).

[105] They could, however, say something slightly different, but it makes forfeiture highly implausible. I will return to this in Chapter VI.

[106] Again, this is McMahan's formulation of the principle. *See* McMahan, *Killing in War* (n 2) 7–15. Frowe speaks of indirect causes, that is, someone whose actions, movements, or presence are causally upstream of the direct threat to Victim. Frowe, 'Self-Defence and the Principle of Non-Combatant Immunity' (n 69) 535–40; Frowe, *Defensive Killing* (n 2) 22, 73–81. Finally, Haque talks about killing 'indirectly'. Indirect killers kill someone by inducing those who they effectively control to directly kill someone else. Haque (n 2) 64–65.

Two issues arise with this way of extending the scope of targets to those who are not posing threats. The first one is particular to McMahan's theory of liability, whereas the second applies more generally.

McMahan's theory of liability has a distributive dimension. Liability is a matter of justice in the *ex-ante* distribution of unavoidable harms.[107] As such, liability attempts to approximate what would be a just distribution of harm in ideal conditions.[108] When harms are indivisible and must be imposed on someone, the nearest approximation of that ideal is for the harm to go to the more responsible party, who is therefore liable.[109] Understood in this way, liability is comparative: whether one is liable to harm depends not only on what one has done but also on what others have done.[110] Even if one bears some responsibility for an unavoidable harm that cannot be divided, one may not be liable to suffer that harm if someone else is more responsible.[111]

But if liability is comparative in this way, it cannot mean that in the example above the sergeant is liable along with the platoon members. It would mean that the sergeant is the *only* liable person, for among the members of her platoon, she is the one most responsible for the harm that just combatants are facing. (And perhaps, not even the sergeant would be liable, as those most responsible for the threat of the unjust war are higher up in the chain of command, but let's leave this aside for a moment.) This seems to imply that just combatants who target the members of the platoon rather than the sergeant are imposing harm on the non-liable, and thus that the members of the platoon could fight back (and perhaps just combatants would, as a result of posing a threat to the non-liable, become liable themselves).

Maybe we would like to say that the comparative dimension of liability is not meant to differentiate between those who are posing an unjust threat. Thus, in the example above, we would compare the just combatant—let's assume there is only one of them, for simplicity—with each of the members of the platoon

[107] McMahan, 'Self-Defense against Justified Threateners' (n 49) 117; Jeff McMahan, 'Self-Defense and the Problem of the Innocent Attacker' (1994) 104 Ethics 252.

[108] McMahan, 'Self-Defense against Justified Threateners' (n 49) 117; McMahan, 'Self-Defense and the Problem of the Innocent Attacker' (n 107). Both Susanne Burri and Jonathan Quong have argued against understanding liability as a matter of distributive justice. *See* Quong, *The Morality of Defensive Force* (n 98); Burri, 'Defensive Liability' (n 90).

[109] Jeff McMahan, 'Duty, Obedience, Desert, and Proportionality in War: A Response' (2011) 122 Ethics 135, 155.

[110] Lazar, *Sparing Civilians* (n 10) 4–5.

[111] McMahan thinks that this idea of the distribution of harms should be understood as a deontological theory. The defensive action against someone is itself just; it is not permissible because it produces a better or more just outcome. McMahan, 'Who Is Morally Liable to Be Killed in War' (n 93) 552; McMahan, 'Self-Defense and the Problem of the Innocent Attacker' (n 107) 260; Lazar, *Sparing Civilians* (n 10).

and the sergeant separately. This seems consistent with how McMahan and others have understood liability justifications, as working by 'pairwise comparisons,'[112] and it would ensure that all of these unjust combatants, considered individually, would be liable in relation to the just combatant. Comparisons between the unjust combatants would be precluded or relevant only pertaining to the relation among themselves. Yet if liability really is meant to capture the most just distribution of harm, it seems that the sergeant must bear all of the harm if that is feasible, making the rest of her platoon non-liable. Otherwise, we would need to explain why just combatants are at liberty of not restricting themselves to the fairest distribution of harm possible and why not doing so would not render them liable. Hence, if liability is comparative in this way, it cannot provide an account for killing in war, nor can it extend the range of permissible targets. What it seems to provide is not only an argument in favour of targeted killing—and this, indeed, is how Michael Walzer interprets McMahan's position[113]—but an account of why targeted killing is demanded by morality.

If we drop the comparative dimension, liability would be able to extend the range of permissible targets in hierarchical relations like the one between the sergeant, who is currently not posing a threat, and the members of her platoon, who are posing the threat. Yet this extension *only* works for hierarchical relations; it doesn't work horizontally. If squad A is resting while squad B is shooting in the front, it is quite difficult to say that squad A is responsible for the threat squad B is posing. They may not even know that squad B is posing a threat, and they certainly didn't order it. Because wars are fought by hundreds of platoons, organized under battalions, divisions, and so on, this extension is still too limited. The problem of non-threatening combatants remains.

Remember the conscientious driver. She was posing an imminent threat to the victim. And this is true of all the examples that McMahan provides, yet it is often not true of war. The problem, then, is to explain why unjust combatants, who are not posing any threats and are not responsible for the threats that others are currently posing, may be liable to harm without abandoning individualism or reductivism. Is there a way to do this?

McMahan has suggested that there is reason to think that the liability of unjust combatants is strict.[114] That is, once their country is waging or planning to

[112] Jeff McMahan, 'Liability, Proportionality, and the Number of Aggressors' in Saba Bazargan and Samuel C Rickless (eds), *The Ethics of War* (Oxford University Press 2017) 5. Kamm and Rodin accept this view too. FM Kamm, *Ethics for Enemies: Terror, Torture, and War* (Oxford University Press, 2011) 133–34. Rodin, 'Justifying Harm' (n 52).

[113] *See* Michael Walzer, 'Response to McMahan's Paper' (2006) 34 Philosophia 43.

[114] McMahan, *Killing in War* (n 2) 183–84.

wage an unjust war, they may be liable, even if they have done nothing objectively wrong, at least not yet.[115] Their liability is grounded on the fact that they made a voluntary choice to join the army, which committed them to obedience.[116] Thus, it is reasonable for them to think that they will be ordered to fight and that they will participate in the unjust war.[117] These two factors—their choice of joining the army and the fact that they will be ordered to fight—make them share moral responsibility for the threat their country is posing.[118] This, in turn, makes them liable to preventive attack when that is necessary to avert the threat posed by their country.[119]

This argument is not committed to individualism, which is probably why, in his more recent work, McMahan argues that unjust combatants who make no contribution to their side's unjust war, and those, such as military lawyers, who restrain or impede the prosecution of the war, are not liable.[120] He has suggested that one cannot infer from the fact that a person is a combatant, or even from the fact that he is an unjust combatant, that he is liable to be killed.[121] And indeed, the argument about strict liability is simply not individualist, but collectivist: what is imminent is only the collective threat of the unjust war. Moreover, responsibility is grounded not on what I do, but on something my collective does (for which I may have none or little moral responsibility, and maybe not even knowledge) together with something I did in the past (join the army) and something I might do in the future (fight in the war).

There are other reasons why the argument is unconvincing. First, the judgment about what is reasonable to believe is unsound. It is not true that it is reasonable for combatants to believe that they will be ordered to fight. Some of them may not participate in the war at all. It is more reasonable to think that not *all* divisions of the army will be asked to participate in the war (countries like to leave armed forces at home, for obvious reasons) and that not all soldiers of each unit will go. Second, the argument seems to imply that if we have reason to believe that civilians will be conscripted to fight in an ongoing unjust war, they may be liable to preventive attack *before* being drafted, if that is necessary to avert the threat. This completely blurs the distinction between civilians and

[115] ibid.
[116] ibid.
[117] ibid.
[118] ibid.
[119] ibid.
[120] Jeff McMahan, 'What Rights May Be Defended by Means of War?' in Cécile Fabre and Seth Lazar (eds), *The Morality of Defensive War* (Oxford University Press 2014) 122–23.
[121] ibid.

combatants, but it also, once again, puts tension on the idea of forfeiture for being disproportionate.

Another solution to the problem of non-threatening combatants is Frowe's. Frowe argues that someone working in an administrative branch of the mafia can be killed in self-defence if shooting him would enable the intended target of a hit to save her life, and there is no less harmful way in which she could do this.[122] She thinks the same is true of any person who intentionally lessens the survival chances of another person, and, according to her, this explains why non-attacking soldiers on the unjust side of the war are legitimate targets. They are indirect threats and, given their responsibility for the unjust threat that they pose, it is permissible to kill them to save one's own life. In a footnote she adds that she is assuming that everyone who signs up for the armed forces is morally responsible for doing so.[123] Thus, she concludes, there is no principled difference between self-defence and national defence, but only contingent differences about the likelihood of being in this sort of situation.[124]

Applied to combatants, the argument has some difficulties. First, many non-attacking soldiers are not actually lessening the survival chances of just combatants. Further, we cannot identify a set of just combatants whose survival chances are lessened by a particular set of unjust combatants. We might say that non-threatening unjust combatants lessen the chances of winning the just war, but it is difficult to argue that any individual unjust combatant is doing so. In a sense, non-attacking unjust combatants are like the conscientious driver *before* she gets into the car.

Intentionality is also diluted, particularly if we want to argue for the forfeiture of rights. Many non-threatening unjust combatants do not intend to lessen other people's chances of survival—they might not even know they are doing so. And such an account of intentionality, whereby a person is morally responsible for endangering you if she intentionally fails to avail herself of a

[122] Frowe, 'Self-Defence and the Principle of Non-Combatant Immunity' (n 69) 542. A similar argument is made by Draper: 'Suppose that, two days ago, Jones hid an IED in a public place and that he plans to detonate it in two days. Suppose further that there is no chance of our discovering where the IED is hidden. Even if we find Jones sleeping in his bed, and even though he poses no threat of harming anyone in the immediate future, I take it to be obvious that if the only way to prevent Jones from detonating the IED and thereby killing innocent persons two days from now is to kill Jones immediately—perhaps we know that if we do not kill him now he will escape and in all probability detonate the IED—then the appeal to defense justifies killing Jones immediately. Nor is there any basis to reach a different conclusion if we alter the case by changing the two days to two weeks or two months or even two years. The critical question in such cases is whether inflicting harm in defense is immediately necessary to eliminate a threat of unjust harm. Whether the unjust aggressor is currently using a weapon or whether the relevant threat is a threat of being harmed in the near or in the distant future is irrelevant.' Kaila Draper, *War and Individual Rights: The Foundations of Just War Theory* (Oxford University Press 2015) 187.

[123] Frowe, 'Self-Defence and the Principle of Non-Combatant Immunity' (n 69) 542.

[124] ibid.

reasonable opportunity to avoid posing an unjust threat, would mean that most of us are liable for our contributions to global warming. Some of us are, indeed, knowingly lessening the survival chances of a lot of people by choosing to eat meat, using electricity, driving cars, travelling by plane, and so on.

Second, just combatants are not only or primarily fighting the war to avert threats to themselves but also to avert the macro-threat. Indeed, some combatants are never in danger; technology allows them to operate remotely. Third, most armed groups fighting in conflicts today rely, in fact, in forced conscription or violence to get people—sometimes children—to fight. Finally, I know of no state whose criminal law of self-defence would justify killing the mafia clerk. Of course, Frowe may think the law is wrong or that there are other reasons why it should be narrower. Indeed, in cases like the mafia one, there are other, less harmful options available, such as calling the police. Thus, maybe she thinks that the establishment of a legal system and the actions of legal institutions in maintaining security and punishing wrongdoers can make it morally impermissible to use violence, except in a very narrow range of circumstances.[125] This is a plausible view, and it would explain why in war the permission to commit violence is broader: there is no monopoly of violence.[126] However, it does stray away from reductivism, as it implies that institutions or their absence can change the moral demands hanging over us and that we can transfer privileges of self-defence to collectives, such as the state. A different view would be to suggest that the existence of certain institutions—like law enforcement—make it more unlikely that it is necessary to kill someone like the mafia clerk. Given the absence of such institutions in war, finding oneself in the situation where killing the mafia clerk is necessary is more likely.

Haque, on the other hand, thinks that combatants who perform combat functions may be targeted even while they are not directly participating in hostilities on grounds of anticipatory self-defence.[127] Although they currently pose no unjust threat, they are very likely to do so in the future.[128] It is objectively permissible to kill such combatants when this will in fact prevent them from posing unjust threats in the future, and it is permissible in an evidence-relative sense to kill such combatants when the attacker has decisive reason to believe that this will prevent them from posing unjust threats in the future.[129]

[125] *See* Mark Greenberg, 'The Moral Impact Theory of Law' (2013) 123 Yale Law Journal 1288, 1311.

[126] This is Fishback's position. *See* Ian Fishback, 'Necessity and Institutions in the Self Defense and War' in Christian Coons and Michael Weber (eds.), *The Ethics of Self-Defense* (Oxford University Press, 2016) 274, 277.

[127] Haque (n 2) 89–90.

[128] ibid.

[129] ibid.

The fact that a combatant performs a combat function is a very strong reason to believe that such combatants will pose unjust threats in the future unless prevented from doing so.[130] This is not true for combatants who perform non-combat functions.[131] I'm sympathetic to Haque's view, insofar as it draws a distinction between combatants with a combat function and those without one, only making the first kind liable. If adopted, it would have significant effects in restraining war and shifting proportionality calculations, all of which I take to be beneficial. But of course, that someone is permitted in an evidence-relative sense to do something—kill combatants with a combat function—does not in fact render those combatants liable.

An additional problem with these views about non-threatening combatants is that they fail to capture the feeling that something wrong is done to them when they are harmed and killed. It is doubtful that we can ground the forfeiture of rights of people who are not posing any threats, even if we correctly think they may pose them in the future. Further, it is also doubtful that we can permissibly act *as if* people have forfeited their rights simply because we believe they might pose threats in the future. Note that requiring certainty will not do. If we required something close to certainty that a certain platoon or division will in fact pose threats against just combatants in the future, the view would collapse into pacifism. But if we only require a mere chance that the threat materializes, then forfeiture seems implausible—even if that chance would in fact have materialized. Perhaps more importantly, it also seems that our reactive attitudes toward cases of non-threatening combatants are inconsistent with forfeiture. We would think that in these cases it is appropriate to feel regret and to apologize to unjust combatants for what has been done to them. I will return to this point in Chapter VI.

C. Killing the Many: Unjust Combatants and the Number of Aggressors

Killing in war does not usually take the form of one-on-one confrontations. Threats are posed by many individuals against many individuals. This feature of war, which I have been ignoring so far, makes self-defence accounts difficult to apply.

130 ibid.
131 ibid.

In particular, the number of participants in war poses two issues for liability-based views of self-defence. The first one relates to the requirement of narrow proportionality and applies to defensively harming unjust combatants posing micro-threats and those who are not. The second one pertains to the requirement of necessity and applies most obviously to defensively harming unjust combatants who are not posing micro-threats.

Though accounts of liability vary, necessity and narrow proportionality are well-accepted requirements of defensive action. These requirements can be internal or external to liability. When necessity and narrow proportionality are internal to liability, unjust combatants cannot, by definition, be liable if harming them or killing them violates those constraints.[132] Killing or harming them would wrong them. When the requirements are external to liability, their absence might make defensive action impermissible, but *not* because such defensive action would wrong the unjust combatant.[133]

Given the number of aggressors and defenders involved in war, these requirements are often not met. Regarding narrow proportionality, many unjust combatants' combinations of degrees and kinds of knowledge, moral responsibility, and causal contributions will be relatively low. This makes it narrowly disproportionate to kill or harm them, even when they are posing micro-threats to other just combatants. If narrow proportionality is internal to liability, this makes unjust combatants non-liable. If narrow proportionality is external to liability, it might make it impermissible to kill them.

Regarding necessity, even though harming unjust combatants who are posing micro-threats might be necessary for just combatants to avoid threats to themselves, this will not be true of unjust combatants who are not posing any micro-threats. If necessity is internal to liability, it implies that a significant number of unjust combatants are not liable to substantial or lethal harm.

1. Unjust combatants and narrow proportionality

The requirement of narrow proportionality sets limits to the amount of harm to which a person is liable, usually as a function of the person's moral responsibility for the threat.[134] In McMahan's understanding, narrow proportionality is

[132] McMahan, *Killing in War* (n 2) 9–21. Narrow proportionality is less controversial as an internal requirement of liability than the necessity requirement. *See* Frowe, *Defensive Killing* (n 2) 89–119.

[133] There are additional requirements. Wide proportionality, which takes into account the number of innocent victims that might be harmed as a side effect of defensive action, is generally accepted as a requirement for defensive action to be permissible—that is, it is external to liability. A success condition has also been explored, though it has been much more neglected in the literature. Daniel Statman, 'On the Success Condition for Legitimate Self-Defense' (2008) 118 Ethics 659.

[134] McMahan, 'Liability, Proportionality, and the Number of Aggressors' (n 112) 4; McMahan, *Killing in War* (n 2) 9–21; McMahan, 'What Rights May Be Defended by Means of War?' (n 120) 123–24.

often a function of moral responsibility: the more responsibility for the threat an individual has, the more harm one can inflict on them to defend oneself.[135] But he has also suggested it may be a function of whether attacks against combatants make only negligible contributions to the achievement of the just cause.[136]

These two issues—moral responsibility and causal contributions—are related. When threats are posed by numerous individuals, each of them might be only morally responsible for their own contribution to the threat. If the number of individuals posing the threat is very high, this may entail that each individual is responsible for a very small threat of harm. In the same vein, Fabre thinks that causal contributions are fundamental in justifying killing in war and that the size and kind of those contributions may have an impact on the degree of harm individuals are liable to.[137]

If narrow proportionality—the degree of harm one is liable to—is a function of the size or type of one's causal contribution to a threat and/or of one's degree of moral responsibility, we are forced to conclude that a significant number of unjust combatants will not be liable to death or substantial harm.[138] Since many unjust combatants will be acting under duress or coercion, their moral responsibility will be diminished. And since most threats in war are posed by many individuals, causal contributions to the threat—and the size of the contribution for which each combatant is morally responsible—will be small or overdetermined. As a result, individuals are unlikely to be liable to substantial harm or death; most likely, they would be liable to lesser degrees of harm, and it would be narrowly disproportionate to kill or harm them in excess of those small degrees of harm.

This view of liability, whereby liability is a function of moral responsibility or causal contributions, is attractive. It can explain the intuition that something wrong has been done to someone who is killed in order to avert a threat when the person had little responsibility for it and/or contributed marginally to it. But this understanding of the narrow proportionality requirement would inevitably lead us to pacifism. Many, if not most, unjust combatants would simply not be liable to substantial harm or death; a just war could not be conducted

135 McMahan, *Killing in War* (n 2) 175–82.

136 Jeff McMahan, 'Necessity and Proportionality in Morality and Law' in Claus Kress and Robert Lawless (eds), *Necessity and Proportionality in International Peace and Security Law* (Oxford University Press 2020) 25.

137 Cécile Fabre, *Cosmopolitan Peace* (Oxford University Press 2016) 6.

138 Others, such as Tadros, think that in most cases, the facts that make causal contribution scalar are morally irrelevant or redundant in determining liability. Tadros also believes that causal contributions are not necessary for liability to be harmed, though they do make a difference to it. Tadros, *To Do, to Die, to Reason Why* (n 4) 203.

permissibly. The view also poses problems in cases where several aggressors threaten one victim. It forces the conclusion that the aggressors would not be liable to substantial or lethal harm.

McMahan's original solution to this problem relied on the idea of there being a 'residual injustice' in cases where someone who is minimally responsible for a lethal threat of harm is killed in order to avert it. In ideal conditions, harm would be distributed according to a threat's moral responsibility. But because victims will often not be able to distribute harm according to their respective degrees of moral responsibility but only to shift it entirely to the aggressor, there will be a residual injustice: the aggressor will have suffered more than his fair share and compensation may be appropriate.[139] In cases where numerous aggressors are involved, and there is a residual injustice in each of their killings, those residues might add up and forbid killing them.

The idea of residual injustice suggested by McMahan, though attractive, seems to contradict the very idea of liability. For it is hard to grasp how that residual injustice is something other than A's being wronged, of him having a legitimate complaint against the defensive imposition of harm. But then, if he has been wronged, whatever we may want to say here, we can't say that he was liable to harm or death. If he was, why should we think there is a residual injustice in how he was treated? And if he wasn't, then why could he be killed?

Jonathan Quong has recently offered a different solution to the problem of the number of aggressors. He argues that the widely accepted view—whereby narrow proportionality is a function of moral responsibility—is wrong,[140] and proposes the stringency principle instead: the correct approach is for narrow proportionality to depend on the stringency of the right that attackers threaten to violate. The more stringent the right, the more harm they are liable to.[141] Narrow proportionality is also insensitive to the degree of causal contribution a person makes to a threat. This allows Quong's view to avoid the problem raised by cases where there are many attackers who together threaten significant harm against one person. Unlike McMahan's view, his view allows for them to be liable insofar agents in a group act together and have sufficient evidence to understand the gravity of the threat that they pose.[142] In this way, the evidence available to the attackers plays a role in determining the type and stringency of the right that they threaten to violate.[143]

139 McMahan, 'Duty, Obedience, Desert, and Proportionality in War' (n 109) 155.
140 Quong, *The Morality of Defensive Force* (n 98) 97.
141 ibid 98.
142 ibid 122–23.
143 ibid.

Quong's stringency principle, thus, is able to explain why several aggressors acting together to threaten a stringent right are liable, even if each of them individually considered contributes little to the threat. Quong's view is, however, an account of defensive action—not an account of killing in war.

Applied to the ethics of war, the view can explain why unjust combatants posing micro-threats together with other combatants may be liable to significant degrees of harm. Since these combatants threaten stringent rights—they pose micro-threats to the lives of just combatants—they might then be liable to significant harm, and even death, if they meet the required threshold of moral responsibility. Yet, this view will struggle to make unjust combatants who are not posing micro-threats liable. It is hard to understand these individuals as acting together in threatening rights that are stringent, unless we say that they can be understood as contributing to the macro-threat posed by the unjust war, which is a threat against stringent rights. But this seems like an artificially broadened version of Quong's view, which would also make many civilians liable.

2. Unjust combatants and necessity

The requirement of necessity entails that if harming a person is unnecessary for the achievement of a relevant type of goal, that person cannot be liable.[144] Not any goal will do. In self-defence, the goal is for the victim to defend herself and avoid harm. Because killing in war also involves killing combatants who are not posing micro-threats against others, the goal must be significantly different. We need a different account of necessity, whereby the goal must be about averting the macro-threat of the unjust war. Enemy combatants are often killed not to avert threats to other combatants, but to achieve some kind of military advantage that might contribute to victory, either in the short or long term. Victory is not always the same as averting the macro-threat posed by the unjust side of the war, as it may include other goals too (such as, presumably, the achievement of a sustainable peace).

But even if we understand necessity in this way—which is difficult to square with a reductivist view about war—if necessity is internal to liability, a significant number of unjust combatants cannot be liable. Individual instances of harming or killing unjust combatants contribute very little—if anything—to averting the macro-threat of the unjust war. More importantly, the killing of one combatant cannot *on its own* achieve the goal of averting the macro-threat, and in some cases, it will also not be enough, *on its own*, to

[144] McMahan, *Killing in War* (n 2) 9. This is related to liability's instrumental character.

achieve a significant military advantage. Only killing *several* unjust combatants plus the destruction of other military objects would achieve that result. But if killing one combatant cannot avert the threat unless it is premised on the condition of killing several other combatants (and, perhaps, destroying some military objects too), then individual unjust combatants cannot be liable to harm. Their liability partly depends on whether it is necessary to harm them but also whether it is necessary to harm them depends on whether others are harmed. But whether others are harmed depends on whether they are liable to be harmed, yet whether they are liable to be harmed partly depends on whether it is necessary to harm them. And so on. In summary, defensive action against each of them fails to meet the necessity requirement because their death, on its own, fails to avert the threat. As such, they are not liable to defensive harm.

The problem of unjust combatants and the number of aggressors regarding narrow proportionality and necessity is then as follows. A liability view, even one with a low threshold of moral responsibility, where necessity and narrow proportionality are internal requirements makes a very significant number of unjust combatants, even those who are posing micro-threats, *not* liable to death or significant degrees of harm.

A liability view where necessity and narrow proportionality are external to liability would make it impermissible to kill or harm them, unless an alternative explanation for harming them could be provided. This explanation cannot, of course, be the just war if we want to stick to reductivism. Moreover, this view—where narrow proportionality and necessity are external to liability—has counterintuitive implications. It implies that harming someone when harming them accomplishes nothing and when the harm imposed on them is disproportionate to the degree of their moral responsibility and causal contributions to a threat does not wrong them. It entails that harming someone in this way leaves no moral residue; that there is no reason to apologize for having done so. This seems implausible to me. But the more plausible view—where narrow proportionality and/or necessity are internal to liability—fails to make most unjust combatants liable.

All said, it is impossible to escape the fact that many unjust combatants will not be liable, even according to a theory that requires a low degree of moral responsibility. As a result of the liability—or lack thereof—of at least some unjust combatants, it turns out that just combatants are *also* partaking in an activity that they know or have reason to know may impose harm on non-liable people: precisely, those non-liable unjust combatants. If we add to this that just combatants also pose risks to non-liable civilians from the unjust side—who are likely to be many[145]—just combatants begin to look more and more like the conscientious driver. In fact, under the responsibility view we must conclude that most combatants, just and unjust alike, are like the conscientious driver. They are all partaking in an activity (war) and thus, they know or have reason to know that they may contribute to imposing harm on other people who are not liable as, in fact, most wars, just and unjust, have historically done so far. That is, it is plausible to conclude that *all* combatants will meet the required threshold of responsibility for liability. And under a moral responsibility account such as McMahan's, just combatants would become liable when the threat against non-liable individuals materializes. This is so even if just combatants are subjectively justified in believing that unjust combatants are liable.[146]

In order to escape this conclusion, McMahan and other philosophers working on self-defence have argued that 'objective justification defeats liability'.[147] That 'objective justification defeats liability' means, in the case of the just combatant, that although he may be responsible for a threat of harm against someone who is not liable, the just combatant does not become liable himself for posing that threat if he has an objective justification to pose that threat (eg a lesser evil one).[148] This solution, however, cannot resort to the just war or the just cause as a justification: doing so would be letting go of reductivism. It thus needs to rely on some other justification for posing harm on the non-liable, such as a lesser evil one. But, as I will argue in Chapter VI, a lesser-evil justification for killing in war is very unlikely to succeed.

[145] As pointed out by Uwe Steinhoff. *See* Uwe Steinhoff, 'Debate: Jeff McMahan on the Moral Inequality of Combatants' (2008) 16 Journal of Political Philosophy 220; Steinhoff (n 53).

[146] McMahan, *Killing in War* (n 2) 164–67. McMahan later introduced a revision to the responsibility account in order to escape the conclusion about reasonably mistaken just combatants. He suggested that if unjust combatants are responsible for making it reasonable for just combatants to believe that their use of force is permissible, just combatants do not become liable by posing unjustified threats towards non-liable combatants. By contrast, unjust combatants themselves become liable just for their responsibility in just combatants' having formed these wrong but reasonable beliefs. This solution, however, is hard to swallow. I readily admit that I am not sure I understand how unjust combatants could be responsible for the wrong beliefs that others form about the moral character of the war they are fighting in. Unjust combatants have, in fact, very little responsibility for the beliefs that others may form about their war. This responsibility is likely to fall on their government or leaders. It also seems implausible that people may forfeit their rights simply because they are responsible for the wrong beliefs of others. *See* McMahan, 'Who Is Morally Liable to Be Killed in War' (n 93).

[147] *See*, eg, Jeff McMahan, 'Debate: Justification and Liability in War' (2008) 16 Journal of Political Philosophy 227; McMahan, *Killing in War* (n 2) 43; Quong, 'Liability to Defensive Harm' (n 50). Against this view *see* Steinhoff (n 52); Adam Hosein, 'Are Justified Aggressors a Threat to the Rights Theory of Self-Defense?' in Helen Frowe and Gerald Lang (eds), *How We Fight* (Oxford University Press 2014); Kai Draper, 'Who Should Die?: The Ethics of Killing in War' in Bradley Jay Strawser, Ryan Jenkins, and Michael Robillard (eds), *Defensive Liability: Four Common Mistakes* (Oxford University Press 2018) 82–84.

[148] McMahan, 'Self-Defense against Justified Threateners' (n 49).

Finally, there is a more general objection we can make against the reductivist individualist view. I believe that at least part of our moral rules for war ought to guide conduct in a meaningful way. That is, they need to be applicable by agents who are engaging in war (as opposed to being applicable by someone watching the war from afar and with perfect knowledge). I think that even if views like the one I described above worked as a retrospective evaluation of a war, they fail to guide agents' behaviour. It is one thing to look at a war after it has happened and try to evaluate who was liable and why, and a very different one to have to make these decisions while fighting one.

The problem is, of course, that we don't *know* who is liable and who is not, and because the view is based on the particularities of each moral agent, we have no way of knowing.[149] A battle in which every combatant is trying to ascertain liability before choosing who to target would be chaos, which is a lot to say for something that is already chaotic in its own right.

One possible response is that, generally, no-one has perfect knowledge of the circumstances when she decides on a particular course of action. This is obviously true. When a victim decides to shoot her aggressor to defend herself, she doesn't know whether the aggressor is liable. She *thinks* he is, and, later, we will judge whether she acted on reasonable beliefs or not by looking backwards. But what the victim needs to ascertain doesn't seem unrealistically demanding. She is minding her own business, and suddenly, someone is running at her with a knife. She doesn't need to form a very high number of reasonable beliefs to decide what to do: she needs to reasonably believe that someone is trying to harm her; that she is innocent or that she is not the aggressor; that the only way to avert the threat is to defend herself; and so on.

Yet even when we only have one unjust combatant attacking a just one, we have the added complexity of the just combatant's having to determine whether he is fighting a just war; whether the enemy combatant is fighting a just war; how both his actions and the actions of the enemy combatant relate to the macro-threat each of their sides is posing; and what degree of moral responsibility the enemy combatant has for his micro-threat and for the macro-threat as well.[150] And this list does not even account for issues related to necessity

[149] Henry Shue and Janina Dill have expressed similar concerns. *See* Henry Shue, 'Do We Need a "Morality of War"?' in David Rodin and Henry Shue (eds), *Just and Unjust Warriors: The Moral and Legal Status of Soldiers* (Oxford University Press 2008); Janina Dill and Henry Shue, 'Limiting the Killing in War: Military Necessity and the St. Petersburg Assumption' (2012) 26 Ethics & International Affairs 311; Jeff McMahan, 'The Morality of War and The Law of War' in David Rodin and Henry Shue (eds), *Just and Unjust Warriors: The Moral and Legal Status of Soldiers* (Oxford University Press 2008).

[150] Moshe Halbertal, 'The Goldstone Illusion' (*New Republic*, 5 November 2009) <https://newrepublic.com/article/70956/the-goldstone-illusion> accessed 18 August 2022. Related to this and pressing

and proportionality, which are complicated by the fact that the primary goal of combatants is not to avert threats to themselves but to avert the macro-threat posed by the collective they are fighting against, and that the death of any individual enemy combatant is likely to contribute little to the overall victory.[151]

These difficulties are related to a larger question regarding how concessive the demands of morality can be to real-world constraints that take place in a variety of circumstances, including war.[152] As it stands, the moral responsibility view oscillates between total war and moral paralysis that corners individuals into pacifism or surrender. Uncertainty in war is so pervasive and knowledge about the liability of individuals is so hard to reach, that in most circumstances revisionism must either demand that individuals do not kill[153] or simply assume the liability of all unjust combatants. This issue might be solved in different ways—such as creating a decision procedure[154]—but so far, it hasn't been.

McMahan is very much aware of these difficulties. He thus suggests that 'most unjust combatants' will be liable to defensive killing because most of them will be, at best, 'partially excused threats'. This fact allows just combatants to operate on the belief that unjust combatants are liable to defensive harm unless they have evidence that proves otherwise.[155] Similarly, Haque acknowledges that some unjust combatants are not liable to defensive killing, but that the harm is permitted in an evidence-relative sense: it is reasonable to believe that they are.[156]

I think the harm will sometimes be permitted in an evidence-relative sense—but not always, and perhaps not even often, precisely due to the arguments I have made so far. And in any case, reasonably believing that someone is liable does not make that person liable. Even if some just combatants are justified in believing that other unjust combatants are liable, this belief does not render unjust combatants liable.

similar points *see* Seth Lazar, 'Necessity in Self-Defense and War' (2012) 40 Philosophy & Public Affairs 3, 27–28.

[151] Lazar, 'Necessity in Self-Defense and War' (n 150) 27–28.

[152] Seth Lazar and Laura Valentini, 'Proxy Battles in Just War Theory' in Peter Vallentyne, David Sobel, and Steven Wall (eds), *Oxford Studies in Political Philosophy*, vol 3 (Oxford University Press 2017) 185–87.

[153] ibid.

[154] ibid.

[155] *See* McMahan, 'Who Is Morally Liable to Be Killed in War' (n 93) 547–49; McMahan, *Killing in War* (n 2) 182–88. Partially excused threats are those who act without a full excuse. McMahan says that excuses may have relevance in the distribution of risk: it may affect what he calls 'narrow' proportionality. ibid 192–98.

[156] Haque (n 2) 95.

IV. Conclusion

I have argued that forfeiture-based views of self-defence that are committed to reductivism and individualism cannot provide a complete theory of the morality of killing in war without abandoning methodological revisionism. Killing in self-defence is significantly different from killing in war, and unjust combatants are significantly different from villainous aggressors.

All said, substantive and methodological revisionism must come apart. In order to make most unjust combatants liable, we would need to adopt a theory of liability with a low threshold of moral responsibility and where narrow proportionality and necessity are external to it. This is a troubling view that, in addition, requires an explanation that makes it permissible to kill liable individuals when the harms imposed on them are not narrowly proportionate nor necessary. If we want to hold on to reductivism, the answer cannot be the just war.

In abandoning reductivism and individualism, forfeiture loses its initial appeal. But if forfeiture doesn't abandon reductivist individualism, it struggles to make many unjust combatants liable without, at the same time, becoming an implausible theory of killing in war. In fact, the most plausible account of forfeiture cannot make many unjust combatants liable. This is what I will argue in Chapter VI.

VI
The Non-liability of Unjust Combatants

I. Introduction

There are significant differences between killing in war and killing in self-defence. As a result, a view committed to both individualism and reductivism struggles to justify most killing in war. But this struggle is also explained—and until now, I have been mostly ignoring this—by the implausibility of a forfeiture-based view to account for most killing in war. Unjust combatants are unlike villainous aggressors. Yet forfeiture theories of self-defence have the most plausibility when applied, precisely, to the case of the villainous aggressor—that is, when the person posing the threat is *culpable* for it. Once we start to move away from the central case of the villainous aggressor, forfeiture loses its appeal.[1]

Even if revisionists can make most unjust combatants liable by setting a low threshold of moral responsibility, by not requiring the threat to be imminent or even posed by the person who is liable, and by making necessity and narrow proportionality external to liability, in doing so, liability becomes difficult to defend as a plausible moral theory. How can it be that people forfeit their rights against significant harm or death on account of negligible degrees of moral responsibility, small causal contributions to threats, and so on? The forfeiture of basic rights against harm—along with its implications regarding counter-defence and the non-wronging nature of the harm—seems disproportionate to these features of the participation of many combatants in war.

In other words, the challenge for forfeiture theories is the following. To explain why conducting a just war is permissible, a forfeiture theory needs to claim that most unjust combatants are liable, at the very least, to significant degrees of harm—that is, that most unjust combatants have forfeited rights against substantial harm and/or death. Forfeiture of rights against trivial harm

[1] On the problem of forfeiture to deal with cases outside of the villainous aggressor, *see* Kimberly Kessler Ferzan, 'Culpable Aggression: The Basis for Moral Liability to Defensive Killing' (2011) 9 Ohio State Journal of Criminal Law 669; Whitley Kaufman, 'Is There a "Right" to Self-Defense?' (2004) 23 Criminal Justice Ethics 20.

The Morality of the Laws of War. Marcela Prieto Rudolphy, Oxford University Press. © Marcela Prieto Rudolphy 2023.
DOI: 10.1093/oso/9780192855473.003.0006

will not do; it will corner the view into pacifism. But the nature of many unjust combatants' participation in war makes it very difficult to argue that they have lost rights against significant harm. In fact, the best possible version of a forfeiture view of self-defence cannot support the conclusion that most unjust combatants are liable to significant harm.

The arguments in this chapter will thus be arguments more directly related to the problems with forfeiture. I will not discuss the relatively more abstract issue of whether forfeiture is a plausible moral theory pertaining to basic moral rights. My point is narrower: forfeiture, I will argue, is not a plausible moral theory to account for most killing in war. Many unjust combatants are not liable to significant harm or death; they are wronged when killed.

Some of the arguments I will make in this chapter have force both for unjust combatants who are posing a micro-threat against just combatants—that is, those who are posing an imminent threat to other(s) just combatants—and for those who are not. For some, the fact that the unjust combatant is posing a micro-threat will be enough to make the unjust combatant liable, so that the arguments presented here will be, at best, able to show that *only* non-threatening unjust combatants are not liable. However, I believe that the most plausible position is that even some unjust combatants who are posing micro-threats are not liable to death or significant degrees of harm. Unjust combatants are unlike villainous aggressors in many ways, even when they are posing micro-threats against just combatants. Nevertheless, a successful argument as to why many non-threatening unjust combatants are not liable to death or significant harm is good enough for my purposes.

Section II argues that the forfeiture of rights against significant harm and death is not a plausible explanation for most killing in war. Many unjust combatants are not liable to significant harm or death. Precisely due to the difficulties that a liability-based view has in justifying killing in war, some have introduced a lesser-evil justification to justify killing and harming someone in excess of what they are liable to. This is known as the 'combined view'. Section III explains what a lesser-evil justification is. Section IV describes the combined view and argues that it fails. Section V concludes.

II. Unjust Combatants and the Problem with Forfeiture

A forfeiture view *can* make most unjust combatants liable. It can abandon individualism, in the sense of abandoning the idea that individuals can only be liable for facts about themselves. It could thus hold a collective account of liability,[2] where liability is grounded in the unjustness of the war and the fact that an individual belongs to a collective that is fighting an unjust war. But such an account would be problematic precisely because it is collective.[3] Though we can understand that soldiers in war are fighting together and in the name of a collective entity, the implications of saying that a class of people forfeit rights against significant harm simply because they belong to the armed forces of a collective who is fighting an unjust war is deeply troubling. Not even collective punishment is plausibly justified. How can the forfeiture of our most basic rights be so?

Thus, I agree with some revisionists that liability—if we are to employ liability at all—should be grounded individually, and not in the mere fact that the war is unjust. But if a plausible account of liability must be individualist, the inevitable conclusion is that many unjust combatants are simply not liable.

An attractive account of forfeiture needs to be grounded in something about the individual who forfeits their rights. But not anything will do. If that something is to be significant enough to ground forfeiture, it must be an action to which the agent is significantly related and for which they are strongly morally responsible (namely, culpable). This is precisely why forfeiture accounts of self-defence usually argue that innocent attackers—such as those who lack agency or moral responsibility—are not liable and that it is impermissible to kill them.[4] But between the innocent attacker and the villainous aggressor there is a vast realm of possibilities: different combinations between different degrees of moral responsibility, knowledge, and causal contributions.

There is no point in exploring all these possibilities here. The justificatory problem for forfeiture, as it arises in war, is that a significant number of combatants are thinly related to the war effort. Yet at the same time, we need to find a way of justifying significant degrees of harm against them by arguing that their thin connection to the war grounds the forfeiture of rights against significant harm—that such a connection can make it the case that harming those combatants does not wrong them, that they cannot defend themselves. If this can't be done, we are cornered into pacifism.

[2] *See*, eg, Saba Bazargan, 'Complicitous Liability in War' (2013) 165 Philosophical Studies 177.

[3] I leave open, however, the possibility that states are liable to attack, if those attacks are well-tailored and narrow ones and which harm only what belongs to the state and no individuals at all. In favour of such liability, *see* François Tanguay-Renaud, 'Can States Be Corporately Liable to Attack in War?' in Saba Bazargan and Samuel C Rickless (eds), *The Ethics of War: Essays* (Oxford University Press 2017) 127.

[4] *See* Ferzan (n 1); Jeff McMahan, 'Self-Defense and the Problem of the Innocent Attacker' (1994) 104 Ethics 252; Michael Otsuka, 'Killing the Innocent in Self-Defense' (1994) 23 Philosophy & Public Affairs 74; David Rodin, *War and Self-Defense* (Clarendon Press 2002).

This issue can be understood as a question about the kind of relationship there must exist between the features of someone's conduct and forfeiture itself. The uneasiness some of us feel about saying that someone like the conscientious driver or an epistemically justified aggressor who turns out to be mistaken have lost their rights against harm comes from a certain kind of dissonance between the features of these individuals' conduct and the forfeiture of rights against significant harm.

But there is also a problem with the intuitions that examples like Conscientious Driver rely on in ascertaining liability. Let us restate the example:

> Conscientious Driver. A person who always keeps her car well maintained and always drives carefully and alertly decides to drive to the cinema. On the way, a freak event that she could not have anticipated occurs that causes her car to veer out of control in the direction of a pedestrian.[5]

This example elicits two intuitions. First, the intuition that it is permissible for the pedestrian to defend himself, which might be less or more appealing to different people. Second, the intuition regarding the reactive attitudes that would be appropriate for the pedestrian to have if he defends himself against Conscientious Driver and kills her. Some of us would say that it would be appropriate for the pedestrian to feel regret for having harmed Conscientious Driver and that he should apologize to her if possible.

The example is meant to help us elucidate the limits of liability. But these intuitions pull in different directions. The intuition about permissibility weighs in favour of the driver's liability, while the intuition about reactive attitudes weighs against the driver's liability. It is unclear why the first kind of intuition (about permissibility) should prevail over the second kind (about appropriate reactive attitudes) in helping us elucidate when someone is liable. After all, what characterizes liability-based explanations is not merely that it is permissible to kill or harm someone in self-defence but more importantly, that the imposition of non-consensual harm is non-wronging, in an agent-relative sense. As a result, certain reactive attitudes towards Conscientious Driver, such as regret, the pressure to apologize, etc, should not be present in liability cases. Yet those are the kind of reactions some of us are likely to have in cases like Conscientious Driver and others that stray away from the case of the villainous aggressor.

[5] Jeff McMahan, *Killing in War* (Oxford University Press 2009) 165.

More generally, and going back to the dissonance between the features of these individuals' conduct and the forfeiture of rights against significant harm, it seems implausible that the driver cannot defend herself, that she is not owed an apology. But these are the appropriate consequences if one is to argue that she is liable, that she is not wronged. However, one would expect the pedestrian to feel precisely that, to understand that he has done something to the driver for which he ought to apologize. If the driver survived, and the pedestrian visited her in the hospital and apologized to her, it would hardly be an appropriate reaction to advise the pedestrian that he had no reason to do that; that there is no reason why he should feel the need to apologize. If that pressure to apologize is not misguided then it seems that the driver has not forfeited her rights. And the reason we might think such pressure exists comes precisely from the dissonance between what the driver has done and what must be done to stop her from posing a threat (ie inflicting significant harm or death on her).

This pressure to apologize exists in other examples as well where the person who poses the threat has not been as diligent as the conscientious driver. And that pressure to apologize may stand in tension with the intuition about what is *permissible* to do. If a reckless pedestrian, absorbed on his phone, simply crossed the street, thus leaving a diligent driver with no choice other than to run him over or die, I don't think we would say the driver ought not to apologize for defending himself against the pedestrian. And it would be little comfort to the driver who, strictly speaking, allowed the pedestrian to be killed, to hear that the pedestrian, in his recklessness, had forfeited his rights against death or significant harm. She would be likely to think he had not.

The pressure to apologize is not only a function of the level of harm that needs to be imposed on the threatening person in order for the victim to defend himself successfully. Suppose someone is walking down the street and receives an urgent call. They are thus distracted and can't see that you need to move past them in order to prevent a car from running over you. If you push the pedestrian aside to accomplish that, surely you ought to apologize for it. But if the pedestrian was intentionally blocking your way to safety, you would not need to do so. In both cases, the harm imposed on the pedestrian is trivial.

This pressure to apologize can be more or less severe depending on the circumstances. But it seems to stem from the dissonance of forfeiture as an explanation of harm-imposing behaviour when the person is not culpable, is not posing an imminent threat, or contributes very little to a threat for which they are not responsible. In war, this dissonance is particularly troubling.

In terms of moral responsibility, the problem is that although unjust combatants are fighting an unjust war, many are not morally responsible for the decision to initiate the war, many do not choose to go to war, and face significant costs if they refuse. Further, even if many combatants are morally responsible for joining the army, once they are part of it they will have no say in whether they may participate or not. And even when the decision to participate in a war is made by individuals, most of the time they don't have much of a say on what they will be doing and where.

Now, regarding the moral responsibility for joining the army, not all combatants choose to do so. About those who do, I readily acknowledge that given that just wars are few and far between, there is a degree of moral risk in doing so, because there is a significant probability of being ordered to fight an unjust war. The degree of moral risk varies considerably depending on the collective the soldier joins. Soldiers of many liberal democratic states, with the exception of the United States, which in the twentieth and twenty-first centuries has fought many wars without a just cause and/or has fought using impermissible means, probably incur in relatively less moral risk when deciding to join the army.[6] Soldiers of authoritarian regimes may incur in more moral risk but they may be under severe epistemic constraints. Still, in both cases, responsibility for putting oneself under moral risk does not ground responsibility for the unjustness of the war, and it seems implausible that it can ground liability to be killed in self-defence. Even if it is true that in deciding to act as someone's agent, we must consider what we will be instructed to do, and that if what we will be instructed to do is to act wrongly, then we ought not to become someone's agent,[7] at least when joining the army in many democratic states, there is no way to know whether we will be instructed to act wrongly, and so, there is no decisive reason not to become the army's agent.

The situation is significantly different for people who choose to join irregular armed groups. These individuals are usually not just putting themselves at moral risk, because the decision to join these groups is more readily determined by whether they are fighting a just war, or at least, whether they are fighting for a just cause. Thus they may have more responsibility, and in some cases, it may be enough to ground liability. However, it is worth remembering that sometimes these combatants will be acting under duress,[8] coercion, or will have been forcefully recruited during their childhood.

[6] And their duty of obedience may be easier to justify. On this, *see* David Estlund, 'On Following Orders in an Unjust War' (2007) 15 Journal of Political Philosophy 213.

[7] Victor Tadros, 'Duress and Duty' in Saba Bazargan-Forward and Samuel C Rickless (eds), *The Ethics of War: Essays* (Oxford University Press 2017) 94, 110.

[8] Starvation, for instance, has made people in Yemen join armed groups so they can feed their families. *See* 'Yemenis Are Left so Poor They Kill Themselves Before Hunger Does' (*The Guardian*, 11 December 2018) <http://theguardian.com/global-development/2018/dec/11/yemen-war-norwegian-refugee-council> accessed 18 August 2022.

Now I accept that an individual has a moral obligation not to participate in what they reasonably believe to be an unjust war, if they can do so at a reasonable cost to themselves and their loved ones. I also accept that an individual can act wrongly by participating in an unjust enterprise they know to be unjust, even if the unjustness is not their responsibility in the first place. And, certainly, I agree that individuals should try to judge by themselves, if possible, whether the wars they are called to fight are just or unjust. But this is different from saying that someone who may not even be threatening anyone forfeits their rights against significant harm or death simply because they have some moral responsibility for their participation in an unjust enterprise. Or at the very least, their degree of moral responsibility is not proportional to the degree of harm that just combatants may supposedly inflict on them without apologizing or wronging them: severe injury and death. This can be understood as a problem of narrow proportionality under the widespread view, but it need not be. It is a problem of the reactive attitudes elicited by imposing significant or lethal harm on people who are not culpable. These reactive attitudes do not seem misguided to me.

It should strike us as quite implausible that most individuals—and perhaps *all* individuals—could ever be liable to the horrific pain and suffering that is often involved in dying while fighting a war. It is not *only* that many unjust combatants' degrees of moral responsibility and causal contributions are disproportionate to being killed or maimed by just combatants.[9] It is also that it is hard to imagine what degree of moral responsibility could possibly be enough for a person to become liable to burning to death while trapped in her vehicle or slowly bleeding to death while lying on the ground alone.[10] What degree of moral responsibility could ever justify a harm like this to the point that the harm does not wrong them? To the point of saying that unjust combatants do not deserve an apology from those who inflict that harm on them? That they cannot defend themselves? That is, to the point of being liable? Most likely, only someone like the villainous aggressor will fit the bill: someone who is culpably

[9] McMahan also points out that some unjust combatants will be partially excused, and that them being so will diminish the amount of harm they will be liable to, but he stops short from concluding that if so, then many unjust combatants cannot be liable to be killed. Jeff McMahan, 'What Rights May Be Defended by Means of War?' in Cécile Fabre and Seth Lazar (eds), *The Morality of Defensive War* (Oxford University Press 2014) 127–28; McMahan, *Killing in War* (n 5).

[10] *See*, eg, accounts on Andrew Carroll (ed), *Operation Homecoming: Iraq, Afghanistan, and the Home Front, in the Words of U.S. Troops and Their Families* (University of Chicago Press 2008); Svetlana A Aleksievič, *Zinky Boys: Soviet Voices from the Afghanistan War* (Julia Whitby and Robin Whitby trs, Norton 1992); Bảo Ninh, *The Sorrow of War: A Novel of North Vietnam* (Frank Palmos tr, Riverhead Books 1996); David Finkel, *The Good Soldiers* (3rd edn, Picador USA 2010); Phil Klay, *Redeployment* (Penguin Books 2015).

and knowingly posing a significant and imminent threat of harm against someone else. Yet many unjust combatants are not like the villainous aggressor, and many non-threatening unjust combatants are hardly like the villainous aggressor *at all.*

In terms of their contributions, many unjust combatants, particularly those not engaged in micro-threats, contribute little or nothing to the macro-threat posed by their side.[11] And even those who contribute more actively—say, by taking a hill or crossing a river—usually make a rather small contribution to victory. Moreover, a significant number of combatants are unwilling to kill, even after rigorous training, or are simply inefficient.[12] Thus, in many cases each individual combatant's contribution to the collective threat will be small, or, at the very least, not big enough to make forfeiture of rights against significant or lethal harm plausible.

Finally, in the knowledge domain, many unjust combatants lack knowledge of whether and how their contributions add up to victory or to posing a macro-threat. It seems implausible that merely being aware of the possibility that they might be contributing to the macro-threat posed by their side, even when they are not posing a threat to any just combatants, is enough to ground forfeiture.

Unjust combatants do not always have or can reach reasonable beliefs about the unjustness of their war. They face similar epistemic constraints to those faced by just combatants in ascertaining the justness of the war. And even if they can reach a reasonable belief that their war has an unjust cause, it may be that the other side is also fighting an unjust war. Thus, knowing that our war is unjust doesn't seem to provide a non-defeasible reason not to fight. The unjust combatant would need to be able to reach a reasonable belief that the other side of the war is just for him to conclude that he may not fight at all. But this is an exceedingly difficult judgment to make. How can morality be so demanding as to make individuals liable to be killed or significantly harmed if they fail to reach correct beliefs that most of us would fail to reach as well?

The combination of these three features of many unjust combatants' participation in war—the dilution of moral responsibility, somewhat small causal contributions that sometimes do not threaten the rights of other combatants, and lack of or limited knowledge—renders forfeiture implausible. For how can small or null contributions added to little or no responsibility, and little to no

[11] *See* Seth Lazar, 'Responsibility, Risk, and Killing in Self-Defense' (2009) 119 Ethics 699.

[12] *See* Dave Grossman, *On Killing. The Psychological Cost of Learning to Kill in War and Society* (Open Road Integrated Media 2014).

knowledge on how they relate to the macro-threat of the war they fight make someone liable to death or significant harm?

Note that I am not claiming that, for example, a gang of villainous aggressors can never be liable to be killed in self-defence if their contributions are small individually considered. Suppose an *N* number of individuals decide to kill B together. Each of them will impose an X degree of harm so small as to not be threatening, but together they will kill the victim. It might be plausible—though still very difficult—to ground liability here, even despite their small contributions to B's death, because they are killing B *together*.[13] They each know and are responsible for their behaviour and for how their behaviour and that of others contribute to the threat. Many unjust combatants are not in this situation. Though some of them—and we don't always know which ones—may make small contributions that together constitute a serious threat, sometimes they are not responsible and do not have knowledge about the connection between their contributions and the macro-threat, for which they may also have no responsibility. And the fact that they are not liable does not necessarily imply that the victim could not defend herself against each individual gang member if responsibility and knowledge were lacking. I am simply making the weaker claim that, in that case, they would not be liable; that they would not forfeit their rights against significant or lethal harm.[14]

Although the threats faced in war are organized and massive in scale, they are posed by individuals who are often unlike 'gangs of villainous aggressors'. Think about the following example: 100 bandits raid a village where 100 persons live. Each of the villagers is eating a bowl of 100 beans. Each bandit eats one bean from each bowl, so they each end up eating a full bowl of 100 beans, but each of them only causes little harm to each villager.[15]

Unjust combatants are like the bandits in the following sense. Though individually considered, each may be posing an insignificant threat or no threat at all to any just combatants,[16] an *N* number of them together pose a very significant threat, not against any particular just combatant, but against the whole side

[13] It is difficult only under the widespread view. Under Jonathan Quong's stringency principle, it is relatively straightforward. *See* Jonathan Quong, *The Morality of Defensive Force* (Oxford University Press 2020).

[14] For a similar example and argument *see* Cécile Fabre, *Cosmopolitan War* (Oxford University Press 2012) 76–78.

[15] Jonathan Glover, 'It Makes No Difference Whether or Not I Do It' (1975) 49 Proceedings of the Aristotelian Society, Supplementary Volumes 171, 174–75.

[16] Those who are posing no present threat to anyone in particular could be understood as posing a threat that is imperceptible, and that contributes to the macro-threat of the unjust war. At some point (we don't know when) it becomes perceptible and dangerous. *See* Derek Parfit, *Reasons and Persons* (Clarendon Press 1992) 75–86.

of the just war (or against victory). But the differences with bandits are many. Except for people in high positions of command, many unjust combatants are not morally responsible for the macro-threat and have no or little knowledge on how their actions or the actions of their units contribute to victory. This is related to combatants being hierarchically organized: some of them are giving orders, and some of them are following them. Additionally, those contributions are usually not equal (eating one bean): some combatants will contribute more, some less, some nothing,[17] and by doing different things.

Given all I have said above, we might think that the problem is with forfeiture itself as an explanation of non-consensual impositions of significant or lethal harm. There is truth to that, but importantly, even if we move past forfeiture, there is the issue of whether the harms that are imposed on unjust combatants during war wrong them, something which is not just connected to forfeiture, and for which the arguments made above still stand.[18]

If the most plausible version of forfeiture results in the non-liability of most or many unjust combatants, we now find ourselves in need of an account of why it might be morally permissible to kill them. Before attempting to do just that, there is an important point to make about these particular qualifications 'most' and 'many'.

There are two ways in which we can understand this. On the one hand, there is a sense in which the claim—that most or many unjust combatants are not liable—depends on factors that will vary from one conflict to the other. Thus, this 'many' might not be a constant 'many'; it might not hold true of every conflict. It is possible that an armed group is mostly formed by people whose degrees of moral responsibility and actions resemble that of the villainous aggressor. In these cases, we could then say that 'most' unjust combatants will be liable, or that their degree of moral responsibility is enough for liability. This could be true of armed groups committed to terrorism and of armed groups or armies fighting genocidal wars,[19] which are important actors in today's conflicts. Consider a US citizen choosing to travel to Syria and join ISIS. Or consider someone choosing to join the Bosnian Serb Army of Republika Srpska just after the Srebrenica massacre. The level of responsibility of these agents is fundamentally different from, and larger than, that of individuals facing a decision to join the army. These individuals are not just putting themselves at

[17] So a theory of collective action needs to be thin enough to account for these differences, like Christopher Kutz's. Christopher Kutz, 'Acting Together' (2000) 61 Philosophical and Phenomenological Research 1.

[18] It is, however, connected to a rights-based account of killing, and that may be the problem.

[19] I thank Joseph Weiler for pushing me on this.

great moral risk—which they are—but are making a decision that is (or should be) primarily determined by the justness of the war they are joining or the moral standing of the organization they want to become members of. This kind of moral responsibility might be enough to ground liability, or at the very least, their degree and kind of moral responsibility very closely resembles that of the villainous aggressor when the organizations they join have evil aims.[20]

If so, this could be a problem for my argument, at least as it applies to today's conflicts. If a significant number of participants in current armed conflicts are like those described above, it is very difficult to still claim that most or many unjust combatants are not liable. The opposite, in fact, may be true. The threats we now face in most armed conflicts might actually be very much like threats posed by gangs of villainous aggressors, engaged in ethnic cleansing, terrorism, or both. And if we are facing gangs of villainous aggressors, it will be difficult to argue that unjust combatants are not engaged in the murder of the innocent and that they should have any kind of legal—let alone moral—privilege to defend themselves.

The easy way out of this is simply to point out that members of non-state armed groups such as these will not actually have any legal privilege to target enemy combatants. They are, in fact, unprivileged combatants and can be prosecuted for their conduct in hostilities.[21] And members of state armed groups engaged in genocide can be prosecuted for such conduct too.

The moral issue is more complicated. Even if it is true that levels of responsibility are higher in joining some non-state armed groups, these organizations—in particular, terrorist groups and non-state armies—are, at least today, usually highly coercive. Thus, a significant number of people will find themselves joining the war under duress, either directly forced by these groups—a common practice in Afghanistan by both the Taliban and the warlords—or by the lack of other choices (in Yemen, for instance, some people had to face the choice of joining one of the armed groups and making money or watching their family starve to death).[22] Most of these groups do not rely

[20] The degree of moral responsibility is still high, but it certainly does not ground liability when those individuals joining non-state organizations are joining just wars.

[21] *See Geneva Convention III Relative to the Treatment of Prisoners of War* (hereinafter, GCIII), 12 August 1949, art 4; *Convention (IV) respecting the Laws and Customs of War on Land and its annex: Regulations concerning the Laws and Customs of War on Land* (hereinafter, The Hague Convention IV), The Hague, 18 October 1907, reg 1; API, art 43; 'Child Recruitment and Use' (*Office of the Special Representative of the Secretary-General of the United Nations for Children and Armed Conflict*) <https://childrenandarmedconflict.un.org/six-grave-violations/child-soldiers/> accessed 18 August 2022.

[22] *See* 'Yemenis Are Left so Poor They Kill Themselves Before Hunger Does' (*The Guardian*, 11 December 2018) <http://theguardian.com/global-development/2018/dec/11/yemen-war-norwegian-refugee-council> accessed 18 August 2022.

entirely on voluntary participation—in fact, most of them don't—which entails that a significant number of combatants will have joined under duress or will be children—indeed, child soldiers have been common in the conflicts taking place in Afghanistan, Chad, the Democratic Republic of Congo, Myanmar, Somalia, South Sudan, and Yemen.

Therefore, it might be true—though subject to empirical confirmation—that many unjust combatants fighting in today's armed conflicts will have a significant degree of moral responsibility. But this is not the only focus: at the same time, because these groups force people to fight under threat of harm and recruit children—who lack any moral responsibility—we can still say that many unjust combatants are not in fact liable—if only contingently.

Nonetheless, and on the other hand, there remains a way in which saying that 'most' or 'many' unjust combatants are not liable doesn't depend on these contingencies. One is to deny, of course, that killing someone can ever be a non-wronging harm. But another one, and the argument that I am making, is to assert that the nature of war itself makes it very difficult for most or many unjust combatants to be liable. Hierarchical organizations and actions performed by numerous people will always tend to the dilution of individual causal contributions, knowledge, and responsibility among those lower down the chain of command. Those lower down in the chain of command are also the most numerous. And the forfeiture of rights against being killed and maimed will most usually be disproportionate to these diluted degrees of individual causal contributions, knowledge, and responsibility.

III. A Lesser-Evil Justification

Troubled by some of the issues discussed above, some revisionists have conceded that some unjust combatants may be liable to degrees of harm that are lower than death or significant harm, and as a result, that they cannot be liable to be killed in self-defence. Unless another justification for killing is provided, the absence of a liability-based justification would make it impermissible to kill them. If we want to stick to reductivism and individualism, the just war itself cannot solve this problem; it cannot act as a *sui generis* moral justification for killing the innocent.

In an effort to solve this issue, a lesser-evil justification has been proposed. Generally speaking, lesser-evil justifications can operate either in conjunction or instead of liability. Operating instead of liability, a lesser-evil justification would justify killing non-liable unjust combatants and harming non-liable civilians as a side effect. Non-liable unjust combatants may be few or many, depending on how low the threshold of moral responsibility required to trigger liability is and on how extensive a theory of liability is in covering those that are not posing imminent threats. Operating in conjunction with liability, lesser evil would justify killing or harming someone in excess to the degree of harm they are liable to. This is known as a combined justification, and it has been defended by Jeff McMahan, Helen Frowe, and Saba Bazargan.[23]

This role for a lesser-evil justification is important, because even the moral responsibility account of killing in war acknowledges that a significant number of unjust combatants are not fully culpable or responsible for the threats they pose or for their participation in the war, mostly due to epistemic constraints and duress. Their diminished culpability may affect the amount of harm to which they are liable, that is, the amount of harm it is narrowly proportionate to inflict on them:[24] the less culpable an unjust combatant is, the less harm he is liable to. Thus, if an unjust combatant is liable to an X degree of harm that does not amount to killing her, just combatants may be justified in killing her if the infliction of that additional harm that surpasses X—and amounts to death—can be justified as the lesser evil.[25] In summary, a combined justification would be able to cover cases where unjust combatants are liable to degrees of harm lower than what needs to be imposed on them.[26]

This solution has many problems, however. It struggles to hold on to reductivism—at some point it becomes necessary to appeal to the just war as a moral justification—and non-consequentialism.

A. What is a Lesser-Evil Justification?

What is a lesser-evil justification? Among justifications to impose harm non-consensually, philosophers usually distinguish between two kinds: lesser-evil ones and liability-based ones.[27]

[23] Saba Bazargan, 'Defensive Liability Without Culpability' in Christian Coons and Michael Weber (eds), *The Ethics of Self-Defense* (Oxford University Press 2016); McMahan, 'What Rights May Be Defended by Means of War?' (n 9); Helen Frowe, *Defensive Killing* (Oxford University Press 2014).

[24] McMahan, 'What Rights May Be Defended by Means of War?' (n 9) 127–28.

[25] ibid 139.

[26] Bazargan, 'Defensive Liability Without Culpability' (n 23); McMahan, 'What Rights May Be Defended by Means of War?' (n 9).

[27] On the distinction between liability and non-liability (or lesser evil) justifications, *see* Adil Ahmad Haque, *Law and Morality at War* (Oxford University Press 2017) 7; David Rodin, 'Justifying Harm' (2011) 122 Ethics 74, 74.

The difference between these justifications is better understood in terms of rights. Liability-based justifications do not infringe or violate any rights: they do not wrong those who are harmed.[28] By contrast, lesser-evil justifications for imposing harm do infringe rights, although they do not violate them. Because they infringe rights, we say that rights have been overridden; that is, that the moral reasons for infringing certain rights against harm have outweighed the moral reasons against doing so.[29] When that is the case, the permissibility of harming someone can be explained as a preference for the lesser evil.[30] This way of thinking about rights is, as noted by Tomlin, controversial; it implies seeing rights as inputs in deciding what is permissible by weighing them in favour or against a course of action, but not necessarily blocking it.[31] Lesser-evil justifications are also a feature of what Walen calls the 'infringement model of rights', whereby rights can be overridden and act as inputs in deciding what is, all things considered, permissible to do.[32]

There are three important features of lesser-evil justifications. First, because lesser-evil justifications do infringe rights, the individual who is harmed *is* wronged. For some, the fact that someone is wronged explains the need to apologize or compensate, and for others, it also entails that the victim is permitted to fight back against that imposition of harm.[33] Thus, if a lesser-evil justification works—though I don't think it does— it might imply that some unjust combatants can defend themselves.

Second, despite initial appearances, the kind of reasoning required by a lesser-evil justification need not be purely consequentialist. It will be purely consequentialist if we only compare states of affairs: X number of people dead, a war being lost, etc. But lesser-evil justifications can also take into account certain non-consequentialist moral disparities, such as the one that exists between inflicting harm and allowing harm.[34] They can also be sensitive to modes of agency,[35] that is, to the level of intentionality or culpability of the agents. In this

[28] The non-wronging nature of the harm can be either due to the forfeiture of certain rights or to the existence of an enforceable duty to bear that harm. On the issue of whether liability to harm is limited to forfeiture, *see* Jeff McMahan, 'Individual Liability in War: A Response to Fabre, Leveringhaus and Tadros' (2012) 24 Utilitas 278; Victor Tadros, 'Duty and Liability' (2012) 24 Utilitas 259.

[29] *See* Haque (n 27) 7; Rodin, 'Justifying Harm' (n 27) 73.

[30] ibid.

[31] Patrick Tomlin, 'Innocence Lost: A Problem for Punishment as Duty' (2017) 36 Law and Philosophy 225, 249.

[32] Alec Walen, *The Mechanics of Claims and Permissible Killing in War* (Oxford University Press 2019).

[33] Though Frowe thinks they cannot do so. *See* Helen Frowe, 'Lesser-Evil Justifications for Harming: Why We're Required to Turn the Trolley' (2018) 68 The Philosophical Quarterly 460.

[34] ibid, 460.

[35] ibid 464.

'non-consequentialist' approach to a lesser-evil justification, we don't only balance consequences or states of affairs but we also assign some value or disvalue to how we achieve those consequences. Thus, if achieving a given outcome requires me to intentionally kill an innocent, the outcome to be achieved needs to be significantly better or more valuable than it would have to be if achieving it only required my letting someone die.[36]

Third, lesser-evil justifications are not universally accepted as justifications for killing or causing grievous harm to other people,[37] especially when that kind of harm is imposed intentionally. The paradigmatic example of a lesser-evil justification is the example where someone breaks into a mountain cabin to save his own life or the life of someone else.[38] In this case, it is not too difficult to figure out the correct balance between the different moral considerations.

IV. Lesser-Evil Justifications, Killing in War, and the Combined Justification

A lesser-evil justification could potentially cover three cases where the reach of liability is limited. The first case is the killing of non-liable unjust combatants. The second one is the harming of non-liable civilians and non-liable combatants as a side effect. And the third one, the imposition of any amount of harm that surpasses what someone is liable to. In the first two cases, lesser evil works *instead* of liability. In the third one, it works *with* liability. I will only talk about the first and third cases (though some of what I say will be relevant to the second one).

[36] This outlook certainly assumes that consequentialist considerations and deontological ones can be weighed against each other, or, in different terms, that they are commensurable. On the difficulties of this position *see* Larry Alexander, 'Deontology at the Threshold' (2000) 37 San Diego Law Review 893, 908–10.

[37] Jeff McMahan, 'The Morality of War and The Law of War' in David Rodin and Henry Shue (eds), *Just and Unjust Warriors: The Moral and Legal Status of Soldiers* (Oxford University Press 2008) 23. *See* Rodin, who says 'that one ought never to kill a person who has done nothing to make themselves liable to be killed has a kind of elementary status in our morality ... Like the existence of Moore's hands in epistemology, it seems that *whatever* else we can know about morality, we can know *these* things.' David Rodin, 'Morality and Law in War' in Hew Strachan and Sibylle Scheipers (eds), *The Changing Character of War* (Oxford University Press 2011) 461.

[38] *See* Mitchell N Berman, 'Punishment and Justification' (2008) 118 Ethics 258, 277; Claire O Finkelstein, 'Duress: A Philosophical Account of the Defense in Law' (1995) 37 Arizona Law Review 251, 281; Bazargan, 'Defensive Liability Without Culpability' (n 23) 77.

A. Lesser Evil Instead of Liability

Operating instead of liability, a lesser-evil justification would morally account for harming non-liable civilians and combatants as a side effect[39] and for the killing of non-liable unjust combatants.[40] Both possibilities have been suggested by Seth Lazar but haven't been developed extensively.

Let's start with non-liable unjust combatants. Unjust combatants may be non-liable for many reasons. Some of them will not be liable because they are not posing any threat at all to other just combatants, or because they are not contributing to the unjust war.[41] Others will not be liable simply because harming or killing them fails to meet the necessity and narrow proportionality requirements or because they do not meet the required threshold of moral responsibility to trigger liability. How many non-liable unjust combatants there will be on the latter case will partly depend on the threshold of responsibility endorsed for liability: the lower the degree of moral responsibility required to trigger liability, the more unjust combatants will be liable (and civilians). The higher the threshold required for liability—such as requiring culpability for a threat, rather than moral responsibility—the fewer unjust combatants will be liable.

In what I have argued is the most plausible view of liability to defensive harm, a significant number—if not most—of unjust combatants are not liable. This makes a lesser-evil justification carry most of the burden of justifying killing in war, thus making the view very close to a consequentialist account of killing in war. I think this is enough reason for discarding it. But it is also difficult to adopt a lesser-evil justification for intentionally killing and harming individuals without abandoning reductivism and individualism.

The problem is that a lesser-evil justification needs to explain why doing or allowing some action X is morally preferable to doing or allowing some other action Y. In simpler words, a lesser-evil justification implies that there is a lesser evil in comparison to *something*. If we want to employ a lesser-evil justification for killing in war, and we want to stay committed both to individualism and reductivism, we have to argue that each killing of each non-liable unjust combatant is the lesser evil to something else. But what could this be?

Think again about the cabin example. Mary breaks into someone else's cabin to save John's life. The balance here would go something like this: intentionally

39 Seth Lazar, 'National Defence, Self-Defence, and the Problem of Political Aggression' in Cécile Fabre and Seth Lazar (eds), *The Morality of Defensive War* (Oxford University Press 2014).
40 Seth Lazar, *Sparing Civilians* (Oxford University Press 2015).
41 As suggested by McMahan, 'What Rights May Be Defended by Means of War?' (n 9) 122–23.

breaking into someone's house is morally forbidden. This moral prohibition—and the harm it causes—needs to be balanced against not being able to save John's life, and thus, against John's dying. John's dying is a greater evil than intentionally breaking into somebody else's house (and the harm it causes). Hence, Mary is justified in doing so.

Let's go back to war. Just combatant A is pondering whether he may kill non-liable unjust combatant B. How would the balance go for our just combatant? Intentionally killing someone is very bad. Intentionally killing someone who is innocent is even worse. This is one of the evils to be balanced. But balanced against what, exactly?

One—rather implausible—possibility is to argue that the killing of one unjust combatant is the lesser evil to losing the just war. This, of course, cannot be right. It is almost never the case that the killing or harming of one individual unjust combatant makes a significant difference—or any difference at all—in whether a just war is lost or won. Thus, the choice that most individual just combatants are in a position to make is not really a choice between the moral wrongness and the harm caused by killing a non-liable unjust combatant and the moral wrongness and the harm caused by losing the war.

Another possibility is for the killing of the innocent unjust combatant to be balanced against just combatant A's being killed, when the former is posing an immediate threat against the latter. How numerous these cases are will depend on how low the threshold of responsibility for liability to be triggered is. In a liability account with a low threshold of responsibility, most unjust combatants posing a micro-threat will be liable; with a high threshold, there will be more cases where a non-liable unjust combatant will be threatening a just combatant. The problem then is to figure out how it could be that the killing of a non-liable unjust combatant is the lesser evil compared to the killing of the just combatant, since it is a choice between two innocents. Because revisionists are committed to individualism and reductivism, we can't argue that the just war is what makes the killing of the non-liable unjust combatant the lesser evil, because that would be the equivalent of making the just war a *sui generis* moral justification.

The possibility that I find most consistent with methodological revisionism is for the balancing between evils to proceed as follows: intentionally killing non-liable unjust combatant B must be balanced against the mere possibility of just combatant A making a (usually small, most usually, of unknown relevance) contribution to winning the just war. Killing the innocent unjust combatant then must be the lesser evil to (perhaps) not being able to make a contribution (of unknown relevance) to winning the just war. The size and importance of that contribution to winning the war will be given by the military gain that those killings achieve. But how can the intentional killing of someone who is innocent be the lesser evil compared to the mere possibility of making this contribution? And even if we have certainty about there being a contribution to victory at all, how can the intentional killing of an innocent be the lesser evil to not being able to make what will usually be a very small contribution to winning the just war?

Imagine that 2 million people are building a road for your town. This road is very important: it will connect your town to another town where there is a hospital. Lives could be saved. Each of you will contribute two bricks. It remains unclear how many bricks you will need to finish the road: it could be anywhere from one million bricks to more than the 4 million bricks there actually will be, in which case no-one knows what will happen to the bridge. Now imagine you are one of these 2 million people. One of your bricks is taken by someone (why, it doesn't matter, but suppose he is non-liable for this example). The only way to get it back is to injure him. Maybe he will lose an arm. Maybe he will lose a leg. Maybe both an arm and a leg. Is intentionally harming this innocent person the lesser evil compared to the other evil in the balance, which is your not being able to contribute your two bricks to this very important road? It seems unlikely.

The most glaring problem is that it doesn't look like harming the innocent is the lesser evil. How can it be that intentionally killing an innocent could be the lesser evil to the mere possibility of not being able to make a very small contribution to some worthy end? How worthy is this end meant to be? And here is how we run into trouble, because the more we think about it, the more *sui generis* the just war justification starts to appear. Because we are now tempted to say that perhaps the road is, after all, not *that* important. Could we change the road for something else? But for what? A hospital? A whole town? The establishment of a democratic government? It doesn't seem to solve the problem, which is that in an individualist reductivist framework, each agent has to balance killing or harming an innocent against the mere possibility of not being able to make a very small contribution to something just or important, whatever that is. And the problem, of course, is that whatever that thing is only begins to, and *merely* begins to, make sense when we think of what is, and should be, at stake in a just war, and how the just war itself provides a special kind of justification. One that cannot be captured by a framework that is committed to reductivism.

The above could be solved if we adopted a different approach to a lesser-evil justification, whereby the deaths of unjust combatants can be aggregated.

So, we could say something like the deaths of 2,000 non-liable unjust combatants are the lesser evil to the consequences of losing the just war. Those consequences need to be able to justify intentional killing. As a result, they need to be significantly worse than the deaths of 2,000 non-liable just combatants and civilians. Some accept a 1:10 ratio, whereby it might be permissible to kill one innocent to save 10 innocents from death. That means that killing 2,000 non-liable unjust combatants could be justified as the lesser evil only if the lives of 20,000 individuals from the just side of the war were at stake. This is, however, barely consistent with non-consequentialism. Other accounts, such as threshold deontology, require something close to a catastrophe in order to justify intentionally killing the innocent. If threshold deontology is correct, the pursuit of most wars, even those with a just cause, would be forbidden.

There are two additional difficulties, and I will focus mostly on the second one. First, it's unclear how the aggregation should work here. Must the death of each unjust combatant save 10 lives, or is it enough that the combined deaths of 2,000 unjust combatants save 20,000 lives? That is, must the 1:10 ratio be preserved at the level of each individual or is it enough that it holds regarding the total of lives lost? I think there are powerful reasons to adopt the first view—that the death of each unjust combatant must save 10 lives. It seems almost impossible to justify intentionally killing someone if their own death, considered independently, does not actually meet the required 1:10 ratio for a lesser-evil justification. If this is correct—though I will not defend it here—then lesser evil seems very likely to fail as a justification of killing in war.

A second difficulty is that no just combatant, on their own, can kill the number of non-liable unjust combatants required to achieve victory. As a result, just combatants, individually considered, would not have a lesser-evil justification. All of them together might, but this would depend on whether they all succeed in killing the required 2,000 unjust combatants. In other words, individual killings of non-liable unjust combatants do not become permissible until 2,000 unjust combatants are killed—they cannot be justified as the lesser evil until 2,000 of them are dead. This entails that individual just combatants are engaging in forbidden behaviour. There is no lesser-evil justification until they have reached the required number of killings. And because each just combatant does not have a justification to inflict harm on the non-liable until the just war has been won, it would make just combatants themselves liable to harm or death under the responsibility view, since some of them would be morally responsible for posing objectively unjustified threats to the non-liable.

One might argue that although the availability of a lesser-evil justification is contingent upon successfully killing all 2,000 unjust combatants, this is only true in a fact-relative sense. In an evidence-relative sense, just combatants would have a lesser-evil justification available if it is reasonable for them to believe that they will succeed in killing all 2,000 unjust combatants. This would allow us to say that although just combatants might be engaged in forbidden behaviour in a fact-relative sense, they would at least be acting permissibly in an evidence-relative sense. Of course, if they fail to kill all 2,000 unjust combatants, each of their killings would turn out to be forbidden in a fact-relative sense.

Although this does not solve the problem of a just combatant's becoming liable under the responsibility view, it gives a plausible explanation of why just combatants might not be engaging in forbidden behaviour in an evidence-relative sense. However, this solution still poses some issues and is, ultimately, unlikely to succeed.

First, just combatants will be acting permissibly in an evidence-relative sense only when it is reasonable for them to believe that killing 2,000 unjust combatants will be successful in winning the (just) war. Given what is at stake—intentional killings of 2,000 non-liable individuals—the probability of success should be quite high for the justification to become available. This might be a difficult threshold to meet.

Second, the judgment about what is evidence-relative permissible to do cannot be limited to this or that mission, but must extend to the entire war. After all, it is winning the just war, and consequently saving X number of people, that renders all those killings permissible as the lesser evil. That is, there is a powerful case for arguing that just combatants should have a reasonable belief regarding the success of the war, and not merely regarding one mission within it. If the war is lost, a lesser-evil justification cannot but fail: each mission or operation within a war is just a contribution to the bigger aim of winning. If victory is not achieved, those missions cannot be justified. To assess them independently from each other is quite artificial.

The above seriously undermines the availability of evidence-relative permissibility concerning lesser-evil justifications in war. A plausible account of evidence-relative permissibility, given what is at stake (innocent lives), should be quite demanding in terms of the quality and quantity of evidence required for the justification to take off, and, in particular, regarding the probability of success of the mission and, further, of the war itself. This standard seems incredibly difficult to meet.

Additionally, the difficulty in meeting this standard seems to grow significantly as we move up the chain of command, which would entail that individuals who are called to make decisions about which missions to pursue and how are seldom justified in an evidence-relative sense. Even if we can accept that those lower down in the chain of command might be justified in an evidence-relative sense because they have good reason to trust their superiors about the probability of success, their superiors are the ones actually in charge of making that assessment. And it seems that very often they will lack sufficiently robust evidence to be justified in an evidence-relative sense. This chain-of-command issue significantly undermines the availability of evidence-relative justifications more generally, in the context of war. I will come back to this in the next chapter.

B. Lesser Evil *with* Liability: The Combined Justification

Now let's turn to the second role for a lesser-evil justification, which is for it to work in conjunction with liability. This is known as the combined justification, and it has been developed in detail by McMahan, Frowe, and Bazargan.

The reason why a lesser-evil justification might need to do some work 'on top' of liability is that under most accounts, liability is subject to an internal constraint known as 'narrow proportionality'.[42] Narrow proportionality means that if an attack would harm someone in excess of the degree of harm to which the person is liable, that excess of harm would wrong him. The degree of harm to which someone is liable depends, partly as we have seen, on their degree of moral responsibility: the less responsible the aggressor is, the more harm the victim must bear to spare him some; the more responsible the aggressor is, the more harm the victim can impose on him to defend herself.[43] Thus, under the combined justification we would find cases where unjust combatants, due to their diminished responsibility, are liable to some X degree of harm that does not amount to killing (or to a specific degree of harming). Here, just combatants would need a lesser-evil justification for the harm that surpasses X.

As Bazargan has argued, the combined justification requires adopting a complex account of liability. In a complex account, the degree of harm to which someone is liable is a function of their moral responsibility for the threat, among other possible factors.[44] Under a simple account, a person's liability

[42] Narrow proportionality is less controversial as an internal requirement of liability than the necessity requirement. *See* Frowe (n 23) 89–119.

[43] McMahan, *Killing in War* (n 5) 175–82. McMahan seems to think there is a residual injustice in the case of the villainous aggressor as well, though a very small one. Jeff McMahan, 'Duty, Obedience, Desert, and Proportionality in War: A Response' (2011) 122 Ethics 135, 155–56.

[44] Bazargan, 'Defensive Liability Without Culpability' (n 23) 72.

is not a function of their moral responsibility. It is enough for liability to be killed that they meet the required threshold of moral responsibility and that they pose a sufficiently serious threat.[45] Under a complex account, we can say that someone is liable to only X degree of harm (X being a function of certain features of the person's conduct), even though in order to successfully avert the threat, X + 1 would be required. That is, we would be able to say that someone like the Conscientious Driver might be liable to having her arm broken, but not to anything else. Under the simple account, if defending yourself from the Conscientious Driver required killing her, the driver would be liable to killing. The simple account is, as its name indicates, simpler. But its implications sit uneasily with some of our intuitions about what is required of the pedestrian—in terms of reactive attitudes or apology—regarding the driver.

If successful, the combined justification would have significant advantages in justifying killing in war. Paired with a high threshold responsibility view of liability to significant or lethal harm—which, I have argued, is the most plausible view of liability—it could explain almost all killing and harming that takes place in war. It would also allow us to make room for the intuition that some unjust combatants are wronged when they are killed in war; that they are owed an apology.[46] Additionally, it might explain why unjust combatants can defend themselves or, at the very least, why unjust combatants may be able to shift the excess of harm back to just combatants.[47]

An additional advantage of the combined view is that it could potentially explain why it is wrong to kill multiple minimally responsible killers (such as the Conscientious Driver) who are independently threatening a single innocent person.[48] Since the rights infringements above the degree of harm to which each of them are liable are additive, but the harm to the victim is held constant, there comes a point where killing multiple minimally responsible killers will no longer be justified as the lesser evil.[49] This would also apply to views where other individuals, different from the Conscientious Driver, are not liable to significant or lethal harm due to a higher threshold of moral responsibility.

The combined justification was first suggested by McMahan.[50] He argued that harm that surpasses what the victim is liable to could be justified as the

[45] ibid 73.

[46] *See* Saba Bazargan, 'Killing Minimally Responsible Threats' (2014) 125 Ethics 114, 134–36; 136, fn 38.

[47] Surprisingly, this might lead to the conclusion that the symmetry thesis is morally right after all. If we wanted to avoid this conclusion without abandoning reductivism, we could not make reference to the just war as a *sui generis* justification.

[48] Bazargan, 'Killing Minimally Responsible Threats' (n 46) 132.

[49] ibid 134.

[50] McMahan, 'What Rights May Be Defended by Means of War?' (n 9) 139.

lesser evil insofar as the excess would be proportionate in the wide sense.[51] Proportionality in the wide sense refers to the idea that defensive action must not inflict disproportionate harm on the innocent as a side-effect.[52] It operates by weighing an act's good effects against the harms it causes to those who are not liable.[53] Although he originally thought it would help solve a number of issues, particularly the problem of the number of aggressors, he ultimately discarded it.[54]

In Bazargan's view, a lesser-evil justification can account for the killing of certain unjust combatants who are similar to minimally responsible killers such as the Conscientious Driver or those who make small contributions to threats, since it is generally better to kill someone who is liable to substantial but sublethal harm than it is to kill someone who is liable to nothing at all.[55] Bazargan's combined justification does require unjust combatants to be liable to substantial harm. It also requires a non-negligible difference in moral responsibility between the attacker and the victim.[56] This already limits the view quite significantly, particularly if what I've said so far about war is correct.

As Bazargan himself acknowledges, a lesser-evil justification has lot of justificatory work to do due to the 1:10 ratio usually accepted in wide proportionality calculations. The 1:10 ratio holds that killing one innocent might be permissible only if it would be necessary to save 10 other innocent people from death.[57] In order to circumvent this problem—which would make the lesser-evil justification ineffective in most cases—he adopts what he calls 'the lesser-evil discounting view'.[58] Under this view, we ought to discount the disvalue of the harm that surpasses what the attacker is liable to relative to the weight of the harm that the attacker would impose on the victim.[59] This sort of discounting is explained by considerations of distributive justice.[60]

Frowe also accepts a combined justification. In her case, the justification is a combination of the duties that an innocent threat has—based on Tadros' earlier views—with a lesser-evil justification for inflicting the harm.[61] An innocent

[51] Jeff McMahan, 'Liability, Proportionality, and the Number of Aggressors' in Saba Bazargan and Samuel C Rickless (eds), *The Ethics of War* (Oxford University Press 2017) 1.
[52] ibid 3.
[53] ibid.
[54] ibid 20.
[55] Bazargan, 'Defensive Liability Without Culpability' (n 23) 74; Bazargan, 'Killing Minimally Responsible Threats' (n 46) 131–32.
[56] Bazargan, 'Killing Minimally Responsible Threats' (n 46) 128.
[57] ibid 127.
[58] ibid 128.
[59] ibid 127.
[60] ibid.
[61] Frowe (n 23) 67.

threat, such as someone who will fall on you and crush you to death (Falling Person), has a duty to prevent herself from harming the innocent. This duty exists because we have special responsibility for what our bodies do.[62] The fact that Falling Person will kill the victim, and the fact that Falling Person would be obliged, if she were able, to avert her body at the cost of some harm to herself, entails that the victim is permitted to inflict that degree of harm on Falling Person since she is merely getting Falling Person to comply with her moral duty not to kill the victim.[63]

In attempting to justify the level of harm that surpasses the harm that Falling Person has a duty to bear, Tadros and Frowe invoke a lesser-evil justification.[64] The harm that Falling Person is obliged to bear as a duty can be discounted.[65] According to Frowe, our duty to avoid non-defensively killing the innocent is more stringent than our duty to avoid defensively killing the innocent.[66] That means that if the victim is killed, she would suffer more morally weighted harm than Falling Person would suffer if Falling Person is killed.[67] As a result, killing Falling Person is the lesser evil.[68] Because this is a lesser-evil justification, Frowe acknowledges that it cannot permit harming as a means or side effect of merely trying to avert a slightly greater harm to oneself; the difference between evils must be significant.[69]

But the combined justification faces a dilemma. If the lesser-evil justification is combined with a view that requires a high threshold of responsibility for liability to significant harm, it will have too much evil to justify and, as a result, it will fail. If it is combined with a view that requires a low threshold of responsibility for liability to significant harm, it renounces part of what had made the combined justification attractive (ie the high threshold of responsibility coupled with the relation between responsibility and narrow proportionality). It also becomes trivial.

The combined justification is also untenable and fails to deliver on its promised advantages. Some of the problems have already been explained by McMahan,[70] Steinhoff,[71] Quong,[72] and Walen.[73]

[62] ibid.
[63] ibid 68.
[64] ibid 69.
[65] ibid.
[66] ibid.
[67] ibid.
[68] ibid.
[69] ibid.
[70] McMahan, 'Liability, Proportionality, and the Number of Aggressors' (n 51).
[71] Uwe Steinhoff, 'Against a "Combined Liability-Lesser-Evil Justification"' (2019) 47 Philosophia 533.
[72] Quong (n 13) 64.
[73] Walen (n 32) 147–50.

Some think that it is a disadvantage of the view that it fails to solve the problem of the number of aggressors—it cannot explain why a victim could kill numerous minimally responsible threats. This failure is because the combined justification is not 'endlessly repeatable': with each additional responsible killer, the more harm there is to be justified as the lesser evil, but although the harm to be justified increases, the harm to be averted against the victim stays constant.[74] At some point, killing the responsible killers will no longer be the lesser evil in comparison to the victim's death.[75] This point is accepted by Bazargan as an advantage of the view. As a theory of killing in war, it poses some difficulties as there might be instances where killing numerous unjust combatants will become impermissible. This will be the case when the harm to be averted on the just side will be the death of one just combatant or other, lesser, evils.

The justification also fails in two-person cases. First, as McMahan notes, it may involve an objectionable form of double counting. Even though the death of the victim is not actually prevented twice, preventing harm to the victim operates twice as a justification.[76] First, in justifying the harm to which the aggressor is liable to, and second, in justifying the harm that is permissible to inflict on the aggressor as the lesser evil.[77]

A more serious objection concerns liability itself. Since the only thing that will prevent the death of the victim is the death of the aggressor, harming the aggressor to any degree less than death is not effective. And if that harm is not effective in preventing the threat, then the aggressor cannot be liable to it—if effectiveness is internal to liability.[78] As a result, the aggressor cannot, by definition, be liable to a lesser degree of harm. There is no liability that the lesser-evil justification can work with.

If effectiveness, necessity, and narrow proportionality were external to liability, the lesser-evil justification would become redundant. The person would already be liable to death or harm. But such a view of liability is unattractive: it would entail that individuals are not wronged when they are harmed disproportionately or gratuitously. This seems hard to accept.

In the two-person case, it is also hard to see how the harm imposed in excess of what the person is liable to can be the lesser evil. Both Steinhoff and

[74] McMahan, 'Liability, Proportionality, and the Number of Aggressors' (n 51) 20.

[75] This might be true in the case of highly culpable threats as well. *See* ibid; Victor Tadros, *To Do, to Die, to Reason Why: Individual Ethics in War* (Oxford University Press 2020) 171.

[76] McMahan, 'Liability, Proportionality, and the Number of Aggressors' (n 51) 21.

[77] He then responds to the objection but does not determine whether it is sufficient to rebut it. ibid.

[78] ibid 22.

McMahan have come to this conclusion.[79] Briefly, the problem is that lesser-evil justifications require a significant difference between harms in order to be able to justify infringing the rights of the innocent. This difference usually requires a 1:10 ratio. But this is not the kind of difference that we encounter in the two-person case. Although Bazargan incorporates the discount view to solve this problem, it's unclear why we should accept the view that imposing harm to which someone is not liable has less disvalue when that person is liable to something else.[80] In war, the discount view is also often not applicable. Because it requires to discount the disvalue of the harm that surpasses what the attacker is liable to relative to the weight of the harm the attacker would impose on the victim, it cannot operate in cases of unjust combatants who are not posing micro-threats, where no harm is threatened against the just combatant.

But without the discount view, the justification fails. In order to stick to the 1:10 ratio, the person should be liable to 90 units of harm when only 100 units of harm would effectively avert the threat. Those 10 additional units of harm would be the lesser evil compared to the death (100 units of harm) of the victim. But this would make the view's usefulness trivial, as it would require endorsing a view whereby the attacker is already liable to something close to 90 per cent of the harm.[81]

This way of thinking about the 1:10 ratio is also controversial. The 1:10 ratio is about killing one innocent individual to save 10 innocent individuals from death. It is not clearly applicable to units of harm accrued by one single individual. If so, we would have to accept that individuals can impose harm on the non-liable any time that it would spare them 10 times the amount of harm directed their way. If A's death brings about 100 units of harm, A could intentionally inflict 10 units of harm on the innocent to spare herself. But 10 units of harm is quite a lot: it's 10 per cent of the harm involved in death. We cannot just stipulate that it merely amounts to pushing or slapping the innocent. It probably amounts to a relatively significant injury.

An additional problem with the claim that the harm that surpasses that to which the threat is liable is the lesser evil is that the harm imposed on the threat is imposed intentionally and not as a side effect of defensive action.[82] This requires an even bigger disparity between the evil averted by killing the threat and the evil averted by the victim letting herself die at the hands of the threat.

[79] Steinhoff (n 71).

[80] Expanding on this objection, *see* Walen (n 32) 146–48.

[81] McMahan argues that it would require the person to be liable to 99 per cent of the harm. McMahan, 'Liability, Proportionality, and the Number of Aggressors' (n 51) 24.

[82] ibid 22.

In other words, the 1:10 ratio I suggested above is way too generous. This is even more so if we consider how combatants are killed in war.

Contrary to what most of the literature seems to think,[83] unjust combatants are not often killed eliminatively. They are killed opportunistically or, at the very least, in a partially opportunistic manner. Opportunistic killing takes place when the killer gains a benefit from the death of the victim that he would otherwise have not enjoyed.[84] Eliminative killing takes place when the goals achieved by killing the victim are goals that the killers could have secured just as easily in the victim's absence.[85] The victim is treated and seen as an obstacle or problem that killing can solve, not as a resource from which to derive benefits.[86] It is widely believed that opportunistic killing is worse than eliminative killing because it treats people as a means to an end.[87] This makes it harder to justify and might require an even bigger disparity between the harms.

Lazar, for example, thinks that soldiers are not killed as tools or means. Rather, they are obstacles or threats that need to be eliminated.[88] If a soldier is running at me with a bayonet, I kill him eliminatively, not opportunistically: I derive no benefit by killing him that I would have derived had he not been there.[89] This seems to be the view of most of the literature.

But modern warfare tells a different story. Often, combatants are killed opportunistically, or at the very least, they are killed partially in an opportunistic manner. For war is not about eliminating threats as they arise; it is often about causing enough damage in order to force the enemy to surrender. Probably the most common strategy to accomplish the surrender of the enemy is to kill or hurt as many enemy combatants as possible.[90] In other words, it is often the case that killing unjust combatants does provide the just party with a benefit: it is a way of forcing surrender on the unjust party. It is not just a matter of eliminating threats. This is even more apparent when unjust combatants are not posing micro-threats. If this is correct, then a combined justification has an almost impossible job to do.

As we have seen, the combined justification fails even in the two-person case, where one partially liable individual poses a threat to a non-liable one.

[83] *See*, eg, Lazar, *Sparing Civilians* (n 40) 57.
[84] ibid.
[85] ibid.
[86] ibid.
[87] *See* Lazar, *Sparing Civilians* (n 40); Tadros, *To Do, to Die, to Reason Why* (n 75) 127.
[88] Lazar, *Sparing Civilians* (n 40) 57.
[89] ibid 60.
[90] Yishai Beer, 'Humanity Considerations Cannot Reduce War's Hazards Alone: Revitalizing the Concept of Military Necessity' (2015) 26 European Journal of International Law 801, 825.

In war, there will be cases where the partially liable individual—the unjust combatant—will not be posing threats to any just combatants. In those cases, the intentional killing of the unjust combatant cannot even be balanced against the death of the just combatant. As a result, killing the unjust combatant is unlikely to be the lesser evil relative to the failure of the just combatant's making a small contribution to victory. In other words, we run into the same issues discussed in section V.B.

Finally, the combined justification is even harder to apply than a pure liability-based account. These difficulties are obviously compounded by the uncertainty that accompanies war. Ultimately, no-one really knows what it will take—how many lives will be lost—to win a just war. Of course, we never have full certainty about how any given course of action will turn out, but it would be absurd to deny how pervasive uncertainty is in war. And even if that level of uncertainty were pervasive in other spheres of action, the nature of the uncertainty in war seems to have some discontinuity to other kinds of uncertainties simply because there are lives at stake. To once again use our cabin example: if Mary enters the cabin and it turns out that John dies anyway, then it is truly unfortunate that Mary had to enter the cabin without the owner's consent in vain. But 'unfortunate' does not even begin to make sense of what happens when thousands of individuals die horrible deaths—on both sides—so that a just war is won, and yet, it is not.

V. Conclusion

In Chapters V and VI, I have pointed out the limitations of the revisionist view in accounting for killing in war. In Chapter V, I argued that a self-defence view committed to reductivist individualism fails to account for most, or a significant amount of, killing in war. In this chapter, I have argued that a plausible forfeiture-based view cannot make many unjust combatants liable to death or significant harm and that the introduction of a lesser-evil justification, whether in conjunction with, or instead of, liability fails.

Although the revisionist view aims to provide a complete account of killing in pursuit of a just war, it cannot plausibly provide anything but an account of why it is justified to kill villainous aggressors who are together posing significant threats to others.

Ultimately, a plausible forfeiture view cannot explain most killing in war. This is so because, to paraphrase Bernard Williams, a forfeiture-based view for killing in war has 'too few thoughts and feelings to match the world as it really is'.[91] This is not to say that the framework lacks complexity or sophistication. On the contrary, it is both complex and sophisticated. The problem is that in the version that allows it to explain most killing in war, a forfeiture view struggles to capture the moral ambivalence and the tragedy that is very often present in killing in war. If morality has a place in human affairs, it cannot simplify what is at stake in war without losing its grip on it. It cannot make it so that it prevents us from recognizing what makes war a tragedy in the first place. A plausible view of killing in war has to leave enough room for that.

[91] Bernard Williams, 'A Critique of Utilitarianism' in JJC Smart and Bernard Williams (eds), *Utilitarianism: For and Against* (Cambridge University Press 1973) 150. Williams here is referring to utilitarianism as 'simple-minded'.

VII
Beyond Self-Defence

I. Introduction

In the past chapters, I argued that the most plausible version of a liability-based view of self-defence cannot justify many instances of killing in war. At its best, this view leaves a vast space of killing in war morally unaccounted for. It is precisely within this vast space where we can begin to deconstruct the analogy between what unjust combatants do and the murder of the innocent, by proposing an alternative account of killing in war.

This account starts from the limitations of the revisionist view. If many unjust combatants are significantly different from the central case of self-defence known as the *villainous aggressor*, and expanding liability to account for these significantly different cases is both morally unattractive and implausible, many unjust combatants are *not* liable. But if many unjust combatants are not liable—and, as we saw, a lesser-evil justification is unlikely to explain the killings of those non-liable—we require an argument that accounts for those killings. Alternatively, we must embrace pacifism.

Perhaps it is striking that we find ourselves in this situation, where many killings in pursuit of a just war are still morally unaccounted for. In a sense, the question about why just combatants can kill in war seems at first like an easy one. Surely, if we know *one* thing about war, it is that just combatants are justified in killing unjust ones. Yet, in the previous chapters, I hope to have shown not only that this is the central issue in justifying the legal privilege to kill but also one of the hardest moral questions to answer. It is partly, of course, because these are matters of life and death.[1] But, also, what seems to make questions about war so perplexing is that we are faced with two kinds of objective reasons clashing against each other: agent-neutral and agent-relative reasons, particularly deontological ones.[2] While we can understand that in an agent-neutral

[1] *See* Moshe Halbertal, 'The Goldstone Illusion' (*New Republic*, 5 November 2009) <https://newrepublic.com/article/70956/the-goldstone-illusion> accessed 18 August 2022.

[2] *See* Thomas Nagel, *The View from Nowhere* (Oxford University Press 1989) 152–85. As explained earlier, according to Nagel, both agent-neutral and agent-relative reasons are objective if they can be understood and affirmed outside the viewpoint of the individual who has them. Neutral reasons

The Morality of the Laws of War. Marcela Prieto Rudolphy, Oxford University Press. © Marcela Prieto Rudolphy 2023.
DOI: 10.1093/oso/9780192855473.003.0007

sense a just war is better than an unjust one, we struggle to grasp what agent-relative (deontological) reasons just combatants may have for inflicting death on other individuals who are innocent in morally significant ways. The clash is between what we think best to happen and what we think someone should do: though it is best that the just war is won, we think we should not do certain things, regardless of their outcome.[3] Indeed, that an outcome A is better than another outcome B does not necessarily mean an agent has a moral permission to bring it about.[4]

This conflict is certainly magnified by the pervasive uncertainty there is in war. But even with full knowledge of the facts (something we can only dream of ever achieving), judging that pursuing a war that will result in the deaths of thousands of people is the correct thing to do should give us pause. The things we are balancing—among them, human lives and deontic moral constraints—are incommensurable in important ways.[5] There is a sense in which what is being balanced cannot simply *be* properly balanced, and in any event, we certainly lack any way of making any sort of precise measurement.[6]

A plausible moral theory of killing in war needs to make significant space for this conflict. It cannot simply wish it away. An attractive account of killing in war, if there is such a thing, must also reflect the perpetual peace ideal; that a just war is second best to peace. A just war is, in many ways, a tragedy. It is a tragedy not just because thousands of civilians and just combatants are killed: unjust combatants' deaths also share in that tragedy. Our attitudes towards some of those who die fighting an unjust war are not the same attitudes we have towards those who are killed in self-defence while villainously aggressing against their victims.[7] We tend to think that what happens to (some) unjust

underlie judgments about what should happen, whereas relative reasons can affect judgments on what people may do. Deontological reasons are a type of agent-relative reason, and they stem from the claims of other persons not to be maltreated in certain ways. They limit what we are permitted to do in the service of impersonal or autonomous values. They are objective because they can be affirmed outside the viewpoint of the individual who has them.

[3] For a similar view, *see* Thomas Nagel, 'War and Massacre' (1972) 1 Philosophy & Public Affairs 123.

[4] This, of course, is not unique to war but a problem for philosophy in general. *See* Nagel, *The View from Nowhere* (n 2); Thomas Nagel, 'The Fragmentation of Value' in *Mortal Questions* (Cambridge University Press 2012).

[5] *See* John Gardner, 'Justification under Authority' (2010) 23 Canadian Journal of Law & Jurisprudence 71.

[6] Samuel Scheffler, *The Rejection of Consequentialism: A Philosophical Investigation of the Considerations Underlying Rival Moral Conceptions* (Clarendon Press 1995) 63.

[7] On the use of the first-person plural pronoun here, I avail myself of Samuel Scheffler's explanation: 'When I talk about *our* attitudes and *we* think or feel, I do not intend to be making strictly universal claims. I do not mean to claim, in other words, that literally everyone is prone to these attitudes. My use of the first-person plural might instead be thought of, to borrow some terminology that David Lewis employed in a related context, as a "wait-and-see" use... To put it in a slightly different way, in characterizing *our* attitudes, I mean to be characterizing my own attitudes and the attitudes of any other

combatants is *also* a tragedy, made even more senseless by the fact that sometimes they have died and killed in vain.

In this chapter, I will propose an account of killing in war that aims to capture the moral ambivalence that is often present in killing in war. I will give an account of the kind of conflict one faces in killing in war and argue that, at best, just combatants have a moral permission, in the *sui generis* sense I will develop, to kill non-liable unjust combatants. The structure of the argument relies on the idea of there being a small difference in weight between the reasons for killing in war and the reasons against it. This feature is what makes the account of permissibility I will develop *sui generis*. But the most difficult hurdle the account must overcome is to explain why killing non-liable unjust combatants could be morally permitted at all in the circumstances of war. If what we encounter is a moral prohibition—in other words, some form of pacifism—the instrumentalist justification I proposed in Chapter III will have to carry almost the full burden of justifying the legal privilege of combatants to kill. It will have to justify not merely the permissibility of targeting just combatants, but also the permissibility of targeting unjust ones. This is not unthinkable. While the idea of permissibility that I will develop has welcome implications and can account for some non-consensual impositions of harm, as a theory of killing in war it leaves several issues unresolved.

This chapter proceeds as follows. In section II, I provide a summary of the morality of war left in place after the past two chapters and expand upon what I have called 'the chain of command issue'. In section III, I attempt to sketch a moral account of killing in war by suggesting that killing (some) non-liable unjust combatants is merely permissible, not justified. Section IV provides a brief account of the (*sui generis*) idea of moral permissibility. In section V, I outline some implications of this view. Section VI deals with some objections. Section VII concludes.

II. Recap and the Chain of Command Issue

The picture we have of killing in war, after the previous chapters, is the following. Some unjust combatants who pose micro-threats might be harmed on the basis of liability-based justifications. Some harm directed against unjust

people who share them however numerous those people happen to be. On the one hand, I don't think that the attitudes are mine alone. On the other hand, I don't wish to claim that they are universally shared . . . .' Samuel Scheffler, *Death and the Afterlife* (Kolodny ed, Oxford University Press 2013) 17–18.

combatants who pose micro-threats might be objectively unjust, but permitted in an evidence-relative sense.[8] When that is the case, unjust combatants would be allowed to defend themselves (and in some accounts, just combatants would become liable for posing objectively unjustified threats against the innocent). Finally, we might posit that just combatants have agent-relative prerogatives—explained by an appeal to the disproportionate weight each of us may attach to our own lives[9]—to defend themselves against non-liable unjust combatants who pose micro-threats.

This account fails to make many unjust combatants, including those who do not pose micro-threats, liable. A lesser-evil justification, as we saw, is likely to be unworkable. There is thus a moral prohibition against harming a significant number of unjust combatants. Just combatants cannot kill them—attempting to do so might make them liable, and unjust combatants would be justified in defending themselves.

An account like this can convey the horrors of war—even just wars. But it also fails to justify most killing in war. In fact, I think it is an argument for pacifism: it puts into question the very permissibility of resorting to war. This is true, as we will see, even in an evidence-relative sense. And this problem concerning evidence-relative permissibility is even more pressing for those who have to make the decision to resort to force, that is, for military and civilian leaders. This is what I have been referring to as the chain of command issue.

In a reductivist individualist account where liability requires a high threshold of moral responsibility, the problem regarding evidence-relative permissibility and the chain of command is as follows. Partly because armed groups are organized hierarchically, the higher the position someone occupies in the chain of command, the more 'bloated' moral decisions become: they include assessments about more information, the lives of more individuals, and higher stakes. This hierarchical organization also has implications for the moral situation of just combatants at the lower end of the chain of command: many of them have no or little moral responsibility concerning the decision to resort to

[8] Haque holds a similar view of killing in war. He rejects the responsibility view of liability and its distributive dimension. On his view, individuals become liable to defensive killing by threatening to unjustly kill directly, jointly with others, or indirectly through others one controls. As a result, he argues that unjust combatants who participate in hostilities or have a combat function are liable to defensive killing. It will also be often evidence-relative permissible to kill them on the basis of anticipatory self-defence and fact-relative permissible to do so when killing them in fact prevents them from posing threats in the future. In contrast, combatants who do not have a combat function are not liable, but it might be evidence-relative permissible to kill them. Haque's view implies that a just war will often inflict objectively unjust harm—though evidence-relative permissible—on many unjust combatants. *See* Adil Ahmad Haque, *Law and Morality at War* (Oxford University Press 2017) 85–95, 257–70.

[9] Jonathan Quong, *The Morality of Defensive Force* (Oxford University Press 2020) 69–70.

force; contribute very little, if at all, to the war; and often do not—and cannot—have first-hand knowledge about the justness of the war. Some of the information is not readily accessible and might be classified, and the *ex-ante* judgment of whether a war is just is complex, among other epistemic obstacles. As a result, a significant number of just combatants may be relying on what they have been told about the justness of the war by their superiors or civilians in charge. This reliance, when justified,[10] can limit the scope of the moral decisions that just combatants at the lower end of the chain of command face, by making them less bloated. These combatants no longer need to assess proportionality issues, effectiveness, or chances of success. Their decision might be right in an evidence-relative sense if they can reasonably trust their superiors.

But while many combatants at the lower end of the chain of command might reasonably rely on the judgment of their superiors, and thus, can make more limited decisions about their participation in war, this ceases to be true as we move up along the chain of command. Those higher up don't face limited decisions about small contributions, nor are they able to rely on someone else's judgment: *someone* has to make the judgment that fighting the war, pursuing mission A or B, is the right thing to do; that the war will be proportionate and successful, that it will be effective, that killing X number of non-liable individuals will save more than 10 times X. This issue reaches its maximum point at the level of those who have to decide whether to resort to force. And given the limits of this version of the reductivist individualist view, it looks like it is impossible to justify their decision to go to war, even in an evidence-relative sense.

First, even if we don't—and can't—know *who* the innocent will be, we know, with a high degree of certainty, that many individuals, both combatants and civilians, do not meet the threshold of moral responsibility for liability. That is, we have enough evidence to be reasonably certain that targeting any number of combatants is likely to involve harming people who are not liable.

Second, if many individuals do not meet the threshold for liability, most killing in war must be justified as the lesser evil. And because killing combatants is done at least partially in an opportunistic manner, one might have to embrace a more demanding ratio than 1:10. This introduces even more uncertainty, and it will be rare to have reasonable grounds to believe that a lesser-evil justification will be available: this would involve having some degree of certainty that the war will succeed and that the 1:10 ratio will hold.

[10] *See* David Estlund, 'On Following Orders in an Unjust War' (2007) 15 Journal of Political Philosophy 213. Moreover, soldiers may have strong feelings of betrayal, or of being 'suckered' when those people turn out to be lying or negligently mistaken. *See* Nancy Sherman, *The Untold War: Inside the Hearts, Minds, and Souls of Our Soldiers* (1st edn, W.W. Norton 2010) 48–63.

If the prohibition against killing the innocent is truly stringent, particularly when killing at least partially in an opportunistic manner, it seems that the only reasonable conclusion for military and political leaders, given the evidence that is often available to them, is that they should *not* resort to force. Things like success and proportionality—which would potentially make lesser-evil justifications available— are uncertain, but the fact that many combatants (and civilians) are not liable is, on the contrary, quite certain. And if the decision to resort to force is often impossible to justify, even in an evidence-relative sense, then just combatants on the lower end of the chain of command have no good reason to trust their superiors when they are ordered to fight. The chain of command issue, then, has implications not only for those higher up in the chain of command but permeates the decisions of those lower in the chain of command too.

Unless one introduces an evidence-relative account that makes it permissible to resort to war when there is considerable uncertainty about what is fact-relative right but reasonable certainty about the likelihood of harming the innocent (which is fact-relative wrong),[11] some form of, if not outright, pacifism about war seems the only tenable alternative.[12]

If this account is true, the justification for the symmetry thesis could only be provided on the basis of the perpetual peace ideal, and only contingently. And the fragility of such justification would *increase*, rather than be attenuated, given that, on this account, although non-liable individuals on both sides might have agent-relative prerogatives to defend themselves, both sides are also likely to be engaged in killing the innocent—if to different degrees.

Although enough to provide a (quite fragile) justification of our legal regime, as a moral account of war, this view seems somewhat unsatisfactory. I have said repeatedly that pacifism about war is a difficult position to resist.

[11] For example, one could posit a presumptive or a practical account of just cause if (1) in a war of self-defence, one reasonably believes one's country did nothing to start the war or (2) in a war of defence of others, one reasonably believes that the members of the third-country being attacked are morally innocent, and, in both cases (3) if one's country does not go to war, innocents will die. Here, I'm following Helen Frowe's account of what a practical account of self-defence would look like. Note that this is a different question to the question of whether someone is liable to defensive harm, which is answered at the fact-relative level. *See* Helen Frowe, 'A Practical Account of Self-Defence' (2010) 29 Law and Philosophy 245.

[12] I leave somewhat open the possibility that an entire war might be justified, in the fact-relative sense, partially on the basis of liability and mostly on the basis of lesser evil (so that the killings of the non-liable save more than ten times the lives of those killed). This seems to me an extremely rare occurrence, and often, something regarding which we are very unlikely to have reasonable evidence when deciding how to proceed.

But pacifism about war is also a troubling position to embrace: it is hard to accept that there is nothing morally permissible to do in the face of aggression or slaughter; that the only morally available alternative on the face of something like Nazi domination is not to resort to force. This is a particularly salient issue when not fighting will result in the unjust side's imposition of a deeply unjust and repressive regime, which will imperil the lives and wellbeing of many individuals.

The decision not to fight can thus have significant moral costs. But these costs—which are often uncertain—are unlikely to solve the chain of command issue or make a lesser-evil justification for most killing in war available. First, because intentionally and opportunistically inflicting harm (fighting the war) is harder to justify than allowing harm to take place (not fighting). And, second, because these costs do not change the 1:10 ratio typically required by lesser-evil justifications.

There are additional reasons why the above reductivist individualist account is unsatisfactory as an account of war. First, relying on agent-relative prerogatives to explain why just combatants may harm threatening (non-liable) unjust combatants falsifies the nature of war. Just combatants' reasons for killing non-liable unjust combatants might have to do with their prioritizing their own lives, but it would be perplexing if those reasons had no relation to the justness of the war itself. Otherwise, (non-liable) just and unjust combatants who pose micro-threats would be in almost completely symmetrical moral positions. Second, it is difficult to accept that when just combatants target non-liable unjust combatants who are not posing micro-threats, they are often breaching a moral prohibition. This view would entail that there are many cases in war where there is nothing normatively significant that distinguishes between just and unjust combatants—nothing that relates to whether they are fighting just or unjust wars. This would be true not only in instances where just combatants are engaged in, say, war crimes, but also when they are targeting non-liable unjust combatants who are not posing micro-threats, where lesser-evil justifications are unlikely to be available. This is far from an uncommon situation in war. I find this conclusion difficult to accept.

These issues with a reductivist individualist view provide good reasons not to embrace absolute pacifism and to try to develop an alternative account of killing in war. That is what I will attempt to do in the rest of this chapter—but I'm not sure it succeeds. Providing such an account and defending it properly would require another book. Here, I can only sketch the form of the argument and some of its implication and objections.

III. Just War and Killing

The account I will propose is not meant to displace the justificatory work of self-defence in killing in war.[13] As I have stated repeatedly, self-defence works partially. What I say next aims to work in the (rather vast) space left by self-defence.[14] That space is vast because self-defence will not be able to justify killing unjust combatants who are not posing a direct and immediate threat (ie a micro-threat) against other just combatants. That space might be even vaster if we think, like I do, that forfeiture of rights against significant or lethal harm has little plausibility outside of the case of the villainous aggressor. A reductivist individualist framework seems incapable of providing a complete account of killing in war.

The challenge for those of us who believe that liability-based views of self-defence are implausible outside of the case of the *villainous aggressor* is precisely to find an alternative explanation to non-consensual impositions of harm in the context of a just war.[15] This is no easy task.

If it is true, as I have argued, that many unjust combatants are innocent in important ways, there might simply not be a good enough moral reason for killing them. And I must admit that despite what I say below, sometimes I find myself more inclined towards a form of pacifism that would make wars morally forbidden until we find ways of sparing those non-liable. If pacifism is true, what I say below is not enough to displace the moral prohibition on killing. Maybe, in the spirit of Seth Lazar's work,[16] it will explain why targeting civilians is morally worse than targeting combatants. But, once again, the fact that something is morally worse than something else does not make what is better morally permissible.

[13] Still, I have my qualms about leaving a forfeiture-based view of self-defence in place simply because of the problems associated to forfeiture theories of self-defence. But we can assume that what is being left in place is not only forfeiture, but whatever justification we think there is for killing in self-defence, like Michelle Dempsey's reasons-based account, which does not rely on forfeiture. I think my account of permissibility is compatible with her theory of self-defence, perhaps even more so that with forfeiture views. *See* Michelle Madden Dempsey, 'Against Liability' in Christian Coons and Michael Weber (eds), *The Ethics of Self-Defense* (Oxford University Press 2016); Quong (n 9).

[14] Interestingly, the Department of the Army, United States of America, ADP 3-0 Operations, Army Publishing Directorate (https://atiam.train.army.mil/catalog/dashboard) states that rules of engagement always recognize a soldier's inherent right to self-defence.

[15] Think about the Second World War, what should have been an intervention to stop the genocide in Rwanda, and the intervention by NATO in Kosovo. Though I acknowledge that the Second World War was not fought justly and there is disagreement about Kosovo. *See*, ie, Paul W Kahn, 'War and Sacrifice in Kosovo' (1999) 13 Philosophical Dimensions of Public Policy 201; Michael Walzer, *Just and Unjust Wars: A Moral Argument with Historical Illustrations* (Basic Books 2006) 255–68.

[16] *See* Seth Lazar, *Sparing Civilians* (1st edn, Oxford University Press 2015).

Still, I think it is important to attempt to provide an alternative account of killing in war that can resist absolute pacifism and, also, make sense of the distinction between combatants and non-combatants, which has significant intuitive force. My attempt is also motivated by a belief that morality is meant to capture and illuminate, rather than obfuscate, our experiences, and meant to guide our actions. I would thus hope that it would have something to say about those combatants who invaded the Normandy beaches that relates to what they were doing: fighting against Nazi domination. And I would also hope that an account of the morality of war would be workable, in the sense of not demanding of us, beings with imperfect knowledge, to measure things we have no way of measuring, both morally and empirically.

But this highly non-ideal approach to morality has significant dangers: it might be servile to, and uncritical of, the *status quo* and, thus, too permissive when it should be more demanding. I will come back to this at the end.

As I said in Chapter III, in a realistic Utopia there would be no wars. And there would be no wars because people would comply with moral demands (ie do not initiate wars). But in the world we live in, people do not actually comply with these moral demands; they go to war. There is thus a question, firmly located in non-ideal theory, about how to address this kind of (massive and deadly) non-compliance: how to stop it and achieve a sustainable peace.

The most plausible reductivist individualist account—where liability requires a high threshold of moral responsibility—answers this question with pacifism. The less plausible reductivist individualist account—where liability requires a low threshold of moral responsibility—answers it with total war. Perhaps, then, providing an alternative to pacifism and total war requires moving away from reductivism.

There is no way to fight a war by only targeting those who are liable. It is not only that we often do not know who those people are, particularly in the lower ends of the chain of command. It is also that those engaged in posing micro-threats or who have combat functions are often those who are less likely to be liable, for different reasons. It is plausible then that there might be no other way to stop this massive form of deadly non-compliance—war—other than by targeting some of those who have not forfeited their rights. That is, we might think that using force against some of them is the only way to stop the unjust war.

This is a difficult argument to support. It partially relies on empirical questions about what is possible and impossible to do in eliminating collective threats, and, particularly, on whether the collective threat posed by the unjust side of a war can only be eliminated through fighting—through killing. That is, the argument assumes that there is a level at which certain kinds of non-compliance with moral demands—collective threats of violence—can only be dealt with through something like war. But why would this be the case? It seems to me that this can only be true contingently and in very rare circumstances. It is also difficult to argue that this impossibility to deal with non-compliance actually obtains, partly because we do not seem even to be trying to find alternative ways of dealing with this kind of non-compliance that do not involve killing. Further, the fact that those on the lower end of the chain of command are more likely to not be liable is not a fixed fact. It is partly a result of our social arrangements and institutions, whereby states are allowed to, or nor legally prevented from, coercing individuals into fighting, and the evidence and information available to individuals regarding the wars their countries fight is limited.[17] I can thus do nothing other than leave this issue—whether there is a point at which violent and collective non-compliance cannot be stopped by targeting only those who are liable—open.

Assuming that this kind of massive non-compliance cannot be dealt with by targeting only those who are liable, and that just wars are actually able to achieve a sustainable peace, we might thus say that intentionally killing (some) non-liable individuals may be exceptionally permitted in such circumstances—in the circumstances of war. This position strays away from reductivism: it posits that there is a permission to target some of those who are not liable that is exceptional to times of war. That permission is exceptional to times of war only to the extent that there is a level at which deadly non-compliance cannot be stopped by targeting only those who are liable. Of course, a non-reductivist view about war is not a novel idea: a just war has been thought to provide a special kind of moral reason for imposing harm since, at least, St Augustine,[18] but it does run the risk of being too permissive. And it requires a fully developed argument as to why what is factually impossible would have an effect on what morality demands or permits.

Nonetheless, the position I will defend is compatible with weaker versions of reductivism, whereby moral properties that are often present in war—but not

[17] This is why I think accounts regarding permissible or reasonable mistakes in the context of self-defence—whereby the rights of the apparent aggressor are not violated—like Bolinger's, are unlikely to succeed in the context of war. They most often will fail the second condition (that the behaviours that license the defender to assume that the apparent aggressor is liable must be reasonably avoidable) because performing such behaviours, in the context of war, is not something that members of the community can typically avoid without undue costs. *See* Renée Jorgensen Bolinger, 'The Moral Grounds of Reasonably Mistaken Self-Defense' (2021) 103 Philosophy and Phenomenological Research 140.

[18] Ep. Ad Bonif. Clxxxix quoted in Thomas Aquinas, *Summa Theologica* (Fathers of the English Dominican Providence tr, Burns Oats & Washbourne 1947).

often present in killing in self-defence—are important in justifying killing, particularly, in justifying killing when faced with collectively posed threats.

So far, the question of who, among the non-liable, are permissible targets in the context of war is still open. In answering this question, I think we should reject consequentialism, for several reasons.

First, consequentialist accounts of killing in war can be significantly over-inclusive with respect to the laws of war: if killing civilians is instrumental to victory, targeting them might be morally permissible, as long as the disvalue in doing so is less than the disvalue of losing the war. Of course, consequentialist accounts have ways of resisting this implication. For example, one might adopt something like a rule-consequentialist account about the morality of war, whereby it would be better in the long term to have some moral rules governing war, including the prohibition on targeting civilians.[19] These moral rules would lead to overall better results, even if in some individual cases they would lead us to doing what is morally wrong. The rule-consequentialist, then, takes a two-level view: the rules are absolutely binding when making decisions in concrete circumstances,[20] but what justifies the rules themselves are consequentialist considerations.[21] The problems of rule-consequentialism are well known and dovetail with the problems of instrumentalist accounts of the law—although, importantly, they are not the same.[22] In any case, conceptions of the morality of war that hold that targeting civilians is morally permissible do not necessarily render the laws of war morally unjustified. They just pose a question of whether it is sometimes permissible to disobey them.

Nonetheless, consequentialist accounts of killing in war remain unconvincing to me. Already in Chapter III, I pointed at the lack of conclusive evidence to support something like the humanitarian view, or in fact, any instrumentalist account of the laws of war. In the case of consequentialist accounts of killing in war the objection also applies, though in a slightly modified way. The objection cannot be that 'if we are going to permit the killing of innocents, we'd better be sure that doing so is going to have good consequences', since consequentialists should deny that killing the innocent is forbidden simply because they are innocent. Thus, the objection is that if we are

[19] Brandt, for example, develops a rule-consequentialist account of the morality of killing in war. *See* Richard B Brandt, 'Utilitarianism and the Rules of War' (1972) 1 Philosophy & Public Affairs 145.

[20] This feature of rule-consequentialism has been criticized as 'rule worship'. *See* JJC Smart, 'An Outline of a System of Utilitarian Ethics' in JJC Smart and Bernard Williams (eds), *Utilitarianism: For and Against* (Cambridge University Press 1973) 10.

[21] Brandt (n 19) 147.

[22] *See* John Gardner, 'Tort Law and Its Theory' in John Tasioulas (ed), *The Cambridge Companion to Philosophy of Law* (Cambridge University Press 2018) 1.

to construct a moral account of what is permissible to do based entirely on the consequences of our actions (or of our moral rules) then we should have powerful evidence that said account would be good at achieving the consequences we are aiming for. Yet, at the moment, we do not.

But perhaps the most glaring problem with consequentialist accounts is that they are unable to capture in any meaningful way how some of us think that targeting civilians is wrong not because of its overall consequences but because of what is done *to them, as civilians.*[23] A consequentialist account has nothing to say about this. If the distinction between combatants and civilians is justified only because of its consequences, it need not be a distinction between combatants and civilians. Any restrictions on targeting that achieved the desired consequences would be justified—even one that restrained targeting to, say, those living in rural areas. Yet, it seems that we would dismiss this arrangement as arbitrary. A non-consequentialist account seems preferable to me.

I will thus sketch a non-consequentialist account of killing in war; one that is able to account for some of the killings of (non-liable) unjust combatants that the revisionist view leaves unaccounted for. This entails providing an answer to the question of who, among the non-liable, are permissible targets in war.

In very general terms, I will explore an account under which just combatants have a *pro tanto* reason to kill (non-liable) unjust combatants because (1) just combatants are fighting a war with a just cause and (2) unjust combatants are *combatants*; they are the ones who more robustly and directly contribute to posing the unjust threat of the war, by fighting it. This *pro tanto* reason is not a justification, but a permission, in the *sui generis* sense I will develop.

A. A Just War

It is not my aim to develop a complete theory of just war regarding resort to force. However, given that my account strays away from strong forms of reductivism, it is important to discuss what a just war is and which causes constitute just causes for war.

[23] Lazar emphasizes the following features about civilians which might make it worse to target them than to target combatants. First, that killing civilians is ineffective and so cannot satisfy a necessity constraint on permissible killing; second, that attacks on civilians might often be opportunistic rather than eliminative, and that opportunistic killings are worse than eliminative ones; third, that civilians are more likely to be innocent than soldiers, and so killings civilians is morally riskier than killing soldiers, which in turn makes it worse; fifth, that civilians are vulnerable and defenceless, and killing those vulnerable and defenceless is worse than killing those who are not. *See* Lazar (n 16).

I said above that just war theory is a form of non-ideal theory that deals with a particular form of collective and deadly non-compliance. Just war theory is meant to provide guidance on when it is morally justified to resort to war, and also on how to pursue war. Thus, the theory has an *ex-ante* dimension. Certainly, the judgment of when to resort to war is primarily for specific authorities to make, but individuals are also meant to undertake it, as an evaluation of whether they should participate in the war.

Assuming that there are forms of collective and deadly non-compliance that cannot be stopped by targeting only those who are liable, there might be a permission to target some of those who are not liable. This account already has implications for what can constitute a just cause for resorting to force: the need, precisely, to stop massive and deadly forms of non-compliance; the need to respond violently because others are already engaged in violence. And the level of non-compliance also needs to reach a certain threshold so that measures short of war are no longer effective. This would overlap with commonly accepted just causes for war: armed response to aggression and humanitarian interventions,[24] and it would exclude other proposed just causes, such as preventive wars,[25] wars as punishment,[26] and subsistence wars.[27]

Thus, on my account 'just cause' is simply a short way to refer to the existence of massive and deadly non-compliance that cannot be dealt with by targeting only those who are liable. The existence of a just cause is, once again, contingent on that impossibility, and the success of the argument depends on it too.

In traditional frameworks of just war theory,[28] there are requirements additional to the just cause. A war is just if (1) it is waged under a sovereign or

[24] Walzer, who is the most prominent representative of the traditionalist view, supports both. *See* Walzer, *Just and Unjust Wars* (n 15) 51–108. But some revisionists also support war against aggression or humanitarian intervention as just causes. *See*, eg, Helen Frowe, 'Can Reductive Individualists Allow Defense Against Political Aggression?' (2015) 1 Oxford Studies in Political Philosophy 173; Cécile Fabre, *Cosmopolitan War* (Oxford University Press 2012) 51–96, 166–207; Jeff McMahan, 'Just Cause for War' (2005) 19 Ethics & International Affairs 1.

[25] *See* Whitley Kaufman, 'What's Wrong with Preventive War? The Moral and Legal Basis for the Preventive Use of Force' (2005) 19 Ethics & International Affairs 23. He concludes that preventive wars can only be legitimately authorized by the United Nations Security Council.

[26] *See* Oliver O'Donovan, *The Just War Revisited* (Cambridge University Press 2003).

[27] *See* Fabre (n 24) 97–129.

[28] Both in its Early Christian form, whose representatives are Augustine, Aquinas, and Vitoria, and in its later form, as exposed by Hugo Grotius and Vattel, among others. *See* Augustine of Hippo, *City of God* (Gerald G Walsh and Daniel J Honan tr, Catholic University of America Press 2008); Francisco de Vitoria, *De Indis et de Iure Belli Relectiones* (John Pawley tr, Carnegie Institution of Washington 1917); Francisco de Vitoria, *Reflection on Homicide & Commentary on Summa Theologiae* (John P Doyle tr, Marquette University Press 1997) IIa-IIae Q. 64; Aquinas (n 18) II-II, Q. 40, art. 1; Emer de Vattel, *The Law of Nations* (Bela Kapossy and Richard Whatmore eds, Thomas Nugent tr, Liberty Fund 2008); Hugo Grotius, *Hugo Grotius on the Law of War and Peace* (Stephen C Neff ed, Cambridge University Press 2012).

legitimate authority; (2) it has a just cause; and (3) the belligerents are animated by the right intentions.[29] The war must also pass four prudential tests: it must be expected to produce a preponderance of good over evil (proportionality); it must have a reasonable chance of success; it must be a last resort (necessity); and it must be expected to have peace as its final outcome.[30]

Whether a war is all things considered just is, as a result, a complex judgment to make. The first problem is determining the object of moral evaluation. Wars evolve, and an initially just war may turn into an unjust one.[31] Moreover, there may be several points within a war where a re-evaluation of whether the war should continue may be necessary.[32] Finally, a single collective may be fighting a war that is partly just and partly unjust.[33] While the initial resort to force is certainly the first object of moral evaluation, this doesn't exclude re-evaluations along the way—and a fully worked out just war theory should say something about this.

Even after the object of moral evaluation is determined, it is far from obvious what the relations between the requirements for resorting to force are. Perhaps some just causes are sufficient to render a war justified, even if the other constraints are not met.[34] It might be that the requirement about there being a reasonable chance of success is best understood as a requirement that even if the war is lost, we could still say that it was a war worth dying and killing for. In other words, the requirement might be best understood as asking that defeat does not render the war banal. And perhaps this also tells us something about what should be at stake in going to war.

But I think such an account might be overly permissive. My account of just cause already contains a constraint: there is no just cause for war if massive non-compliance can be dealt with by targeting only those who are liable. And for a war to be just all things considered, in addition to having a just cause, it should meet what are described sometimes as prudential tests. It should

[29] *See* Joachim Von Elbe, 'The Evolution of the Concept of the Just War in International Law' (1939) 33 American Journal of International Law 665, 669; James Turner Johnson, 'The Broken Tradition' (1996) 45 The National Interest 27, 29; Francisco de Vitoria, *De Indis et de Iure* (n 28) 420–30, 467.

[30] *See* Johnson (n 29) 28; Nigel Biggar, *In Defence of War* (Oxford University Press 2013) 3; O'Donovan (n 26) 48–62.

[31] *See* Saba Bazargan, 'Morally Heterogeneous Wars' (2013) 41 Philosophia 959.

[32] ibid.

[33] ibid.

[34] I am thinking here about the plausible judgment that, as suggested by Walzer, human losses involved in something like a Nazi victory would not have been simply losses of life, and that fighting against them with little chance of success or knowing that the losses of life would be higher than not fighting at all, would not have rendered the war unjust Michael Walzer, 'World War II: Why Was This War Different?' (1971) 1 Philosophy & Public Affairs 3.

meet a proportionality constraint, be necessary, succeed, and achieve peace.[35] The content of these constraints is partially given by liability-based justifications: success, proportionality, and necessity are, as we saw, internal to some accounts of liability. But if my account of permissibly killing in war succeeds, some of these constraints would be external to it, and proportionality, in particular, should be understood differently. I will come back to this in sections V and VI.

The conditions listed above are very demanding. If we could prove that all wars, no matter how just their cause, fail in bringing peace, then wars cannot be just nor permissible. This is not implausible. Peace imposed by force may lack moral authority, which in turn can make it fragile.[36] And if there are other feasible alternatives to war, that is, if there are ways of dealing with massive and deadly non-compliance by targeting only those who are liable, the permission to impose harm on non-liable combatants cannot be grounded and there can be no just causes for war. That is, we should embrace pacifism.

Although this is enough for the moment, it is still very imprecise. There are definitional issues, such as what constitutes aggression; what forms of violence justify humanitarian interventions; and when collectively posed threats become impossible to eliminate by targeting only those who are liable (ie when collectively posed threats become 'war'). There are also borderline cases (such as the problem of 'bloodless invasion');[37] questions about whether states or the international community have a duty to intervene in cases like the genocide in Rwanda; and about the relation between the UN Charter's prohibition on the use of force and just causes of war. I leave these aside, not because they are not important but because what I will say, if not relevant to hard cases, will hopefully be so to the central instances of wars with a just cause.

[35] Customary international law considers necessity and proportionality to be requirements of the *jus ad bellum*. *See*, eg, Yoram Dinstein, *War, Aggression, and Self-Defence* (5th edn, Cambridge University Press 2012).

[36] Jonathan Glover, *Humanity: A Moral History of the Twentieth Century* (2nd edn, Yale University Press 2012) xii. Again, I leave open the possibility that some wars may be worth fighting even when there is little chance of success, as suggested by Michael Walzer regarding the fight against the Nazi regime. *See* Walzer, 'World War II' (n 34).

[37] The argument from bloodless invasion is a challenge to views of war based on self-defence, and it aims to show that having the end of defending the lives of persons is not sufficient to support wars in defence of aggression: some acts of aggression may in fact threaten the lives of no citizens of the victim state. *See* David Rodin, *War and Self-Defense* (Clarendon Press 2002) 130–40; Frowe (n 24); Jeff McMahan, 'What Rights May Be Defended by Means of War?' in Cécile Fabre and Seth Lazar (eds), *The Morality of Defensive War* (Oxford University Press 2014) 115.

B. Just War and Unjust Combatants

Even if the goals of a just war are worth pursuing, the fact that something (killing) is a means to an end (winning the just war) does not make it morally permissible. The question that remains then is who, among the non-liable, are permissible targets in a war with a just cause. Or, put differently, why would a just cause provide a reason for harming unjust combatants who, I have argued, have not forfeited their rights?

As we have seen, a just war will be conducted to accomplish certain goals. Those goals are not victory for its own sake but stopping aggression or slaughter and, more importantly, achieving peace (and we may wonder whether peace can be achieved through war at all). The accomplishment of those goals will require killing. The killings are instrumental because they are not a goal in themselves; they are means to an end: in order to do X (stop aggression by A), we will have to do Y (use force against A, which will require killing some of those fighting for A).

Now war is not about randomly killing the highest possible number of people. The aim of war is to win by forcing the enemy to surrender; to stop non-compliance. If the war is one of self-defence, the goal will be to stop aggression, and if it is a humanitarian intervention, it will be to stop the slaughter.[38] Traditionally, strategies to win wars have been planned on the basis of killing or hurting as many enemy combatants as possible: as Beer notes, 'the name of the game, from Napoleon's campaigns to Vietnam' has been body count.[39] In other words, the accomplishment of the goals pursued through war may involve targeting combatants, or simply, targeting military objectives which will happen to involve killing combatants.[40] But 'may' is a bit dishonest. We *know* war will involve killing, even though those killings will only be instrumental: they are not the goal of the war.

If we believe that morality is concerned not only with results but also with how we should treat people, we should look for something *about* the non-liable that could have normative significance when deciding if it is permissible to kill them. As Nagel writes, 'hostility or aggression should be directed at the person

[38] As Moshe Halbertal notes, in an ordinary (non-asymmetrical) conflict, the collapse of the enemy's army is a more or less clear event to aim for, and what makes asymmetrical war so vexing is that victory seems to never be final. *See* Halbertal (n 1).

[39] Yishai Beer, 'Humanity Considerations Cannot Reduce War's Hazards Alone: Revitalizing the Concept of Military Necessity' (2015) 26 European Journal of International Law 801, 825.

[40] And also civilians, but this is a different issue.

or persons who provoke it and ... should aim more specifically at what is provocative about them'.[41]

The thought then is that, in the context of war, we can answer the question of who— among the non-liable—are permissible targets in a way that is responsive to facts about them and not just to the goal of stopping deadly and massive non-compliance. And this answer, thus, might be given by a distinction between those who contribute more directly and robustly to those forms of deadly and massive non-compliance and those who don't; a distinction between those who are actively engaged in those forms of non-compliance and those who are less so.

Roughly, this would translate into a difference between those who belong to armed groups and civilians. After all, armies and armed groups are the entities through which states and non-state actors fight wars. Put simply, their job is to fight wars, which is not the job of civilians.[42] Fighting often constitutes a relatively more direct and robust contribution to the unjust war or to posing a collective (unjust) threat than other, non-combat contributions. It is, after all, combatants who directly pose threats to other individuals, and posing threats to other individuals is a particularly morally risky behaviour.

There are two ways in which we could understand this. On the one hand, it can be understood as a comparative assessment: combatants often contribute more directly and robustly to war than civilians because *fighting* is often a more direct and robust contribution to war than the kind of contributions that civilians make. On the other hand, it can be understood as a necessary condition for someone to be a permissible target in war: fighting is a robust and direct contribution to the war of a special kind that can explain the permissibility of targeting combatants. By contrast, non-combat contributions by non-liable individuals cannot support a permission to harm them, no matter how direct or robust.

Both interpretations present problems. The first one would allow harming non-liable civilians who make contributions that are more robust and direct than fighting. This is a troubling implication, even if one might argue that in an evidence-relative sense, it will be often not permissible to target those civilians: we are unlikely to know who they are. The second interpretation requires an argument as to why fighting would have special normative significance in explaining killing in war. This is a difficult argument to make—because it is

41 *See* Nagel, 'War and Massacre' (n 3) 135.

42 Again, this is not true in asymmetrical warfare: usually, there are no clear and defined military institutions; no clear and defined division of labour. *See* Halbertal (n 1).

difficult to resist pacifism—but it is not impossible. It is quite plausible that posing threats of harm against others is the only kind of contribution to war that is robust and direct enough to make it permissible to kill non-liable individuals—that violence should be directed at those who pose threats, and not against others. Further, posing threats to others—or having the job of posing threats to others— is particularly morally risky, unlike other kinds of jobs/functions civilians have.

I think we should reject the first, comparative, view and embrace the second, non-comparative, one. Making civilians permissible targets seems to me a sufficient reason to discard the first view. If this non-comparative version is right, (non-liable) combatants who do not perform combat functions cannot be permissible targets: they do not contribute in a sufficiently robust and direct way to the collective threat. And non-liable combatants who perform combat functions can be permissible targets, in the evidence-relative sense, even if they are not presently posing threats to others, when it is reasonable to believe that they will do so in the future.[43] If that belief is correct, then it will also be fact-relative permissible to target them.

Of course, there are differences in the ways in which people end up fighting wars. Some join voluntarily, some are drafted; in some cases, the background conditions of social injustice might render their consent meaningless. Although I think this will have an impact on their moral responsibility and thus might make them non-liable, it might still be permissible to kill them.

This distinction between those who contribute to the war in a sufficiently direct and robust manner and those who don't still leaves room for difficult cases, even in the non-comparative version. I don't think it could be otherwise. It is often true that fighting in a war is a sufficiently direct and robust contribution, unlike, say, sewing uniforms or planting crops. But it is possible that some unjust combatants, even those with combat roles, never pose micro-threats to others nor contribute to the war in a significant way; it is also possible that some civilians from the unjust side contribute directly and robustly to the war, by fighting it or, more rarely, by making other more direct and robust contributions to posing threats.

As a result, the relatively bright distinction between civilians and combatants drawn by our laws is somewhat artificial or conventional, in the sense that it partially rests on our legal and institutional arrangements. Although it might be generally true that unjust combatants contribute in a sufficiently direct and

[43] Haque has a similar view, but he believes that combatants are liable. I reject that; I think they are permissible targets, in the sense I will develop later. *See* Haque (n 8).

robust manner to posing threats in the context of an unjust war, it is possible that some civilians (often few) also do so and, occasionally, to a larger extent. And it is also possible that some unjust combatants do not contribute to the threat at all. But the general moral command of 'targeting those who directly and robustly contribute to posing the threat of the unjust side' is a somewhat difficult moral norm to manoeuvre around, slightly similar to the more indeterminate moral norm of 'driving carefully'.[44] The laws of war and the current arrangements of modern states make the moral norm more determinate, similarly to the way in which setting up traffic lights, road markings, and speed limits makes the obligation to drive carefully more determinate.[45] In doing so, the legal rules are helping to constitute the moral norm,[46] which previously said 'target only those who contribute directly and robustly to the threat'. Through the laws of war and certain institutional arrangements, we can draw a bright line between those who more directly and robustly contribute to the war and those who don't by isolating a certain sector of the population as those whose job it is to contribute to the war by fighting it.

A similar role is played by the existence of standing and professional armed forces,[47] that is, people professionally dedicated to the business of waging war, if it comes to it. But we could have organized things differently. Although war will always—or at least in the near future—require people to wage war, it doesn't really require the existence of standing and professional armies. It just requires people willing or forced to fight. Having standing armies is a reasonably good way of isolating a certain sector of the population so that they can be readily identified as those who contribute to and wage war. Like the legal distinction between combatants and civilians, it helps to create a bright line that distinguishes between those who robustly contribute to the war and those who don't (and this might well be an advantage—though not necessarily a decisive one—of the existence of standing armies).

Importantly, it is the *brightness* of the line between civilians and combatants that is artificial—the line or distinction itself is not. It is a distinction between those who directly and robustly contribute to a collectively posed and unjust

[44] In the following paragraphs I follow closely Gardner's argument in John Gardner, 'What Is Tort Law for? Part 1. The Place of Corrective Justice' (2011) 30 Law and Philosophy 1, 19–20.

[45] ibid 19.

[46] ibid.

[47] Approximately 30 countries (out of 193) do not have standing armies, so that 84.5 per cent of current countries have standing armies. The percentage may be even higher given than some of the 30 countries in the list have defence agreements or rely on other nations for their defence. *See* Amanda Macias, 'From Aruba to Iceland, These 36 Nations Have No Standing Military' (*CNBC* 3 April 2018) <https://www.cnbc.com/2018/04/03/countries-that-do-not-have-a-standing-army-according-to-cia-world-factbook.html> accessed 19 August 2022.

threat and those who don't. We should thus not exaggerate the role that the law and our institutions are playing in determining our moral obligation here. The moral norm 'target those who directly and robustly contribute to posing the threat' is not as indeterminate as the moral norm 'drive carefully' is. There are, of course, difficult cases at the border. But even without the legal rules in place and without armies, it is not terribly difficult to realize that people who fight the war—combatants—are those who most directly and robustly contribute to posing the collective threat. As I said before, the brightness of the line might be artificial or conventional but the distinction that it is trying to draw is not.

That the distinction between combatants and civilians might be conventional in the above sense, and that we could have other ways of organizing it (and that there have been other ways of organizing it) doesn't mean that it is arbitrary. The ways in which we draw the line and, particularly, the ways in which we make it bright may constitute moral progress (or not), as judged from the perspective of the underlying morality and the perpetual peace ideal.[48]

Ultimately, I don't think that the just cause and what we find about unjust combatants—that they are those who fight the unjust war, and thus, often contribute more directly and robustly to it—are reasons weighty enough to ground a moral justification for killing them. It certainly doesn't make it '*all right*' that they are killed.[49] A just war is, after all, morally fragile. This moral fragility comes from many of the features that make war what it is. The pervasive uncertainty. The fact that a just war draws most of its value from the goals it might be able to achieve—which are not the just war itself but stopping aggression or slaughter and re-establishing peace. That the judgment which we are called upon to make requires us to balance things such as peace, stopping aggression, and so on, against the killings of many people, who, I have argued, are innocent in important ways.[50] I will thus suggest that what we find about unjust combatants can, at most, ground a moral *permission* to kill them—not a justification. This idea of moral permissibility is *sui generis*, and thus, I will develop it in the next section.

[48] Discussing the arbitrary and interdependence of interests of conventions, *see* Jeremy Waldron, *Torture, Terror, and Trade-Offs: Philosophy for the White House* (Oxford University Press 2010) 102–03.

[49] Nagel, 'War and Massacre' (n 3).

[50] We might even believe that we are balancing incommensurables. *See* Gardner, 'Justification under Authority' (n 5).

IV. Permissibility in Defensive Harm

The idea of permissibility I will develop is concerned with individual non-consensual impositions of lethal harm. I will argue, tentatively, that it might provide an account of why it is permissible to kill non-liable unjust combatants in pursuit of a just war and that it plays a role in explaining agent-relative prerogatives to inflict defensive harm on the non-liable. These agent-relative prerogatives, articulated by Jonathan Quong, are limited by the means principle.[51] Thus, they do not provide extensive permissions to inflict harm defensively on the non-liable to advance our own interests or goals. They are quite restricted.

But while I am fairly certain that the account I will propose succeeds in its second role—giving a formal account of agent-relative prerogatives to impose defensive harm—I am not convinced it can do so as an explanation of killing in war. I will thus focus most of the argument on the latter, as I think it is a much more difficult case to make.

One way to start thinking about what this *sui generis* account of moral permissibility is trying to capture—but, as we will see, not the best way—is to consider where it would stand in relation to the distinction between justifications and excuses. Excused acts are generally described as acts that remain wrong even though the agent is not fully culpable or blameworthy.[52] By contrast, we say that an act is morally justified when not only is it permissible to perform it, but there is a positive moral reason to do so.[53] In a narrower account of justification, justified acts are impervious to moral criticism or no-one is wronged by them.[54] If we would like to put this definition in terms of reasons, a justified action is one supported by a reason that is not outweighed nor excluded. In other words, a justification is an undefeated reason.[55] To justify an action, in the relevant sense, is to make it permissible when it would otherwise have been impermissible.[56]

In the ordinary sense in which permissibility is used, permissible actions include justified actions.[57] But permissible actions also include actions we have

[51] *See* Quong (n 9).

[52] David Estlund, 'Human Nature and the Limits (If Any) of Political Philosophy' (2011) 39 Philosophy & Public Affairs 207, 231.

[53] Jeff McMahan, *Killing in War* (Oxford University Press 2009) 43.

[54] Ryan C Jenkins, Michael Robillard, and Bradley Jay Strawser, *Who Should Die?: The Ethics of Killing in War* (Oxford University Press 2018) 4. This definition implies that lesser-evil justifications where rights are overridden and infringed (and so, where someone *is* wronged) are not in fact justifications.

[55] Gardner, 'Justification under Authority' (n 5) 79. On reasons, *see* Joseph Raz, *Practical Reason and Norms* (Oxford University Press 2002).

[56] Joshua Gert, 'The Distinction between Justifying and Requiring: Nothing to Fear' in Errol Lord and Barry Maguire (eds), *Weighing Reasons* (Oxford University Press 2016) 157.

[57] Raz, *Practical Reason and Norms* (n 55) 89.

no duty not to perform[58] or actions whose performance is morally indifferent (such as choosing between two different pairs of shoes): there are no positive reasons to perform them and no negative reasons against performing them.[59] They are neither required nor prohibited by morality.[60]

The idea of moral permissibility I have in mind is not the idea of permissibility as in what is morally indifferent to do. It is also not a liability-based justification nor a lesser-evil one. It is not even a justification. Although not a justification, it is close to that notion, if we take the role of a moral justification to consist of making an act permissible when it would otherwise have been impermissible.[61] As opposed to excuses, permissibility—like justifications—is less concerned with the blame of the agent and focuses on features of the action that make it right or wrong.

The notion of permissibility I will try to develop is, hence, *sui generis*. Maybe the term 'permissibility' is confusing, and we could find a better one such as a 'weak', 'austere', or 'fragile' justification. Or we could reserve the term 'permissibility' for the account I develop and employ the terms 'morally indifferent actions' for what are usually described as permissible actions. Nothing turns on the name itself.

We thus come to the second—and better way—in which we can think of permissibility. For the distinction between justifications and excuses is not the scale we should be concerned with. The idea of permissibility arises from the need 'to warn against a familiar tendency to think that reason has only two voices, namely the raised voice of requirement and the conspiratorial whisper of permission'.[62] Permissibility aims to put a wedge between acts that are prohibited and acts that are justified by introducing something in between that does not simply amount to an act's being morally indifferent.[63]

The point of this strand of permissibility is to capture the moral fragility of certain impositions of non-consensual harm. In the case of agent-relative prerogatives to inflict harm, it can explain the moral fragility of harming someone

[58] John Gardner and François Tanguay-Renaud, 'Desert and Avoidability in Self-Defense' (2011) 122 Ethics 111, 112–13; Raz, *Practical Reason and Norms* (n 55) 89.

[59] It is worth noting here that the moral indifference of actions will depend on the context, since sometimes choosing between different shoes will be important. *See* Samuel Scheffler, *Human Morality* (Oxford University Press 1993) 19–25.

[60] Seth Lazar, Saba Bazargan, and Samuel C Rickless, 'War's Endings and the Structure of Just War Theory' in *The Ethics of War: Essays* (Oxford University Press 2017) 233–34.

[61] Gert (n 56) 157.

[62] John Gardner and Timothy Macklem, 'Reasons' in Jules Coleman and Scott Shapiro (eds), *The Oxford Handbook of Jurisprudence and Philosophy of Law* (Oxford University Press 2002) 440, 468.

[63] This scale may be more varied if we also include acts that are required. To require an action, in the relevant sense, is to make it impermissible not to perform it when its omission would otherwise have been permissible. Gert (n 56) 157.

due to a preference for one's own life. In the case of war, it could explain the moral fragility of killing non-liable unjust combatants in pursuit of a just war. In particular, permissibility is meant to encompass three features about certain impositions of harm. First, killing implies acting against reasons that correspond to important values, and thus leaves an enormous moral residue. Second, killing non-liable individuals wrongs them grievously. And third, the reasons in favour of killing these non-liable individuals are barely weightier than the reasons against doing so. Overall, this will result in a view that has a tendency to collapse into pacifism, *not* into total war. This is part of what, I hope, makes it appealing.

A. Moral Residue: Acting Against Important Values

Let us start with the idea of moral residue.[64] Deciding whether to kill, whether in pursuit of a just war or in self-defence, is a situation of rational conflict, where reasons pull us in two or more incompatible directions.[65] On the one hand, the fact that war involves killing people is a reason not to pursue war.[66] Even if the just cause, in conjunction with the relation between the combatant and the war, can defeat this reason by outweighing it—and as I said before, maybe they simply can't—they cannot make reasons not to kill disappear. We always have reasons not to kill people. In other words, it would always be better to do as every reason would have one do, defeated or otherwise, because every reason corresponds to a value.[67] In this situation of conflict, in deciding whether to kill or not, it is impossible to realize the value in every reason that needs to be weighed. But it would certainly be better if 'by some miraculous change in circumstances' one could do what all the reasons would have one do.[68] After all,

[64] The idea of a 'moral residue' is employed frequently in the discussion of moral dilemmas. The way I use it here has some similarities—regret and remorse also play a role—but it is not the same. On moral dilemmas and moral residue *see* Terrance McConnell, 'Moral Dilemmas' (*Stanford Encyclopedia of Philosophy*, 15 April 2002) <https://plato.stanford.edu/archives/fall2018/entries/moral-dilemmas/> accessed 19 August 2022. On self-defence, *see* Dempsey (n 13).

[65] The distinctive feature of conflicts is the impossibility of complete conformity with reason. *See* Joseph Raz, 'Personal Practical Conflicts' in Peter Baumann and Monika Betzler (eds), *Practical Conflicts: New Philosophical Essays* (Cambridge University Press 2004) 187.

[66] I leave aside in this explanation (temporarily), the fact that we can think of reasons for not killing people as a special kind of reason: duties. *See* Gardner and Macklem (n 62) 465.

[67] ibid 444, 464; Joseph Raz, 'Value and the Weight of Practical Reasons' in Errol Lord and Barry Maguire (eds), *Weighing Reasons* (Oxford University Press 2016) 141, 145.

[68] Gardner and Macklem (n 62) 444, 464.

> [e]very time one does not do what *any* reason would have one do, be that reason mandatory or otherwise, a trace is left in one's life. This is always in principle a matter of regret, for as we saw reasons do not lose their force as reasons merely because they are defeated.[69]

Part of what makes decisions to kill so difficult is precisely that we are forced to choose between things that we value greatly. In the case of agent-relative prerogatives to inflict defensive harm, and in the case of war, we are thinking about inflicting serious harm, suffering, and death on other human beings. Thus, even if we conclude that we ought to choose one course of action over another one—to kill—the course of action we ought to choose will still involve great moral loss. Or in the less technical language of Private Edward Schiff, who fought in the Second World War:

> Other memories are much darker... the shock when almost everyone I knew was killed or wounded, some right next to me. The worst was the time I fired my machine gun into a German patrol that was so close I could see their faces. I never found out what happened to them, which lets me think there are some old men in Germany today, sitting around telling their grandchildren of the American machine gunner who scared them to death. At other times I realize the chances are good that I killed at least some of the men in that patrol. They were probably teenagers, as I was, with their whole lives ahead of them, until I squeezed that trigger. All had relatives and sweethearts whose lives were shattered and who may be grieving even now.[70]

The idea that acting against an important value will leave an important moral residue can be accommodated by a justification. However, the existence of a justification will not tell us anything about the moral residue: we can't generally say that acts that leave little moral residue are justified. If I kill the villainous aggressor in self-defence, most would agree this action is justified. Yet the moral residue is not insignificant: in an agent-neutral sense, there is the disvalue of a life lost, even if in the agent-relative sense, the rights of the aggressor have not been infringed. If a mother kills an aggressor who is not morally responsible for the threat that they pose[71] in defence of her child, we might think this action is justified—even required, yet the moral residue seems enormous. But if I skip

[69] ibid 467.

[70] Stephen E Ambrose, *Citizen Soldiers: The U.S. Army from the Normandy Beaches to the Bulge to the Surrender of Germany, June 7, 1944–May 7, 1945* (Simon & Schuster 1998) 474–76.

[71] This might be due to mental illness, intoxication, or young age.

our lunch date with you because I saved a child from drowning, surely I had a justification for doing so, and yet the moral residue of breaking my promise seems negligible.

What is driving the significance of the moral residue in the examples above are the values that correspond to the reasons which were not acted upon or that were acted against. Anytime we act against reasons or fail to act for reasons that correspond to an important value—such as the value of human life—our actions will leave a significant moral residue.

Whether an action leaves a significant moral residue is important but it does not exhaust the idea of permissibility that I have in mind. Actions that are best described as justified may also leave a significant moral residue. This will be true any time there are important values at stake, such as when an agent decides to rescue six individuals over rescuing three, lesser-evil justifications for inflicting harm on the non-liable, or killing the villainous aggressor in self-defence.

B. Small Differences in the Weight of Reasons

The second feature meant to be captured by permissibility is the somewhat unusual thought that there might be a small difference between the reasons tipping the scales in favour of one course of action.

In the case of killing in war, the substantive argument is that the fact that the (non-liable) unjust combatant contributes more directly to the war by fighting it seems *barely* significant enough to explain why she may be killed. This argument requires more elaboration, and I will come back to it in section IV.

In the case of agent-relative prerogatives to inflict defensive harm, the substantive argument is (briefly) that one might place more value on one's life than on someone else's, or that one might prefer one's life over someone else's,[72] but not by a significant margin. To abandon the agent-neutral point of view in such a way—by placing much more value on one's own life— would be deeply disrespectful to others.[73]

Formally, the idea of moral permissibility is thus attempting to distinguish between positive reasons for doing something that are not weighty enough to

[72] For a complete substantive argument in favour of agent-relative prerogatives, see Quong (n 9).

[73] Susanne Burri, for example, thinks that placing more value on one's life in the context of defensive killing is always disrespectful to others. *See* Susanne Burri, 'The Toss-up Between a Profiting, Innocent Threat and His Victim' (2015) 23 Journal of Political Philosophy 146.

be a justification, but at the same time weighty enough to grant permissibility and move away from prohibition.

Perhaps it is useful to use numbers to understand what needs to be done. As described above, there will be a justification to do X every time there is a positive reason for doing X that is not outweighed nor excluded by other reasons. There is, certainly, something curious, to put it mildly, in figuring out what it takes for a reason to outweigh another one, since reasons are not quantities, and nothing indicates that they would or should behave as such. Or, in different terms, it is unclear whether and how we can assign weight to different reasons.[74] Nevertheless, in principle, a justification would look like any of the following alternatives:

$3 < 3.00000000000000000 1$

$3 < 4$

$3 < 1{,}000{,}000{,}000$

In the above examples, let us assume, for simplicity, that there is only one reason on each side, each of them with different weights. In these three cases, the higher number is acting as the weightier reason, and thus, outweighs the lower one and supports an X course of action by making it justified. Thus, in all three cases there is a difference between the two pairs, no matter how small, that will allow us to conclude that reason (3) is outweighed, and as a result, that the course of action indicated by the weightier reason is justified.[75]

If we look at these comparisons, it doesn't seem so farfetched to suppose that there should be a difference between what is happening in the first two pairs and what is happening in the last one. On the contrary, it might seem more farfetched to suppose that there would be *no* significant difference at all between them. The idea of permissibility is meant precisely to capture those cases, like the first two pairs, where although there is a positive reason in favour of a given course of action, that reason outweighs the other one by a very small margin. Since reasons are not really things that we can weigh against each other, and since we are often balancing things that are, in many ways, incommensurable, this idea concerning small margins might have even more plausibility applied to reasons than applied to the numbers above. After all,

[74] Michael Philips, 'Weighing Moral Reasons' (1987) 96 Mind 367, 368.

[75] I am assuming for all purposes that no reasons are excluded here. On overriding and the strength of reasons, *see* Raz, *Practical Reason and Norms* (n 55) 26.

determining what differences and margins are too small will be a matter of substantive argument—for the moment, this will have to wait.

C. Wronging the Non-Liable

So far I have been ignoring that the moral conflict in killing non-liable individuals, including non-liable unjust combatants, is not the kind of moral conflict where weighing reasons is straightforwardly the way to proceed. The conflict has a different formal structure (although, as we will see, the idea of there being little difference in the weight of the reasons that move us away from prohibition will remain). The moral conflict is not a conflict between reasons of identical type. And it is not best understood as a conflict between first-order reasons. Let us examine this considering the case of non-liable unjust combatants.

Conflicts between first-order reasons are resolved by considering the weight of the conflicting reasons involved and how they are affected by various strength-affecting considerations.[76] Yet the conflict involved in killing the non-liable unjust combatant in pursuit of a just war is not a conflict that is best understood in terms of first-order reasons. It is a conflict between, on the one hand, reasons in favour of stopping an unjust war (which are first-order reasons) and reasons against killing the non-liable, which take the form of a duty.

Duties are protected by exclusionary reasons (reasons not to act for certain reasons).[77] Duties are also categorical and mandatory. They are categorical because they do not succumb to the personal goals of the agent to whom they apply:[78] you cannot break your promise to have lunch with me just because you'd rather stay home and watch TV. And they are mandatory reasons because they are reasons not to break our duty even when fulfilling it becomes more disruptive than we imagined initially.[79] The presence of a duty, at least in a non-consequentialist theory, implies that one ought not to act on the balance of reasons, even when the balance of reasons counsels for breaking the duty.[80] The stronger reason might in fact be excluded from consideration. In simpler terms, you are not supposed to kill David if killing David will provide you with

[76] ibid 46.
[77] ibid 183.
[78] Gardner and Macklem (n 62) 465.
[79] ibid.
[80] Raz, *Practical Reason and Norms* (n 55) 41.

fresh organs to save five individuals. The duty not to kill David is protected by an exclusionary reason that excludes that reason from consideration.

As Gardner and Macklem explain, the force of duties has then two dimensions: first, duties have ordinary weight as reasons for action.[81] Second, they have exclusionary force that protects them against conflicting reasons for action, independent of the ordinary weight that these opponent reasons have.[82] Reasons can, however, be mandatory, as duties are, and still have little weight;[83] that is, 'they can be good at defeating by exclusion but poor at defeating by weight'.[84]

The moral conflict I am concerned with—that between killing the non-liable and stopping the unjust war, losing one's life or killing the non-liable defensively—is best understood as a conflict between first-order reasons (winning the just war; saving oneself) and a duty (the duty not to kill non-liable individuals), which is a protected reason. This kind of conflict has a different formal structure from conflicts between first-order reasons, where reasons can simply be weighed against each other. Conflicts between duties and first-order reasons involve a first-order reason for action and an exclusionary second-order reason to the effect that the first-order reason should not be acted on.[85] These conflicts are not resolved by the strength of the competing reasons but by a general principle of practical reasoning that determines that exclusionary reasons always prevail (unless they are cancelled by a cancelling condition).[86] We cannot just proceed to weigh the reasons against each other.

There are three possible (formal) ways out of this problem that will get us to the weighing stage. First, we can focus on the scope of the exclusionary reason. Exclusionary reasons vary in scope: they may exclude all or only some of the first-order reasons that apply to certain practical problems.[87] As Raz puts it,

> just as any reason has an intrinsic strength which can be affected by strength-affecting reasons so every second-order reason has, as well as a strength, an intrinsic scope which can be affected by scope-affecting reasons.[88]

[81] Gardner and Macklem (n 62) 466.
[82] ibid.
[83] ibid.
[84] ibid.
[85] Raz, *Practical Reason and Norms* (n 55) 40, 46.
[86] ibid.
[87] ibid.
[88] ibid 46.

In other words, some reasons against performance of the duty remain available to be acted upon: the exclusionary force of duties, even the duty not to kill, need not be absolute.[89] Thus we could say that the scope of the exclusionary reason is limited, and, as a result, it does not exclude first-order reasons for killing those who more directly contribute to a collectively posed threat that cannot be eliminated otherwise. At this point—where we have determined that the scope of the exclusionary reason is limited—everything goes back to depend on the weight of the reasons left in play, including the duty, which is now conceived as an ordinary reason for action.[90] The duty stands bare, divested of its exclusionary force, and we can now weigh all these first-order reasons against each other. The sketch of moral permissibility I have developed will take place, precisely, at this level.

The above explains how the exclusionary force of the duty not to kill can be discounted. By discounting the exclusionary force of the duty, we can weigh the reasons against each other. It is important to emphasize that once in the weighing stage, the reason not to kill subsists, if only as a first-order reason. Its exclusionary dimension—its being a duty—will have been cast aside. But casting aside its exclusionary dimension only takes us so far, because reasons not to kill are not only good at defeating by exclusion but also at defeating by weight. That is, first-order reasons not to kill are still very weighty reasons, even when divested of their exclusionary dimension—even when standing alone. Both these features—casting aside the exclusionary dimension and the weight of the first-order reasons that persist—will be important in understanding why unjust combatants are wronged and why they are wronged grievously. I will come back to this later.

The second alternative is to see the conflict as one between two duties: the first one, a duty not to kill the non-liable, and the second one, a duty to stop the unjust war. If that is the case, these duties can be weighed against each other, and so the argument about the slight difference in weights also plays out.

The third alternative would be for a just war to be a cancelling condition of the exclusionary reason.[91] This alternative, however, is unable to grant permissibility on its own, because cancelling conditions are not reasons for action. They only remove an obstacle; the reason for action must be found somewhere else.[92] A typical cancelling condition is someone releasing you from a promise;[93] if so, then you *don't* have to fulfil your promise, and if you don't,

[89] Gardner and Macklem (n 62) 466.
[90] ibid.
[91] Raz, *Practical Reason and Norms* (n 55) 46.
[92] ibid 187.
[93] ibid 27.

there is no breach. But say the promise is to do something that has value beyond merely being a promise. Then there are other reasons that count in favour of fulfilling your promise, but the mere fact of your having promised doesn't provide one anymore. Thus, if a just war can operate as a cancelling condition, there might be other positive first-order reasons—the fact that combatants contribute more directly to the war—for killing the non-liable unjust combatant.

This third alternative—which might be the best way to account formally for forfeiture—is unable to explain why the harm against unjust combatants wrongs them. By contrast, the first two alternatives can better explain what is cast aside, morally speaking.[94] In addition to the moral residue, which is given by there being two great values at stake, there is the breach of a duty. The two first alternatives are thus able to explain why killing the non-liable unjust combatant wrongs *him*, and not only why it is bad that a human life is lost. However, the second alternative is less plausible than the first one because it requires an argument for there being a duty to fight a just war, which seems unlikely in most cases.

Whichever of the two first alternatives I have proposed is correct, they will ultimately lead us to weighing and to the idea of there being a small difference in weight between the reason counselling for permission and the reason counselling for prohibition.[95]

Now that the issues about the formal structure of permissibility have been dealt with,[96] it is time to think about the substantive arguments that might explain *why* the difference in weight between the reasons at play is small.

[94] In fairness, although the first two alternatives can explain better the non-wronging nature of the harm than cancelling conditions, they also struggle to accommodate the idea that rights are infringed. One might say that if the exclusionary reason is limited, and as a result, does not exclude first-order reasons for killing those who more directly contribute to a collectively posed threat that cannot be eliminated otherwise, then the individual cannot be wronged (there is no duty not to kill them in these limited set of circumstances). However, given that it is the duty not to kill that is limited in these circumstances, but not the right to life, one can plausibly say that they are still wronged.

[95] There may be other ways to get to the weighing stage. For example, what Gardner calls 'cancelling permissions' and what Raz calls 'exclusionary permissions' could also fulfil the same role, insofar as they are able to cancel or discount the exclusionary force of the duty not to kill the non-liable. *See* Raz, *Practical Reason and Norms* (n 55) 90–92; John Gardner, *Offences and Defences: Selected Essays in the Philosophy of Criminal Law* (Oxford University Press 2007) 106–07, 155.

[96] There is an additional formal question regarding whether permissibility can be understood under Hohfeld's scheme of jural relations. Similarly to lesser-evil justifications, it cannot. Hohfeld saw himself as positing a scheme to better understand *legal* concepts, not moral ones. Not every moral concept is thus reducible to Hohfeld's scheme. *See* Wesley Newcomb Hohfeld, 'Some Fundamental Legal Conceptions as Applied in Judicial Reasoning' (1913) 23 Yale Law Journal 16; Wesley Newcomb Hohfeld, 'Fundamental Legal Conceptions as Applied in Judicial Reasoning' (1917) 26 Yale Law Journal 710.

D. Back to Small Differences in the Weight of Reasons

We can shed light on this idea by thinking about some examples. These examples have a similar formal structure: they are not conflicts between first-order reasons but conflicts where there is, on the one hand, a duty, and on the other hand, a first-order reason. Their formal structure is important: it allows us to explain why individuals are wronged, and in the case of war, why non-liable unjust combatants are wronged—why their rights are infringed—when they are killed in pursuit of a just war.

As we will see in section V, the breach of a duty generates certain pressure for regret or similar reactive attitudes. It also helps us understand why killing unjust combatants is not just bad in an agent-neutral way but also in what is done *to them*. Yet even though these examples share a similar formal structure—and in all of them there might be space for these feelings, attitudes, and pressure—they also seem fundamentally different, morally speaking. In some of these examples, the pressure to apologize is very small and the space for feeling regret or remorse barely exists at all. In others, they seem like essential features of the conflict. The idea of moral permissibility, its focus on the small difference between the weight of reasons, is precisely trying to account for these differences.

Placed at one extreme is our first example, already familiar at this point: you promise to have lunch with me tomorrow at noon to discuss my latest paper. On your way to lunch, you see a child drowning. Breaking your promise—we only had until 1 o'clock in the afternoon—you save the child. You don't make it to lunch.

The formal structure of this case is similar to the one that concerns us: there is a first-order reason to save the child conflicting with a mandatory reason or duty (given by the promise). When the time comes to weigh the promise and the first-order reason (most likely because saving children-related reasons are not excluded by the promise to have lunch), the promise seems to be, as Williams suggests, 'clearly and unanswerably overridden'.[97] It is not that the promise has ceased to exist, as Williams correctly points out:[98] you did, indeed, break your promise. And even though you may have a duty to explain to me why the promise was broken, it seems unacceptable for me to voice any complaints.[99] It would be utterly unreasonable, as Williams says, to doubt your decision or even to feel uneasy about having made it.[100] The decision is so

[97] Bernard Williams, 'Politics and Moral Character' in Stuart Hampshire (ed), *Public & Private Morality* (Cambridge University Press 1978) 60–61.
[98] ibid.
[99] ibid.
[100] ibid.

easy: the value of saving the child clearly overrides the value of fulfilling your promise.[101]

Not all cases, unfortunately, are like that, 'of the straight overriding kind'.[102] If we look away from cases where some reasons are 'clearly and unanswerably overridden', we can find what Williams refers to as 'the most extreme cases of moral conflict, tragic cases':[103]

> —Somebody betrayed us ... The Germans found out where the camp of our partisan unit was. They cordoned off the forest and the approaches to it on all sides. We hid in the wild thickets, we were saved by the swamps where the punitive forces didn't go. A quagmire. It sucked in equipment and people for good. For days, for weeks, we stood up to our necks in water. Our radio operator was a woman who had recently given birth. The baby was hungry ... It had to be nursed ... But the mother herself was hungry and had no milk. The baby cried. The punitive forces were close ... With dogs ... If the dogs heard it, we'd all be killed. The whole group—thirty of us ... You understand?
>
> The commander makes a decision ...
>
> Nobody can bring himself to give the mother his order, but she figures it out herself. She lowers the swaddled baby into the water and holds it there for a long time ... The baby doesn't cry anymore ... Not a sound ... And we can't raise our eyes. Neither to the mother nor to each other ....[104]

This tragic example also has a similar formal structure. A first order-reason (saving 30 people) conflicts with, in this case, two duties: of not killing the innocent and of protecting your child. Macklem and Gardner suggest that we may think of these cases as involving some sort of incommensurability, where incommensurability 'denies us any justified way forward'.[105] These are special cases, they say, in which there is a conflict between two or more reasons that are protected against each other: they each exclude the other from consideration.[106] And neither defeats the other because they are both incommensurable.[107] This is the core of what is known as a moral dilemma.[108] Although

[101] ibid.

[102] ibid.

[103] Bernard Williams, 'Ethical Consistency' (1965) 103 Proceedings of the Aristotelian Society, 110–11.

[104] Svetlana Aleksievič, *The Unwomanly Face of War* (Reprint Edition, Random House Trade Paperbacks 2018) xxxii–xxxiv.

[105] Gardner and Macklem (n 62) 470.

[106] ibid.

[107] ibid.

[108] ibid.

many things referred to as moral dilemmas do not reach this level of tragedy, they suggest, all carry 'intimations of it'.[109] Moral dilemmas have the ability of imperilling our rationality by 'locking us into a situation in which, whatever our virtues, we are doomed to act against the basic maxims of rationality'.[110]

In such a case, Williams suggests that the notion of 'acting for the best' simply loses its content; there is no better thing to be done.[111] Agents who have gone through situations like this will lie awake at night, Williams says, not because of a doubt but because of a certainty[112]—a certainty about the wrongness of what they have done, even if it was the best they could do, given the circumstances. And whoever reproached them for being irrational in doing so would seem a 'glib moralist'.[113] None of the individuals saved by the actions of this woman dare look her in the eye: they cannot thank her; they cannot criticize her or praise her. And it would be wrong of her, almost irrational, *not* to worry about it; not to regret her decision. Although there is no-one to apologize to, she would, she *ought* to, if she could.

It is quite astounding that these examples, which share a similar formal structure, are at the same time so different. The idea of moral permissibility I am interested in involves cases that do not quite reach the 'pitch of moral tragedy'[114] of the example above (which is a moral dilemma, not a case of permissibility). They are thus not incommensurable. But they are also far from conflicts of the 'straight overriding kind'. War, killing in war, even in pursuit of a just one, does carry intimations of moral tragedy. It is not unacceptable for the non-liable unjust combatant to voice complaints. And it is perfectly understandable why the just combatant may lie awake at night.

Where do these intimations of moral tragedy come from? Of course, they partly arise out of the values corresponding to each reason: in this case, from the value of human life. But they also arise out of something else: out of the fact that the harm against non-liable unjust combatants wrongs *them*—which arises out of the formal structure of the conflict, as seen in section IV.C—and of the fact that it wrongs them in a *particular way*; that it wrongs them grievously. That particular way is given precisely by the idea that we are not in front of a case of the straight overriding kind, but in front of something else: in front of reasons in favour of killing weighing little more than the reasons against it.

[109] ibid.
[110] ibid 473.
[111] Williams (n 103) 110–11.
[112] ibid.
[113] ibid.
[114] Gardner and Macklem (n 62) 470.

Now, why would we think that the difference in weight is small?

I have spoken above of 'morally significant reasons' and the 'weight of reasons'. This distinction is important. Some reasons might always be morally significant or important when deciding what to do; that is, they will always merit consideration. Yet these morally significant reasons may present themselves bearing different weights. That is, reasons have a dimension of strength, and some reasons are stronger or weightier than others.[115] For example, whether someone is culpable of an action is a morally significant reason in deciding whether they should be punished. But how weighty this reason will be might vary with their degree of culpability.

In the case of killing in war, what I have tried to do in the past pages is to explain why many unjust combatants do not forfeit their rights against lethal or substantial harm. I have argued that their moral responsibility and knowledge about certain aspects of the unjust war are often diluted or non-existent, that their contributions to it are often negligible individually considered, and that sometimes they are not posing immediate threats to just combatants. If we think that these reasons are generally morally significant when attempting to justify killing individuals then we can also say that in the case of many unjust combatants these morally significant reasons will be barely present and, sometimes, simply absent. In other words, the argument is that although these reasons are generally morally significant, their *weight* will be small if we consider each unjust combatant's situation individually.

The absence or dilution of the above reasons—that is, their reduced weight—could imply, as I have said before, that we cannot ground permissibility and should thus embrace pacifism. After all, the first-order reason against killing non-liable individuals has significant weight, even when standing unprotected. It is a difficult reason to defeat.

The argument from permissibility claims that because fighting is a more robust and direct contribution to the unjust war than other activities, it allows us to move away from prohibition. If collective unjust threats cannot be eliminated by only targeting those who are liable, this difference between individuals—the difference between those who contribute robustly and directly and those who don't—might be a normatively significant reason that explains why, in the context of a just war, one might target (non-liable) combatants. But the reason in question seems barely enough to move away from prohibition: it cannot, it does not, clearly and unanswerably override the duty not to kill. It does not,

[115] Raz, *Practical Reason and Norms* (n 55) 25.

however, present itself as a case of incommensurability: there is something to be done which seems right, if barely so.

This account focuses primarily on the kind of relationship that exists between combatants and the wars they fight; on the fact that they contribute to the war by fighting it. In that way, it presupposes that the extent to which one is contributing to a collectively posed threat by fighting—and not, say, by planting crops that will later be fed to the armed group—is a sufficiently robust and direct contribution to a collective threat, and it should thus determine what is permissible in war, even if it is not enough to ground the forfeiture of rights against significant or lethal (defensive) harm. In other words, for permissibility to be successful as an account of killing in war, we require a fully developed argument as to why fighting is a sufficient or more robust and direct contribution to a threat, and why this fact is normatively significant in a way that allows us to move away from pacifism and the prohibition against killing. I have only sketched such an argument—most of this chapter has been devoted to its structure (ie permissibility).

Some would perhaps prefer the focus to be on moral responsibility. Although moral responsibility is not irrelevant in deciding whom to kill in war—in fact, sometimes liability-based justifications will be available in the context of war—an exclusive focus on moral responsibility quickly turns into an argument for pacifism when the threshold of moral responsibility is high, and into an argument for total war when it is low.

No single focus—except a collective focus on the unjust/just war—can overcome the dilution of moral responsibility, knowledge, and causal contributions that comes with participating alongside many people in the same enterprise. But a focus on the quality of contributions to an unjust war might, perhaps, make it permissible to target those who sufficiently or more robustly and directly contribute to it by fighting.

It is tempting to think that the requirement of a small difference between the weight of reasons is unnecessary for the idea of permissibility to take off; that it is enough that the reason acted against has great value—such as the value of human life—and that in acting against that reason, the harm that is imposed on someone wrongs her (which is given by the formal structure of the conflict and the existence of a duty). Yet I think the idea of the small difference in weights is important: it goes back to the idea of certain reasons being 'unanswerably overridden', and others that are not.

Perhaps more examples can illuminate the point. Suppose an aggressor who is morally irresponsible for his threat will kill your child unless you defend your child by killing the aggressor. In this example, the harm against the aggressor seems to clearly wrong him and infringe his rights: it is implausible for forfeiture to take place without culpability; it is likely that the aggressor may defend himself and may complain about the harm that is inflicted on him. The reason that is acted against also has great value because it corresponds to the value of human life. We may thus think that the act is merely permissible. Yet if a parent failed to even try to defend their child it would seem morally puzzling, even morally suspect. From an agent-neutral standpoint, of course, there is no reason to think that the life of the child has more value than the life of the aggressor. But from the agent-relative standpoint, it is plain to see why the child would matter so greatly to their parents and why the parents might have a duty to protect the child; it seems inconceivable that the parent would not think of their action, at least in some sense, as clearly what they had to do, even if what they had to do was a terrible thing. In other words, the duty of the parent to protect their child has significantly more weight than the reason against killing the aggressor.

Such a case seems different from the case of many non-liable unjust combatants: it would not be unthinkable, nor morally suspect, for a just combatant to hesitate to kill a non-liable unjust combatant who is minding his own business, taking a bath, perhaps smoking a cigarette, guarding the post.[116] They may even hesitate to kill a group of unjust combatants who are advancing or marching, if they are piloting a plane, targeting from a place of safety. It seems best not to even think about it. It would be strange for us to criticize them for not having done so, even strange to criticize them for refusing to kill the unjust combatant, for turning away. This reluctance will sometimes arise due to uncertainty. But even when uncertainty is not present, such reluctance can be explained by the small difference in weight between the reasons in favour and against killing the unjust combatant: from the agent-neutral standpoint, killing this unjust combatant has so little impact on winning the just war; it seems so insignificant. From the agent-relative standpoint, the just combatant, whose life might not even be in danger, has no reason to kill him. And from the agent-relative standpoint, there are so many reasons *not* to kill the unjust combatant: he is not liable, he might not be posing a threat, he may have not chosen to be there; he is barely doing anything at all. He is a combatant, his job is to fight the unjust war, and for that reason, he may be killed. But this reason seems hardly to matter at all, hardly to matter enough.

[116] Michael Walzer discusses some of these cases as cases of 'the naked soldier'. *See* Walzer, *Just and Unjust Wars* (n 15) 138–44.

Still, I am uncertain about whether permissibility, conceived as I have suggested, can be successful.[117] It seems to require an account of what makes certain features of human action (or reasons) generally morally significant and what makes them weightier and to what extent. It is difficult to do this in a general manner, and I don't know whether some reasons will be always and equally morally weighty in different contexts, or whether they will behave differently depending on the circumstances, and if so, whether they will do so in a systematic way.[118]

The sketch of permissibility I have provided here is also quite rough, and I worry it has a tendency to collapse into either a justification or a moral prohibition and be rendered incoherent. After all, one might think that the just war, in addition to the relationship between combatants, the army, and the war, does provide a positive reason for killing unjust combatants, one that is weightier than the reasons for not doing so. If so, one might simply conclude that the just war itself acts as a *sui generis* justification for killing non-liable unjust combatants, even if we think that the decision is fragile and leaves a great moral residue. Or, alternatively, we might simply think that a small difference in weight cannot defeat the prohibition on killing and, as a result, it is simply impermissible to target non-liable unjust combatants: we should embrace pacifism.

Ultimately, even if its coherence remains to be proven, the advantages of this *sui generis* account of permissibility seem to me quite significant. This is especially so due to permissibility's ability to (formally) account for cases which have been traditionally thought of as morally justified, such as agent-relative prerogatives to inflict defensive harm.

In the case of killing in war, I think permissibility has welcome implications (section V), but it also has some serious issues, which might make it untenable as an explanation (section VI). Some of these issues are related to uncertainty, but others are about the argument itself: one might plausibly say that the fact that unjust combatants contribute more directly and robustly to the unjust war is simply not a weighty enough reason for displacing the very stringent moral prohibition against harming them. I will start by addressing the implications

[117] Although Gert, for instance, has drawn a distinction between justifying and requiring, so the idea of drawing a distinction between justification and permissibility is not outright implausible. *See* Gert (n 56).

[118] Philips, for instance, thinks that the weight of moral reasons varies from context to context in certain systematic ways. Nevertheless, as Philips notes, the problem of the constancy of weights might as well make us doubt how useful it is to speak of reasons at all, since if it cannot be solved, we seem to be thrown back to relying on mere intuitions. *See* Philips (n 74) 368.

and advantages of permissibility and then move on to address some of the issues that imperil it.

V. Permissibility: Implications and Advantages

Permissibility has what I think are important advantages over the idea of a justification, both in the context of agent-relative prerogatives to inflict defensive harm and in the context of war.

In the context of agent-relative prerogatives to inflict defensive harm, permissibility can give a better account of the kind of conflict one faces. It also gives a better account of why remorse, regret, and a certain pressure to apologize are appropriate. And it can explain the reluctance one ought to feel in choosing one's life over someone else's.

In the case of war, it can capture why killing in war, even in pursuit of a just one, is a moral tragedy, and why pacifism regarding war is such an attractive position, while, at the same time, preserving a morally permissible course of action in fighting a just war. It can explain why just combatants are not simply breaching a moral prohibition when they target (non-liable) unjust combatants, thus dispelling the notion that (non-liable) just and unjust combatants are in a morally symmetrical position. And it can also provide an explanation of war that has to do with the war itself and not only with individual instances of killing in self-defence.

This account can also (partially) solve the 'chain of command issue' because it is no longer reasonably certain that fighting a war with a just cause will involve breaching a moral prohibition. Although the question of non-liable civilians, who are killed as a side effect, remains, killing many non-liable unjust combatants is, under the account I have suggested, permissible. Although significant uncertainty remains regarding proportionality, success, and the achievement of peace, coupled with the uncertainty about the costs of not fighting, the evidence available to individuals looks now a little bit different: there is significant uncertainty about the courses of actions fight/not fight, but individuals can ascertain whether there is a just cause for fighting *and* they can conclude, with reasonable certainty, that in the case of (non-liable) unjust combatants, they will not be breaching a moral prohibition. If to this we add the fact that some killings of unjust combatants might be justified on the basis of liability, and that some just combatants will have agent-relative prerogatives to defend themselves, at the evidence relative level the decision to fight seems now permissible. Whether the war will be, in the end, all things considered just, is, however, left open by this account.

Permissibility might also sit less uneasily with feelings of remorse some soldiers have about killing enemy combatants:[119]

> I can't stop thinking about the fact that these folks have lost something that they will never get back—innocence (and a life free of guilt). My heart hurts for them.[120]
>
> Each time I sit there with not a clue what to say ... offering tissues ... saying I'm sorry ... trying to normalize ... trying to say, 'It was not your fault that so and so died' and 'If you could have done differently, you would have' and 'You had a right to be scared.' And even worse, 'You had to shoot back' and 'Yes you killed someone, and you still deserve to go back to your family and live your life.'[121]
>
> Next time you are hanging out with a friend, think about what you would do if he turned to you and said, 'My boss made me kill someone, and I know I'm going to hell for it so why bother?'[122]

I don't think these feelings of grief, regret, and guilt that combatants experience are misplaced or irrational. And I don't think we can easily dismiss them as a response to uncertainty about having done the right thing or about having had complete information. I believe they are better characterized as an appropriate response to having done the right thing simply because the right thing here involves great moral loss[123] and seems barely like the right thing. This is so because, as I have said repeatedly, it is not a conflict of the straight overriding kind.

[119] *See* Dave Grossman, *On Killing. The Psychological Cost of Learning to Kill in War and Society* (Open Road Integrated Media 2014) 86–92; Sherman (n 10) 90–91.

[120] Lisa R Blackman, 'Alarm Red' in Andrew Carroll (ed), *Operation Homecoming: Iraq, Afghanistan, and the Home Front, in the Words of U.S. Troops and Their Families* (University of Chicago Press 2008) 310, 311.

[121] ibid 308.

[122] ibid.

[123] As Scheffler points out, 'it is a conspicuous feature of our actual moral lives that such feelings are not always responses to uncertainty or lack of information, or symptoms of wavering conviction. They are often present even on occasions when we have all the relevant information and are confident that we are doing the right thing. What makes these feelings appropriate, when they are, is the evident fact that one can face situations of moral conflict in which there is no act available that is morally fortunate in every respect. In such situations even the right act may involve pain, or loss, or harm, or dishonour, or costs of some other kind. Given that feelings of ambivalence and regret can be appropriate even when we are confident we are doing the right thing, there is nothing in the idea of a moral theory that requires one to be intolerant of such feelings.' Scheffler, *Human Morality* (n 59) 44–45. *See* also Williams (n 103) 113.

Permissibility thus can provide an account of the moral conflict there is in killing non-liable unjust combatants that better does justice to these feelings of regret and (moral) uneasiness. It can better embody what Williams calls a 'sensibility to moral costs'.[124] 'Oughts' that have not been acted upon are not simply eliminated from the scene.[125] Or, as Macklem and Gardner put it, where one fails to do as a mandatory reason would have one do, such failure is not easily dismissed.[126] This is especially true when the mandatory reason is also categorical—such as a duty[127]—because there is a larger moral cost attached to letting a right be overridden,[128] and specially so if it is *barely* overridden.

Doing without this idea of permissibility runs the risk of 'falsifying moral conflict'.[129] This becomes an even more pressing consideration if what remains is only the logic of forfeiture, which seems to demand that in a conflict one of the conflicting 'oughts' must be totally rejected.[130] It is not a mistake, as Williams points out, to think that both oughts—do not kill, stop the unjust war—actually apply to the situation; that these are two things that I ought to do. Thinking that most of the killing in war is the killing of individuals who have forfeited their rights is a logic that 'cancels one of the oughts'.[131] Yet this is not the experience we have of war: it paints a picture with too few colours. A just war involves wronging individuals, even if what we have done is the best thing that could be done, given the circumstances; such a decision leaves pain behind, and should do so.[132] As Walzer puts it, the existence of these sort of conflicts is a crucial feature of our moral lives.[133]

Permissibility is thus better able to account for these feelings of regret, remorse, and grief. And it can also explain why it might be appropriate to share these feelings with another person, particularly if the defeated reason is a relational one, addressed towards someone[134]—as it is in this case. It might also make sense to compensate that other person. Compensation, as Raz says, is a reaction aimed at mitigating the consequences of the failure to conform fully to a reason, of having taken the second-best course of action.[135] Raz says

[124] Williams (n 97) 65.
[125] Williams (n 103) 113.
[126] Gardner and Macklem (n 62) 467. Not being able to conform with reason completely is, as Raz puts it, a matter of regret. Raz, 'Personal Practical Conflicts' (n 65) 189.
[127] Gardner and Macklem (n 62) 467.
[128] Williams (n 97) 63.
[129] Williams (n 103) 122.
[130] ibid.
[131] ibid.
[132] Michael Walzer, 'Political Action: The Problem of Dirty Hands' (1973) 2 Philosophy & Public Affairs 160, 171, 174.
[133] ibid 171.
[134] Raz, 'Personal Practical Conflicts' (n 65) 189.
[135] ibid 191.

that this might be so even if the agent was not responsible or at fault for the failure to fully conform with reason,[136] which might be the case with many just combatants.

The sense that something discreditable has been done will also be shared by the victims—the non-liable unjust combatants—who will have a complaint of having been wronged.[137] We would not think that they would be wrong or misguided to complain, or that they are unreasonable in doing so.[138] It may be that when explained to them, they will understand,[139] if barely. It is thus not only a bad thing that non-liable unjust combatants die (in an agent-neutral sense). It is also a wrong done *to* them; it relates to them, and it wrongs *them*.

Permissibility can also explain the reluctance to kill unjust combatants, which is a correct reaction to the case, because the case does involve a genuine moral cost.[140] But this is not the kind of reaction one would expect from someone thinking about whether they should break their promise to save the child. That would be the wrong attitude to take entirely; it is almost obscene, morally speaking.

All of the above—the formal structure of the conflict, the small difference between the weight of reasons, and the values at stake—give rise to certain attitudes, feelings, and to certain moral pressure. Thus, the moral residue and the fragility underlying permissibility are not normatively inert. They create pressure, and sometimes—but not always—reasons for action, such as remedial and purgatory reasons, like apology or penance.[141] They also create rational pressure to avoid the conflict as much as possible and to avoid for it to take place in the first place. Permissibility also opens up space for unjust combatants to defend themselves (to which I return in Chapter VIII).

Given that killing in war involves a failure to conform to all the relevantly applicable reasons, the pressure to avoid the existence of the conflict in the first place may establish a moral duty for states and those in commanding positions to target combatants as a last resort and to previously offer the opportunity to surrender. This needs to be fleshed out in more detail and hopefully included in our laws of war, to whatever extent possible. At the moment, we are quite far from this.[142]

[136] ibid 192–93.
[137] Williams (n 97) 61.
[138] ibid.
[139] ibid 62.
[140] ibid 65.
[141] Gardner and Macklem (n 62) 467; Williams (n 103) 110.
[142] *See* Haque (n 8) 95–105.

My account of permissibility might also have positive implications regarding proportionality and what has been referred to as 'the dispensable lives of soldiers',[143] at least when compared to an account whereby most unjust combatants are liable.

Standard accounts of the ethics of harm include a requirement known as wide proportionality.[144] Wide proportionality concerns itself with the number of innocent bystanders harmed as a side effect of defensive action.[145] It weighs an act's good effects against the harms it causes to victims who are not liable to those harms.[146] Wide proportionality operates thus as a constraint on lesser-evil justifications.[147] Defensive action that harms the innocent is widely disproportionate, and as a result, forbidden, while harms that are justified as the lesser evil are widely proportionate.[148]

Permissibility is not a lesser-evil justification. Still, there is a question concerning the relationship between permissibility and wide proportionality. Under the standard account, harm imposed on the non-liable is permissible only if it is widely proportionate. Under my account, harm imposed on the non-liable is not explained by the existence of a lesser-evil justification, so that ordinary accounts of wide proportionality are not obviously applicable. This poses a question regarding permissibility and its relationship with proportionality. If it is permissible to kill non-liable unjust combatants, is there a limit to how many non-liable unjust combatants may be killed? And if there is, where does that limit come from?

Answering these questions would require developing a new account of proportionality. Consider the following accounts of killing in war. Suppose that in all these cases, for the just war to succeed, it must kill 10,000 combatants.[149]

(1) *Lesser evil and liability with a low threshold of moral responsibility.* Consider a just war where 9,000 intentional killings are justified on the basis of liability and 1,000 intentional killings are justified as the lesser evil. Those 1,000 intentional killings must be widely proportional for a

[143] *See* Gabriella Blum, 'The Dispensable Lives of Soldiers' (2010) 2 Journal of Legal Analysis 115.

[144] Jeff McMahan, 'Necessity and Proportionality in Morality and Law' in Claus Kress and Robert Lawless (eds), *Necessity and Proportionality in International Peace and Security Law* (Oxford University Press 2020) 3–4.

[145] ibid 17–18.

[146] Jeff McMahan, 'Liability, Proportionality, and the Number of Aggressors' in Saba Bazargan and Samuel C Rickless (eds), *The Ethics of War* (Oxford University Press 2017) 3.

[147] McMahan, 'What Rights May Be Defended by Means of War?' (n 37) 124.

[148] McMahan, 'Necessity and Proportionality in Morality and Law' (n 144) 15–16; Quong (n 9) 14.

[149] Technically, accounts (1) and (2) cannot meaningfully distinguish between civilians and combatants, but given that my concern in this section is to deal with the issue of the number of combatants that might be killed, I will ignore this.

lesser-evil justification to be available. Accepting the 1:10 ratio (which might not be demanding enough, given that combatants are often killed at least partially in an opportunistic manner), those 1,000 killings must save 10,000 lives (or, under a more stringent view, each of those 1,000 killings must save 10 lives, for a total of 10,000 lives saved.)

(2) *Lesser evil and liability with a high threshold of moral responsibility.* Consider a just war where 1,000 intentional killings are justified on the basis of liability and 9,000 intentional killings are justified as the lesser evil. Those 9,000 intentional killings must be widely proportional for a lesser-evil justification to be available. Again, accepting the 1:10 ratio, those 9,000 killings must save 90,000 lives.

I have already rejected alternative (1) and pointed out the many issues that arise in account (2). Regarding the number of combatants that might be killed, because in account (1) most unjust combatants are liable, it is exceedingly difficult to place limits on the number of combatants that can be killed.[150]

If we believe that most unjust combatants are liable, or, in other words, if we believe that most unjust combatants have forfeited their rights against significant or lethal harm, it is unclear whether there could be a limit to how many unjust combatants the just side of the war may permissibly kill. Liability, in and of itself, seems unable to set a limit without, at the same time, unravelling.[151] It seems impossible to construct a limit that is internal to liability, and so, we are left with the choice of coming up with a limit that is external to it, and that makes it impermissible to act *despite* people having forfeited their rights.

However, it is very difficult to set an external limit to liability that is able to set a plausible constraint on the number of liable people who can be killed. For the first obvious choice would be to set a limit consistent with the number of human lives that are lost. Yet, the consequence of such a limit is that anytime the aggressors consist of one more individual than the victims, it will not be permissible to kill them. In other words, one single person ought to make the difference. However, most people tend to believe that a victim may defend herself against, say, five villainous aggressors as she may do so against one. So we

[150] Discussing this difficulty with liability, *see* Jeff McMahan, 'Duty, Obedience, Desert, and Proportionality in War: A Response' (2011) 122 Ethics 135; David Rodin, 'The Lesser Evil Obligation' in Saba Bazargan and Samuel C Rickless (eds), *The Ethics of War* (Oxford University Press 2017); McMahan, 'Liability, Proportionality, and the Number of Aggressors' (n 146); Kerah Gordon-Solmon, 'Self-Defence Against Multiple Threats' (2017) 14 Journal of Moral Philosophy 125.

[151] On this point, *see* McMahan, 'Duty, Obedience, Desert, and Proportionality in War' (n 150); McMahan, 'Liability, Proportionality, and the Number of Aggressors' (n 146) 3; Rodin, 'The Lesser Evil Obligation' (n 150) 30.

are left with the choice of coming up with a hybrid explanation that is somewhat internal and somewhat external to liability, whereby we say that there is a degree to which every liable person is not really liable, that there is always a small residue in defensively killing someone who is liable.[152] The existence of this residue implies that when villainous aggressors or less responsible aggressors are numerous enough, the addition of those residues becomes large enough so that it prohibits killing them.[153] However, this argument seems to imply that no-one is really liable in the first place, as it looks like everyone is, after all, wronged, if to varying degrees. Perhaps because of this, the argument, originally defended in different versions by Jeff McMahan[154] and David Rodin,[155] has been abandoned by both.

McMahan now thinks that a better way to set limits to the number of liable aggressors that can be killed is given by a new form of proportionality, distinct from narrow and wide proportionality, which he calls 'proportionality in the aggregate':[156]

> I suggest that, when we consider the harm that would be involved in killing a thousand responsible killers, taking into account that each is only minimally responsible for the threat he poses, and compare that harm to the threatened harm to the victim, taking into account that she bears no responsibility for the fact that killing is unavoidable in the circumstance, we can judge intuitively that killing all thousand would be disproportionate... This is a judgment of proportionality that, like narrow proportionality but unlike wide proportionality, takes account of the moral responsibility of those who would be harmed. But, like wide proportionality and unlike narrow proportionality, it also takes account of the number who would be harmed. We can refer to this form of proportionality as proportionality in the aggregate.[157]

[152] And some think that the size of that residue should vary according to the moral responsibility of the liable individual, so that there is a very small residue in villainous aggressors and a larger one in less responsible aggressors. *See* Rodin, 'The Lesser Evil Obligation' (n 150); McMahan, 'Liability, Proportionality, and the Number of Aggressors' (n 146); McMahan, 'Duty, Obedience, Desert, and Proportionality in War' (n 150); Gordon-Solmon (n 150).

[153] Reaching a high enough residue might require a higher number of villainous aggressors than what is required for less responsible aggressors. This is meant to capture the intuition that killing a very high number of individuals to defend yourself is impermissible, but that this number is lower when aggressors are minimally responsible for the threats they pose. *See* Rodin, 'The Lesser Evil Obligation' (n 150); McMahan, 'Duty, Obedience, Desert, and Proportionality in War' (n 150); McMahan, 'Liability, Proportionality, and the Number of Aggressors' (n 146).

[154] *See* McMahan, 'Liability, Proportionality, and the Number of Aggressors' (n 146). In earlier work, McMahan thought that harms to which people are liable were simply discounted by any proportionality calculations.

[155] *See* David Rodin, 'Justifying Harm' (2011) 122 Ethics 74.

[156] *See* McMahan, 'Liability, Proportionality, and the Number of Aggressors' (n 146).

[157] ibid 25.

This constraint, says McMahan, is independent of whether any individual is wronged.[158] Gordon-Solmon has suggested that the limit to the number of minimally responsible threats (and not to the number of villainous aggressors) that can be killed comes from the fact that by killing all those liable, the victim makes each of them worse off as a side-effect by bringing it about that (1) each is liable to be killed, and (2) that consequently she kills each.[159] Also attempting to provide a solution to this issue, Rodin has proposed a 'lesser-evil obligation', which would arise when competing and independent moral considerations make it the case that a right bearer has an all-things-considered obligation not to exercise his own rights.[160] This is why in some cases it may be impermissible to defensively kill a large number of liable individuals: the putative defender has the liberty to kill all of a large number of liable aggressors, but it is a right he morally ought not to act on because of a requirement to realize competing goods or values.[161]

Independently of whether these solutions are successful, an account of killing in war whereby most unjust combatants are liable (account (1)) struggles to accommodate constraints on the number of liable individuals it is permissible to kill. Account (2), where most unjust combatants are not liable, relies mostly on a lesser-evil justification, but because wars will rarely meet the 1:10 constraint, it collapses into pacifism. If we don't want to embrace pacifism, we should reject this view.

By contrast, my account of permissibility would be able to set a more constraining framework without collapsing into pacifism. Consider:

(3) *Permissibility and liability with a high threshold of moral responsibility (my view).* Suppose 1,000 intentional killings of combatants are justified on the basis of liability and 9,000 intentional killings of combatants are now permitted, in the *sui generis* sense I have developed.

We can now build an account of proportionality regarding those 9,000 killings. Traditional accounts of just war theory incorporate, as a prudential test, that the war brings a preponderance of good over evil. I think this is too permissive: we must reconstruct it as a constraint, so that a war with a just cause that brings about more evil than good is not all things considered justified (it is not, thus, a just war). But the question of how much good over evil must be brought

158 *See* McMahan, 'Duty, Obedience, Desert, and Proportionality in War' (n 150).
159 Gordon-Solmon (n 150) 130.
160 Rodin, 'The Lesser Evil Obligation' (n 150) 28.
161 ibid.

about is still not answered; the question of how many non-liable unjust combatants can be killed is still left open.

Permissibly killing non-liable unjust combatants infringes their rights, and it does so grievously, leaving a significant moral residue. Thus, this account leaves open a plausible way to explain why a proportionality constraint should be more demanding than whatever constraint we can set under a liability-based justification. My account then, like account (1), can explain all killing in war but compared to account (1), it is significantly less permissive pertaining to the number of combatants that can be killed.

I do not think, however, that it could be *as* constraining or *more* constraining than the 1:10 ratio demanded by a lesser-evil justification. If it were as constraining as the 1:10 ratio and internal to permissibility, it would corner the view into pacifism. If it were as constraining as the 1:10 ratio and external to permissibility, it would render permissibility superfluous: a lesser-evil justification would be able to account for those killings. Permissibility would not be required.

I leave open how constraining that ratio is. But, at the very least, my account of permissibility provides us with powerful reasons to incorporate constraints into the laws of war regarding the lives of combatants. At present, those constraints are almost non-existent. Combatants are permissible targets when they do not pose imminent threats, with very few exceptions,[162] and even when means less extreme than targeting them are available. Although it is forbidden to use methods of warfare that cause superfluous injury or unnecessary suffering,[163] this prohibition does not mean that the weapons cannot be lethal.[164] International humanitarian law tells us nothing or little about whether states have an obligation to capture rather than to kill,[165] or whether they ought to

[162] *See* Chapter I. In general, the following are not permissible targets: prisoners of war; persons that are or should be recognized as *hors de combat*; persons parachuting from an aircraft in distress, only during the time of their descent; members of the armed forces and other persons who are shipwrecked; and members of the armed forces and military units assigned to civil defence organizations, unless their protection ceases for the reasons established in article 65 of Additional Protocol I and provided that they comply with the criteria established in article 67.

[163] *Protocol Additional to the Geneva Conventions of 12 August 1949, and relating to the Protection of Victims of International Armed Conflicts (Protocol I)* (hereinafter, API) 8 June 1977, 1125 UNTS 3, art 35.2; *Convention (IV) respecting the Laws and Customs of War on Land and its annex: Regulations concerning the Laws and Customs of War on Land (Hague Convention IV)* (hereinafter, The Hague Convention IV), The Hague, 18 October 1907, reg 23.e; *Declaration Renouncing the Use, in Time of War, of Explosive Projectiles Under 400 Grammes Weight* (hereinafter, 1868 St Petersburg Declaration) 11 December 1868, preamble. The other restriction regarding weapons is related to the principle of distinction. Thus, weapons that by their very nature or design cannot possibly distinguish between civilians and combatants are forbidden. *See* Yoram Dinstein, *The Conduct of Hostilities under the Law of International Armed Conflict* (Cambridge University Press 2004)62–63.

[164] Blinding lasers are not lethal, and yet, forbidden. *See* ibid, 65.

[165] *See* Ryan Goodman, 'The Power to Kill or Capture Enemy Combatants' (2013) 24 European Journal of International Law 819; Michael N Schmitt, 'Wound, Capture, or Kill: A Reply to Ryan

choose military strategies that spare the lives of the highest number of enemy combatants. If I am right, there is a powerful moral case for introducing such constraints.

First, we should strive for, at the very least, reducing permissible targets, according to a functional test such as the ones proposed by Haque, Beer, and Blum. Haque has proposed that only members of the armed forces who perform combat functions should be permissible targets,[166] and Blum has made a similar argument.[167] Beer has suggested that a 'unit-based functional criterion' should be incorporated to the rules on targeting so that we can draw a distinction between subclasses of soldiers based upon their importance and classification in a given campaign's circumstances.[168]

Second, there must be a moral obligation to minimize military casualties, even for the side of the war that is fighting a just war. A plausible account of the relationship between proportionality and permissibility would ensure that. The obligation looks obviously different for the side fighting an unjust war, though it would be difficult to make this asymmetry part of our legal rules.

There is a difficult question here about how this plays out when we need to balance the general obligation to minimize military casualties and the obligation that armies have towards their own people. Sometimes commanders might need to choose between different tactics of attack: some of them will be riskier for their own soldiers but will produce fewer military casualties on the other side; some of them will be less risky for their own soldiers but will produce major military casualties on the other side. To be clear, I am not referring here to the problem of tactics of attack that are less risky for soldiers but riskier for civilians.[169] I am speaking of military casualties on both sides. I suppose there will always be a reason for commanders to privilege the safety of their own soldiers, yet there must be a limit. There is probably no clear answer as to where that limit is but it is still a question worth asking.

This obligation to minimize casualties should also exist within armies and states towards their own soldiers. We should try to place limits on using people

Goodman's "The Power to Kill or Capture Enemy Combatants"' (2013) 24 European Journal of International Law 855.

[166] Haque (n 8) 89–92.

[167] *See* Blum (n 143).

[168] *See* Yishai Beer, *Military Professionalism and Humanitarian Law: The Struggle to Reduce the Hazards of War* (Oxford University Press 2018) 48–49.

[169] *See* Michael Walzer and Avishai Margalit, 'Israel: Civilians & Combatants' (*The New York Review of Books*, 2009) <http://www.nybooks.com/articles/archives/2009/may/14/israel-civilians-combatants/> accessed 28 October 2015.

as cannon fodder. And we should strive for armies and armed groups to take care of their soldiers.

I hope the obligation to minimize casualties can eventually become part of our rules of war, perhaps similarly to how our current rules require 'constant care' to be taken in the conduct of military operations to spare the civilian population and civilian objects,[170] so that constant care is employed in sparing the lives of combatants as much as possible. Perhaps we could even get to a point where the law forbids or places some limits on military strategies that Beer describes as focused entirely on body count, that is, which aim at hurting as many enemy soldiers as possible.[171]

Particularly interesting is Beer's suggestion that the law does not concern itself with military strategy in general nor with the determination of a war's aim in particular, which are fundamental in shaping a state's or armed group's targeting policy. As Beer points out, this disregard has the price of accepting, by default, the Clausewitzian paradigm of total or 'all-out' war strategy, that is, war aimed at the decisive and total destruction of the adversary's military and the extinction of its will to fight.[172] Beer thinks that requiring adherence in targeting to strict and definite war aims, as strategically selected by a belligerent, might substantially affect the scope of lawful targets, especially in limited wars: the law should demand a connection between the strategic aim of the war and its actual application, as reflected in the desired operational and tactical military advantages.[173]

It is a sad feature of our current framework that more thought is given to which *objects* may be targeted than to the lives of combatants, and that their lives do not count for much in assessments of whether a particular war is justifiably initiated and conducted.[174] It is also worth wondering to what extent our laws of armed conflict have shaped our social attitudes towards the deaths of combatants in considering them 'fair game'.

I think these implications are an advantage of the view of killing in war I have defended thus far,[175] even if—precisely because—they entail the imposition of

[170] *Protocol Additional to the Geneva Conventions of 12 August 1949, and relating to the Protection of Victims of International Armed Conflicts (Protocol I)* (hereinafter, API) 8 June 1977, 1125 UNTS 3, art 57.

[171] Beer (n 39) 825.

[172] Beer (n 168) 117–18.

[173] ibid 120.

[174] *See* Blum (n 143) 154; Beer (n 168) 40; Larry May, 'Humanity, Necessity, and the Rights of Soldiers' in Jens David Ohlin, Larry May, and Claire Finkelstein (eds), *Weighing Lives in War* (Oxford University Press 2017).

[175] Even if my account of permissibility fails—that is, even if killing non-liable unjust combatants is impermissible—these implications also follow from a view whereby most unjust combatants are not liable to significant or lethal harm, and even more forcefully. But in that case, the view turns into an argument for pacifism.

several constraints on war, and plausibly make most wars impermissible and harder to justify. My account can also partly capture why war is such a horrible human practice: it is not only that the lives of just combatants are lost; it is also that the lives of all those fighting for the unjust side are lost. And I'm not sure some of these implications are available to those who hold the liability view of self-defence whereby most unjust combatants are liable. Such an account seems to obscure what makes war morally fragile and struggles to set constraints on the number of combatants that may be killed.

Finally, this account of permissibility leaves enough conceptual space to explain why some cases of defensive killing might be symmetrical, that is, why both the threat and the victim might be allowed to defend themselves. This symmetry has been defended as a formal feature of 'traditional' permissibility—that is, morally indifferent actions—whereby permissibility covers those actions that entail a privilege to do X in relation to B, but not a duty of B to submit to X being done to her,[176] but it is also present in agent-relative prerogatives to inflict harm.[177] Permissibility thus leaves conceptual space to explain why non-liable unjust combatants may be able to defend themselves in certain circumstances. This conceptual space must be filled with a normative argument for the existence of symmetrical privileges to kill, which is one of the goals of Chapter VIII.

VI. Permissibility: Objections

Despite these advantages, the sketch of permissibility I have provided leaves several issues unresolved and is subject to some important objections.

A first objection pertains to the small difference in the weight of reasons. There is a paradoxical quality to this feature of my account of permissibility. Several moral theories about rights accept that rights may be justifiably overridden.[178] But in all these cases, rights are overridden by a large margin—not by a small one. In other words, the weight of the reasons in favour of infringing a right must be significantly higher than the weight of the reasons against doing so. Some even think that it takes something like 'catastrophic moral horror' to override certain rights, as Nozick would call it;[179] a 'supreme emergency', in

[176] *See* Jeff McMahan, 'On the Moral Equality of Combatants' (2006) 14 Journal of Political Philosophy 377.

[177] Quong (n 9) 87–90.

[178] *See* ibid, 14; Lazar (n 16) 5. But Walen argues against this model of rights. *See* Alec Walen, *The Mechanics of Claims and Permissible Killing in War* (Oxford University Press 2019).

[179] Robert Nozick, *Anarchy, Utopia and State* (Basil Blackwell, Oxford 1974) fn 30.

Walzer's words.[180] Yet my account of permissibility takes a seemingly opposite stance: the prohibition against killing non-liable unjust combatants is defeated not by overwhelmingly weightier reasons, but by reasons that are barely weightier enough. How can this be so?

Two issues should trouble us. First, given the importance that rights are supposed to have in moral theory, it seems problematic that they could be overridden by a very small margin, as my idea of permissibility suggests. Second, shouldn't what is at stake in a just war lead us precisely in the direction of there being a large difference in weights? That is, shouldn't the very idea of a just war suggest that rights are overwhelmingly overridden by the catastrophic moral horror that the unjust war threatens?

It is true that what is (or should be) at stake in a just war is often close to catastrophic moral horror. But war, even a just one, also *brings about* something close to catastrophic moral horror. This is, I believe, the tragedy of war, and it puts into question its very permissibility. And even if we want to say that the reasons in favour of fighting a just war are much weightier than the reasons against doing so, those reasons must include reasons in favour of killing individuals who are thinly connected—for the reasons I have pointed out repeatedly—to the unjust war. In these individual cases of non-liable unjust combatants, it is simply implausible to argue that the reasons against killing the non-liable are clearly and unanswerably overridden by the mere fact of their being unjust combatants. For the death of one combatant is a small contribution to victory—and perhaps not even that. Yet death and killing are final, the value of the combatant's life is enormous, and reasons against killing have great weight. I see no way in which this conflict could be anything *but* paradoxical. This is one of those instances—but certainly not the only one—where the perplexing nature and the horrors of war can be seen most clearly.

War would present much less of a moral dilemma if reasons against killing were clearly and unanswerably overridden; if killing unjust combatants could be justified as easily as we can explain breaking a promise to have lunch in order to save a child. A moral theory where these paradoxes were eliminated—say, by arguing that most unjust combatants forfeit their rights—would undoubtedly be simpler, perhaps even more elegant. It would make a just war much more comfortable, morally speaking. But a just war *is not* morally comfortable. Any theory that fails to convey this will result in an impoverished picture of warfare—it cannot grasp what makes war so morally puzzling, why pacifism

[180] Walzer, *Just and Unjust Wars* (n 15) 250–67.

about war is attractive. The paradoxical nature of permissibility can be seen as a defect in the account, but it is also, I think, a virtue of it.

There is an additional advantage to the idea of the small difference in weights: the theory of war it supports has less of a tendency to collapse into total war than a view in which rights are overwhelmingly overridden. If killing non-liable combatants is not even justified, but merely permissible, arguments in favour of killing civilians cannot but fail.

There is still the question of what to make of a theory that allows for moral rights to be overridden by a small margin. This should strike us as an unattractive feature of a theory about rights (unless we would want to embrace consequentialism). Indeed, traditional lesser-evil justifications—whereby rights can be overridden only by a significant margin—have a consequentialist dimension, even when integrated into deontological theories that try to incorporate the disvalue of infringing or violating a right. These justifications usually operate by weighing the consequences of contravening a duty against the duty itself.[181]

My account also struggles with some of the issues of what Walen has called 'the infringement model of rights',[182] which is the model of rights most revisionists work with. According to the infringement model, the rights that people have do not settle what may be done to them.[183] It is sometimes permissible to infringe someone's rights (eg lesser-evil justifications and agent-relative prerogatives to inflict harm), even if they haven't lost or waived them.[184] In those cases, the infringement model claims that the right holder has been wronged, even if the action on account of which they were wronged was permissible.[185] Walen believes the infringement model is incorrect. He rejects the idea that someone can be wronged due to a *permissible* action. If an action is permissible then by definition it cannot wrong anyone; it does not infringe rights.

However, my account of permissibility does not claim that the rights of non-liable unjust combatants are overridden because the consequences in favour of infringing their rights are barely weightier than the reasons against doing so. It claims that when there is a just cause for war, it is permissible to target those who more robustly and directly contribute to the unjust war by fighting it. This is the reason—their direct and robust contributions to posing the

[181] *See*, eg, Quong (n 9) 14; Haque (n 8) 7; Rodin, 'Justifying Harm' (n 155) 74.
[182] Walen (n 178) xv.
[183] ibid.
[184] ibid.
[185] ibid.

unjust threat—which barely defeats the first-order reason against killing them. Permissibility does not proceed by comparing states of affairs.[186]

The consequences of war—states of affairs—are not irrelevant, but they are external to this account of permissibility. That is, the success of the war, proportionality, and effectiveness are important constraints on war—not just prudential tests. After all, a just war draws at least some, if not most, of its value from the goals it is able to achieve: stopping aggression or slaughter, re-establishing peace. But these consequences are not internal to permissibility: it is not the consequences that override the duty not to kill non-liable unjust combatants. The account aims to partially isolate the consequences from the question of who are permissible targets, so it does not collapse into consequentialism.

This entails that it will be sometimes permissible to target combatants, but wrong, all things considered, to do so. This will be the case when killing them is ineffective, gratuitous, disproportionate, and so on. Because under my account unjust combatants are wronged grievously regardless of that outcome, the fact that these constraints are external to permissibility poses less significant issues than when they are external to liability-based justifications. And because under my account combatants are wronged grievously, one should have good reason to believe that killing them will be effective, necessary, and proportionate before targeting them.

I think the main objections along these lines are, first, that it remains unclear why we should accept that an action such as killing the non-liable would become permissible on the basis of a small difference in weight, and why that small difference in weight is given by the quality of one's contributions to a collective threat. I have not, in fact, given a fully developed argument as to why sufficiently direct and robust contributions to a collective threat made by non-liable individuals are enough to move away from prohibition. And though the small difference in weight has attractive features, it is also troubling.

Second, I have not provided a complete argument as to why the factual impossibility of dealing with deadly non-compliance by only targeting those who are liable would have an effect on the nature and stringency of our moral duties not to kill the innocent. Perhaps the answer is that the moral demands persist unchanged, and we must embrace pacifism regarding war.

[186] Thomson, for example, considers conflicts where rights are overridden due to differences in their stringency. *See* Judith Jarvis Thomson, *Rights, Restitution, and Risk: Essays, in Moral Theory* (William Parent ed, Harvard University Press 1986) 41–43, 51, 67, 145, 253. And Jonathan Quong argues that we have agent-relative prerogatives to impose harm on the non-liable and thereby infringe their rights. These prerogatives are not lesser-evil justifications; they do not rely on the consequences of our actions. *See* Quong (n 9) 122.

And, finally, I worry that the account, though plausibly able to resist consequentialism, can be subject to abuse. That is, someone might want to argue that targeting civilians is, in a certain case and given certain circumstances, *permissible*, because the reasons for doing so are barely weightier than the reasons against doing so. I would respond to that by saying that if civilians are contributing directly and robustly to the war by fighting it, then yes, it would be permissible to target them. It seems that in those circumstances, they simply would cease to be civilians (this is why I think the non-comparative version of permissibility is preferable).[187] But the danger I see comes from (ab)using the notion of permissibility to argue that targeting (non-liable) civilians is permissible due to *other* reasons (say, to gain a slight advantage in the war) that outweigh the prohibition, but only barely so. I reject this conclusion. My account does not allow for comparing states of affairs or outcomes and then deciding which one is better, if only slightly so. Slightly better outcomes can never be a reason in favour of killing non-liable individuals. And civilians are almost never permissible targets because they almost never contribute to the war robustly and directly. If the law is to progress in the right direction, the answer is to limit the killing of combatants—not to loosen the prohibition on targeting civilians.

A second objection pertains to uncertainty.[188] Recall that there are two versions of the account I have suggested: a comparative one, whereby targeting non-liable combatants is permissible because fighting is often a more robust and direct contribution to war than other contributions, and a non-comparative one, whereby fighting is a special kind of contribution to war that is sufficient to move away from prohibition, unlike non-combat contributions.

The first version makes room for significant uncertainty. Although the law distinguishes between combatants and civilians, this distinction, I said, is somewhat artificial: it draws a bright line between those who more often contribute directly and robustly to the threat (combatants) and those who don't. The law is thus fulfilling a particular kind of role: we might think it constitutes a previously relatively undetermined moral obligation, or we might think that it is enshrining an evidence-relative permission. That is, given what we know about combatants, it is reasonable to believe that they are those who more directly and robustly contribute to posing the threat of war. This gives us powerful

[187] In the comparative account of permissibility, there might be cases of non-liable civilians who make more direct and robust contributions to the war, which would make them permissible targets. This is why I think we should reject that account.

[188] In what follows, I am deeply indebted to Susanne Burri, who first suggested to me that permissibility might be better located at the evidence-relative level.

reasons not to disobey the prohibition on targeting civilians and to enshrine such prohibition in the law. But as a matter of morality, it is still the case that, when deciding to harm unjust combatants, just combatants will often have no way of knowing for certain whether they are targeting someone who contributes to the war in a way that is sufficiently direct and robust so that it makes them permissible targets, partly because they will not know for certain whether combatants will pose threats against others in the future.

We could do away with this issue if we adopted the second, non-comparative, version, whereby fighting is a direct and robust contribution to the war in a way that is sufficient to move away from prohibition, unlike other, non-combat contributions. This would give us a more determinate moral norm: it would tell us that only targeting combatants is permissible—and barely so. This version also has to deal with some uncertainty, this time regarding whether unjust combatants will pose threats in the future.

Ultimately, eliminating pervasive uncertainty in war by making the moral norm too determinate sits unwell with providing a proper account of war, where uncertainty is, in fact, pervasive. Legal rules, of course, have to manoeuvre around that uncertainty and give action-guiding norms that are easy to apply, and that is, in fact, what I've said they do in this case. Morality might simply be more fine-grained than the content of the legal rule, but if it is, then there are powerful reasons for us to try to gain more knowledge about what is fact-relative right to do.[189]

Third, I have left somewhat open how far away from reductivism the account is. If my account applies only to the context of war—and is, thus, strongly non-reductivist—then it requires a definition that allows us to distinguish between war and peace. If my account applies to collectively imposed threats in both contexts—and is, thus, strongly reductivist—and it is permissible to kill those who more directly and robustly contribute to the threat even if they are non-liable, it seems exceedingly permissive.

I think that the best account we can give here is the following. The question of when collectively posed threats become war—and thus, when killing non-liable combatants becomes permissible—is context-dependent and difficult to answer generally: it is contingent upon the factual impossibility of eliminating collectively posed threats without harming those non-liable. We thus need to rely on law to establish a bright line, and this line might be well or

[189] See Derek Parfit, *On What Matters*, vol 1 (Oxford University Press 2013) 150–74.

badly drawn.[190] And, second, because the discontinuity between the 'morality of peace' and the 'morality of war' is contingent upon the factual impossibility of eliminating collectively posed threats without harming those non-liable, we should strive to eliminate the necessity for 'the morality of war' entirely, so that we come to be governed by a less permissive morality, whereby only liability-based justifications—with a high threshold of moral responsibility—and lesser-evil justifications are available.

Finally, and I think this is the most important objection against the account I have developed—and one that underlies some of the objections above—I worry that it sanitizes war where it shouldn't. Perhaps at the level of fact-relative morality, it is simply true that many unjust combatants are not liable, lesser-evil justifications are rarely available, and just wars involve inflicting evidence-relative prohibited harms on many individuals that might only become fact-relative permissible in the rarest of circumstances. Given what I have argued so far and the 1:10 ratio that lesser-evil justifications demand—which might not be sufficiently demanding, since war involves killing many people at least in a partially opportunistic manner—it will most often be the case that lesser-evil justifications fail to account for wars as a whole. At the very least, it is difficult to have reasonable evidence to the contrary, even in an *ex-post* assessment of any given war.

If permissibility fails and the above account of killing in war is true, the chain of command issue comes back with full force, and agents will often not be evidence-relative permitted to go to war, or we might need a considerably less demanding account of when the evidence one has rises to the level of reasonableness—and, further, perhaps my account would work better as an evidence-relative account of killing in war. But if this cannot be done, it means that the decision to go to war will often be a sort of leap into a (moral) abyss, a moral abyss from which we will rarely come out successfully. If this latter account is true, we should embrace pacifism. If my account is true, we should

[190] Additional Protocol II, for instance, provides for a negative threshold by stating that the laws of war do not apply to 'situations of internal disturbances and tensions, such as riots, isolated and sporadic acts of violence and other acts of a similar nature, as not being armed conflicts.' And the International Criminal Tribunal for the former Yugoslavia (ICTY) has also talked, in defining armed conflict, about 'protracted armed violence' between governmental authorities and organized armed groups or between such groups within a state. *Protocol Additional to the Geneva Conventions of 12 August 1949, and relating to the Protection of Victims of Non-International Armed Conflicts (Protocol II)* (hereinafter, APII) 8 June 1977, 1125 UNTS 609, art 1.2; ICTY, *Prosecutor v Dusko Tadić*, IT-94-1-A, Appeals Chamber, Decision on the Defense Motion for Interlocutory Appeal on Jurisdiction, 2 October 1995, para 70.

still embrace the pacifist option, not regarding all wars, but certainly regarding most of them.

VII. Conclusion

Most of what I have said in this chapter provides very strong reasons against fighting a war—even a just one—and, going one step further, provides powerful reasons in favour of pacifism.[191] It conveys that we find ourselves thinking about war because we are simply unable to stop war from happening in the first place.

Despite its fragility, the account I have developed in this chapter—if successful—allows us to move beyond pacifism and to provide a non-instrumentalist justification for the just combatant's legal privilege to kill in war. This justification relies partly on those parts of the revisionist morality that have not been displaced by my account, and partly on the idea of permissibility that I have developed. Thus, permissibility will not apply to *all* killings in pursuit of a just war. First, and most obviously, because sometimes—particularly any time unjust combatants most closely resemble the villainous aggressor—there will be a (liability-based) justification for killing. Second, because permissibility, as I have conceived it, is 'empty' on the just combatant's side: just combatants may have *other* reasons, besides the one that dominates my account of permissibility, for killing non-liable unjust combatants. They may be defending themselves, defending their loved ones, and so on. Those just combatants might be justified if all those reasons together are able to support a significant difference in weight between the two courses of action. However, even if the weight of the reasons supports a justification, these killings will leave a significant moral residue—due to the values at stake—and often wrong unjust combatants, even if not grievously. Some of these cases may, indeed, be closer to cases where reasons are clearly and unanswerably overridden, but they are still quite morally fragile. They are far from being cases where people have forfeited their rights. And some just combatants—particularly those initiating offensive actions, in charge of aerial bombardment, and in charge of making

[191] Shue thinks something similar: 'War is especially not a mechanism that can allot the evils of death and destruction only, or even mainly, to those who have done wrong and protect the lives and goods that belong to those who have done right. This is a profoundly important reason not to fight wars, especially—it must be said in the wake of the disastrous invasion of Iraq in 2003 by the United States and the UK—discretionary wars, or "wars of choice". Henry Shue, 'Do We Need a "Morality of War"?' in David Rodin and Henry Shue (eds), *Just and Unjust Warriors: The Moral and Legal Status of Soldiers* (Oxford University Press 2008) 103.

decisions about whom to target and when—will most likely only have a permission to kill non-liable unjust combatants.

With this account of moral permissibility in place, we can now begin to understand the reasons that unjust combatants may have for participating in an unjust war. In other words, moral permissibility opens a path where we can deconstruct the analogy between what unjust combatants do and murdering the innocent. This will be the focus of Chapter VIII.

VIII
Unjust Combatants

I. Introduction

Are all unjust combatants engaged in murdering the innocent? Or do they have reasons for fighting and killing in an unjust war? Consider the experiences of Corporal Hans Herbst and Murray Shapiro, as depicted by Stephen Ambrose in his book about the Second World War:

> On several occasions, the Americans and German veterans associations have held joint ceremonies to honor their fallen comrades. But more often individuals have met by chance on some former battlefield. During the ensuing conversation ... they find that they have a lot in common ...
>
> Cpl. Hans Herbst had obviously had one such conversation about his entry into Belgium in late October 1944. 'I was in my armored personnel carrier about 17 or 18 vehicles behind the lead tank on a small dirt track through the forest.' The convoy came to a bridge. 'There were two American anti-tank positions built there. We saw them. The first of our tanks opened fire. My friend, Murray Shapiro, one of the Americans in the position across from us, told me [later] that that first shot killed all but one of his men in the first two positions. The survivors returned fire upon us, blowing up four of the Panther tanks headed down the narrow track. We pulled back.'
>
> 'I only want to add,' said Herbst with a chuckle, 'that today when I am with Americans, we are friends. They tell me I am lucky to be alive. I tell them I'm lucky that they are such poor shots.'[1]

Is Murray Shapiro friends with the man who murdered his friends? In the past chapters, I argued that many unjust combatants are not liable to significant or lethal harm. They are wronged when killed, and it is only permissible to kill them. These arguments open up a path to understand why and how what some

[1] Stephen E Ambrose, *Citizen Soldiers: The U.S. Army from the Normandy Beaches to the Bulge to the Surrender of Germany, June 7, 1944–May 7, 1945* (Simon & Schuster 1998) 492.

The Morality of the Laws of War. Marcela Prieto Rudolphy, Oxford University Press. © Marcela Prieto Rudolphy 2023.
DOI: 10.1093/oso/9780192855473.003.0008

unjust combatants do is not simply like murdering the innocent—they open up a path to deconstruct the analogy between murder and unjust combatants' killings.

There are different ways in which the analogy between murder and what unjust combatants do can be deconstructed. In its strongest form, we would deny all similarities between murder and unjust combatants' killings. In a weaker version, the argument would point out that unjust combatants' killings of just combatants are *somewhat* different from murdering the innocent, and that just combatants might not be *simply* morally innocent.

Justifying the symmetry thesis does not require the strongest version of the dissimilarity between murder and what unjust combatants do. It is not necessary to claim that unjust combatants are generally morally justified in killing just ones, that just combatants are not morally innocent, or that killing just combatants is *never* like murder. Weaker claims can be enough, such as saying that *sometimes* unjust combatants are morally permitted to kill just ones; that *sometimes*, just combatants are not morally innocent, and that *sometimes* killing just combatants is not really like murder.

This chapter will thus provide a partial, non-instrumentalist justification of the symmetry thesis, which relies on two arguments. First, as I anticipated in Chapter VII, the fact that many unjust combatants are not liable makes it possible to argue that they are permitted to defend themselves when they are under attack. This argument is quite modest. The legal equality of combatants has a much broader scope: combatants are not only allowed to defend themselves when under attack, they are also allowed to participate in (unjust) wars, which involve offensive military action. Thus, the second part of the justification of the symmetry thesis involves providing a (limited) argument as to when unjust combatants have *pro tanto* reasons to participate in an unjust war.

These two arguments do not succeed in entirely doing away with the analogy between the actions of some unjust combatants and the murder of the innocent. The fact that we are not able to fully contest the moral asymmetry between just and unjust combatants implies, of course, that the instrumentalist justification I proposed in Chapter III will still play an important role in justifying our laws of war. It could not be otherwise. Just and unjust wars are not morally equal.

In fact, most of the reasons unjust combatants have for killing and fighting are not universal. They are not universal in the sense that not every unjust combatant will have reasons for fighting or killing available. The more closely a combatant resembles the villainous aggressor, the less—relative—weight reasons for fighting and killing will have until they are unable to ground permissibility.

This will be particularly true for those who make greater contributions to their side, who are higher up in the chain of command, and who have more knowledge and moral responsibility for the collective enterprise of the unjust war.

And some of the reasons unjust combatants have are, unlike the reasons that just combatants have for fighting and killing, agent-relative prerogatives to inflict harm or to participate in war. That is, they are reasons for action only for a particular agent, or an agent in a particular position, to do something. They are not reasons for *anyone* to do something.[2]

But to say that a reason is agent-relative is not to say that it is unreal, subjective, a mere appearance, or asocial in character.[3] Agent-relative reasons are universal for all agents in relevantly similar positions: they can be understood and respected not only by the people who have them, but also by people in other positions and against whom they are exercised,[4] even if they cannot be *shared*, or acted upon, by those who are not in the relevantly similar position. If we think that there are no agent-relative reasons and that all reasons are agent-neutral, then the arguments in this chapter fail to grant any permissions (though they might be understood as excuses).

Although there is no conceptual incoherence in the existence of conflicting or symmetrical permissions to inflict harm[5] or to participate in opposing sides of a war, the existence of such permissions requires a normative argument. This is what most of the chapter is devoted to. My aim in this chapter is not to provide a full account of each of the *pro tanto* reasons that combatants may have for fighting and killing in an unjust war, but to sketch them, the arguments in their favour, and their limitations.

Some of the reasons in this chapter are better understood as reasons for fighting or joining the war. Others are better understood as reasons for killing just combatants. Some may be understood as both, though more rarely. Each of these *pro tanto* reasons will not always be able to ground, on their own, permissibility for unjust combatants' actions, but several of them together might, depending on the circumstances. To my mind, this is more likely to be the case for combatants at the lower end of the chain of command, when they face significant costs if they refuse to fight the war, are likely to contribute negligibly to

[2] Christine M Korsgaard, 'The Reasons We Can Share: An Attack on the Distinction between Agent-Relative and Agent-Neutral Values' (1993) 10 Social Philosophy and Policy 24.

[3] Samuel Scheffler, 'Projects, Relationships, and Reasons' in R Jay Wallace (ed), *Reason and Value: Themes from the Moral Philosophy of Joseph Raz* (Clarendon Press 2004) 251–52.

[4] ibid.

[5] *See* Jonathan Quong, *The Morality of Defensive Force* (Oxford University Press 2020).

the war and in non-combat functions, and are uncertain regarding the justness of their and the enemy's war. This is not an exceptional situation.

I don't think it's important to determine whether these *pro tanto* reasons ultimately result in unjust combatants being justified, permitted (on account of agent-relative prerogatives) or permitted in an evidence-relative sense (on the basis of the evidence available to them) to participate or kill in an unjust war. Any of these are enough to partially dispel the analogy between what unjust combatants do and the murder of the innocent.

The general form of the argument is as follows. Generally speaking, the decision to participate in an unjust war will often, but not always, involve giving preference to our interests or the interests of our loved ones[6] in circumstances in which the alternatives (refusing to fight, surrendering) might be exceedingly costly and achieve very little. The reasons unjust combatants have to fight and kill arise then out of certain features of war, and out of certain features of our institutional arrangements. Section II focuses on duress and uncertainty. Section III focuses on the collective nature of war. First, it argues that when wars with a just cause are fought unjustly, unjust combatants will have a reason to participate in the war and to inflict harm. Second, it argues that if unjust combatants have reason to believe that they will make only small contributions to their side of the war, this fact will provide no reason not to join when refusing to fight is exceedingly costly for agents.

Section IV focuses on the fact that war is risky for those who fight it. First, it argues that non-liable unjust combatants have agent-relative prerogatives to defend themselves and other, non-liable individuals, when they are under attack. Second, it argues that, under certain circumstances, unjust combatants will be permitted to join an unjust war because refusing to do so would deflect unjustified harm onto other, non-liable, unjust combatants.

Section V concludes by making what is perhaps the most important argument in this chapter. As we will see, many combatants inhabit a small moral world:[7] surrendering and refusing to fight are almost always exceedingly costly for agents and almost always achieve very little. But although the size of this

[6] *See* Thomas Nagel, *The View from Nowhere* (Oxford University Press 1989); Samuel Scheffler, *The Rejection of Consequentialism: A Philosophical Investigation of the Considerations Underlying Rival Moral Conceptions* (Clarendon Press 1995).

[7] I take the idea of a 'small moral world' where the alternatives are constrained from Korsgaard, although she is using this idea to illustrate a different point: '*Very* roughly speaking, you are not treating him as a mere means if he consents to what you are doing. Of course, the Indian does not in general consent to be shot, and his gesture does not mean that after all he has not been wronged. In the larger moral world he has. But if you and the Indians are forced to regard Pedro and the captain as mere forces of nature, as you are in this case, then there is a smaller moral world within which the issue is between you and them, and in that world this Indian consents.' *See* Korsgaard (n 2) 46.

moral world is partly a result of certain features of warfare, it is also the result of our institutional arrangements.[8] The fact that combatants face high costs when they refuse to fight and the fact that their access to evidence is limited are not simply natural facts of war. The size of this moral world is thus partially the result of our institutions; it is a human construction, and we can change it—we can make it larger.

In important ways, the reasons explored in this chapter are reasons for fighting and killing *because* of how warfare is, because of how our institutional arrangements operate. Conflicting permissions thus exist partly due to our institutional arrangements, which fail to provide combatants with less costly alternatives and more information. Different institutional arrangements, although unable to modify certain features of war, might be able to change what is required of combatants by making surrender and refusal less costly and by providing more evidence regarding when it is permissible to participate in war. But until that is the case, unjust combatants will sometimes have reasons to fight and kill.

II. A Background of Duress and Uncertainty

Unjust combatants will often face decisions where reasons pull them in opposite directions. The decision to participate in an unjust war or to defend oneself or others will usually involve giving preference to one's interests or the interests of one's loved ones in circumstances in which the alternatives—refusing to fight, surrendering—might be excessively costly and might achieve very little.

That the alternatives that combatants face are costly and achieve very little is partly related to the nature of war itself and partly related to our current institutional arrangements.

Regarding our institutional arrangements, or lack thereof, it is not unusual for soldiers to join the war under duress, particularly, but not only, within the context of authoritarian states and non-state armed groups. Soldiers might risk severe punishment and even their lives and the lives of their family members if they refuse to participate in war. Hitler's generals continued fighting even when they knew that they were going to lose the war and that fighting any longer was pointless. Some of them did it out of fanaticism; others did it out of a misguided sense of duty; yet others were moved by fear of execution and reprisals

[8] *See* Samuel Scheffler, *Human Morality* (Oxford University Press 1993) 145.

against their families.[9] German soldiers were shot if they refused to fight. In Afghanistan, the Taliban would threaten families and take their young men to fight for them against the warlords. The warlords did the same.[10] In Iraq, the picture is depressingly familiar, as narrated by Lt Colonel Terry F Moorer:

> I also interviewed and actually felt great respect for several Iraqi officers who were prisoners of war. Most of the captured officers had surrendered their men with little to no resistance. Quit and surrender are *not* U.S. Army words. Officers in every military are responsible for accomplishing the mission first and seeing to the welfare of the command second. But when resistance is futile, the difficult but correct moral choice is surrender. Lee's capitulation at Appomattox is a prime example of such moral courage.
>
> Unlike Lee, however, the Iraqi officers knew that surrender meant they were betting the lives of their families. During the early phases of the war, Saddam inserted Baath party officers into units to report the names of deserters ... The Iraqi officers at Camp Bucca made the difficult but correct moral choice to surrender rather than offer futile resistance.[11]

Although duress is often an excuse, under certain circumstances, threats of harm against non-liable unjust combatants or their loved ones might be able to confer permissions to participate in war. When the harm threatened against the unjust combatant considerably exceeds the harm that the unjust combatant would inflict by participating in the unjust war, unjust combatants might have lesser-evil justifications available. This is most likely to be the case for (non-liable) unjust combatants who participate in roles that do not involve imposing harm on other individuals. As a result, when unjust combatants have reason to believe that they will not impose harm on other individuals, but they are certain that refusing to participate will inflict harm on themselves or others, they might have an evidence-relative permission to fight.

Our institutional and legal arrangements are also insufficient in dispelling uncertainty. Some uncertainty regarding war, is, of course, inevitable: even when there is a just cause for war, uncertainty is pervasive. But some combatants live in societies where reliable information regarding decisions to

[9] Ian Kershaw, *The End: The Defiance and Destruction of Hitler's Germany, 1944–1945* (Penguin Press 2011).

[10] Dexter Filkins, *The Forever War* (Vintage Books 2009).

[11] Terry F. Moorer, 'JAG in the Sandbox' in Andrew Carroll (ed), *Operation Homecoming: Iraq, Afghanistan, and the Home Front, in the Words of U.S. Troops and Their Families* (University of Chicago Press 2008) 110.

resort to force is unavailable and/or propaganda is rampant. Information about the other side of the war—whether they have a just cause or are fighting permissibly—can also be difficult to find. In some (rare) cases, this will imply that unjust combatants might have evidence-relative permissions to participate in the war, due to their reasonably believing that their own war has a just cause or that the other side lacks it.

III. Collective Enterprises and Their Moral Heterogeneity

War is a collective enterprise, fought by hundreds, and often thousands, of individuals. The collective nature of war has implications in shaping and determining what the right thing to do is. Sometimes it may provide reasons to fight and to kill in an unjust war; sometimes it might make other alternatives, such as surrendering or refusing to fight, unattractive due to their being excessively demanding and achieving very little.

Bazargan has referred to war as 'morally heterogenous'. By this he means that wars 'include both just and unjust aims, where the achievement of the unjust aims is not instrumental to the achievement of the just aims'.[12] Here, I mean to use the expression more broadly because wars are 'morally heterogenous' in many ways: some actions in a just war will be morally forbidden (section III.A); some actions in war will contribute little or nothing to victory (section III.B.1; and the collective nature of war will entail that individual acts of surrender will achieve nothing or very little (section III.B.2). I take each of these in turn.

A. Just Wars: Fighting Unjustly

There is an obvious way in which unjust combatants may have reasons to fight and kill. History does not paint a kind picture of war, even just ones:

> We took prisoners, brought them to the detachment... We didn't shoot them, that was too easy a death for them; we stuck them with ramrods like pigs, we cut them to pieces...

[12] Saba Bazargan, 'Morally Heterogeneous Wars' (2013) 41 Philosophia 959.

> What do you know about it?! They burned my mother and little sisters on a bonfire in the middle of our village. (A Russian soldier, who fought in the Second World War)[13]

A war may have a just cause but be disproportionate or unnecessary. A war may begin as a just one but evolve into an unjust one by expanding the scope of its goals. A just war may be generally or overall conducted unjustly by, for instance, targeting civilians or taking no prisoners. Indeed, war crimes and crimes against humanity have often—if not always—been committed in war, no matter how just its cause. The highly idealized just combatant fighting in a highly idealized just war exists mostly in writing. There is rarely such thing as the unsullied just combatant; no such thing as a totally just war.[14] Think about the Red Army fighting the Nazis on the Eastern Front,[15] about the attacks against Hiroshima and Nagasaki during the Second World War. Or about the revolt led by Ismail Khan in Afghanistan, during which more than a dozen Soviet political advisers were slaughtered, as well as their wives and children.[16]

A brief look at the history of warfare is enough to realize that it is not an exception for just combatants—nor for unjust combatants, of course—to engage in horrific moral acts, which we now refer to as war crimes, crimes against humanity, and genocide. Just combatants might execute prisoners of war or *hors de combat*, they might torture prisoners, hold them in terrible conditions, and intentionally target civilians. And all modern wars are likely to involve the deaths of hundreds of (non-liable) civilians as collateral damage, which, although not a war crime, has an uncertain moral status, to say the least. All of these acts will provide a reason for unjust combatants to participate in an unjust war and for killing just combatants engaged in them. Here we have cases where some just combatants will resemble the villainous aggressor.

Some might want to say that in effect these reasons turn just combatants into unjust ones and vice-versa. Some, like Frowe,[17] might think that this shows that the judgment of whether a war is just or unjust is not morally interesting: what matters are the actions of individuals. Whatever the case, it does turn some of

[13] Svetlana Alexievič, *The Unwomanly Face of War* (Random House Trade Paperbacks 2018) xxxiv.

[14] One cannot help but wonder, for instance, if the sense of crusade against the Nazis in the Second World War may partly explain the atrocities that were committed by the Allies: the bombing of cities and the attacks on Hiroshima and Nagasaki immediately come to mind.

[15] *See* Alexievič (n 13); Kershaw (n 9).

[16] Steve Coll, *Ghost Wars: The Secret History of the CIA, Afghanistan, and Bin Laden, from the Soviet Invasion to September 10, 2001* (Penguin Books 2004).

[17] *See* Helen Frowe, 'The Just War Framework' in Seth Lazar and Helen Frowe (eds), *The Oxford Handbook of Ethics of War* (Oxford University Press 2015).

the unjust combatants' actions into permissible ones and, sometimes, into justified ones.

These reasons are, of course, contingent, if to different degrees. After all, if pacifism is false, a just war *could* be conducted justly—there is no use in denying the theoretical possibility. Nevertheless, it seems unlikely that we could ever eliminate this particular form of moral heterogeneity in war, if war is to remain an enterprise that is fought by thousands of individuals. As much as we strive for every single action within war to be justified or morally correct, it is impossible to ensure it. Human beings make mistakes, they have biases in favour of their loved ones and themselves. They can also be incredibly cruel to one another. War is just another instance where these flaws and tendencies manifest themselves.

B. The Problem of Small Contributions

Arneson has suggested that unjust combatants might truly say that 'it makes no difference' whether they fight or not, since they will do no harm that would not have come about anyway.[18] This argument—if successful—has two parts. The first one is that 'if I don't do it, someone else will', and the second one is that 'one person makes no difference'.[19] I will return to the first part of the argument—if I don't do it, someone else will—in section IV.B. The second part of the argument—that one person makes no difference—is important, but applicable to a limited set of cases. It is a problem of collective action, which is, of course, not unique to war but a general problem in political and moral philosophy.[20]

1. It makes no difference whether or not I fight

The 'one person makes no difference' argument is plausible only when applied to soldiers who contribute to war in non-combat roles and who are unlikely to harm anyone. Given the destructive potential of a single well-armed individual, one person can actually make a significant difference in the life of someone else, even if not in the overall outcome of the war.

Although wars are a matter of collective harm in some sense, they are not *only* a matter of collective harm. The issue of collective harm poses a serious

[18] Richard Arneson, 'Resolving the Responsibility Dilemma' in Saba Bazargan and Samuel C Rickless (eds), *The Ethics of War* (Oxford University Press 2017) 78. But he is sceptical about this reason.
[19] Jonathan Glover, 'It Makes No Difference Whether or Not I Do It' (1975) 49 Proceedings of the Aristotelian Society 171, 172.
[20] *See*, eg, Derek Parfit, *Reasons and Persons* (Clarendon 1992).

dilemma partly because each individual contribution does not harm anyone—and thus, it is difficult to understand why each individual contribution is wrong—whereas the sum of many contributions does cause substantial harm (think global warming).[21] War is only like this in the sense that there are many contributions, each of them small compared to the overall effect of the war. However, individual contributions might not be small in themselves. Even if killing one person makes a small—or null—contribution to winning a war, it already causes enormous harm:

> Because while the news in Rustamiyah on September 4 was all about three dead soldiers and a fourth who had lost both legs, and a fifth who had lost both legs and an arm and most of his other arm and been severely burned over what remained of him, that wasn't the news in the United States. In the United States, the news was all macro rather than micro. It was about President Bush arriving in Australia that morning, where the deputy prime minister asked him how the war was going and he answered, 'We are kicking ass.'[22]

It is true, nonetheless, that in some cases unjust combatants contribute to the war in ways that make no difference because their behaviour contributes nothing or very little to defeating the just side and it does not harm anyone. Some combatants are in charge of recovering bodies or damaged pieces of equipment, detecting mines, supplying the front line, and so on. Some combatants save lives. This is the case of those soldiers who are, for example, dedicated to mine-clearing and IED (improvised explosive device) clearing. And in the US army, every soldier is trained in combat medicine.[23] These actions, certainly, constitute small contributions to the soldier's own side of the war, since they aim at preserving the quantity of forces, yet these actions are not aggressive nor defensive in any traditional sense of those terms. In these cases, the argument about not making a difference will have more force.

However, although the argument might hold true of some actions individually considered, it will not hold true of all of them considered together. Here, again, we face the problem of collective harm. That is, although the contribution of each soldier might be small, the contribution of a unit or a regiment to victory might be very big. As Parfit argues, even if an act harms no-one, that

[21] ibid.
[22] David Finkel, *The Good Soldiers* (3rd edn, Picador USA 2010) 140–41.
[23] Sebastian Junger, *War* (Twelve 2011) 57.

act may be wrong because it is one of a set of acts that together harm other people.[24]

Ultimately, at its best, this argument may be able to explain why there might be no reasons *against* participating in an unjust war in certain roles. It does not, however, provide reasons in favour of doing so, unless it is accompanied by some of the other reasons explored in this chapter, such as the fact that surrendering and refusing to fight might be exceedingly costly for agents.

2. It makes no difference whether or not I surrender

Choices in war are constrained by its collective nature in a particular way: although surrendering or refusing to fight might be the correct thing to do if everyone also did it, usually we know that if one person surrenders, *nothing* will happen. The war will rage on. For surrender to work in stopping the war or making victory for the just side easier to accomplish, it needs to be massive surrender. Given the different positions that individuals occupy along the chain of command and thus, the different implications of their actions, surrendering will have different effects. Although a couple of generals refusing to wage war might be able to impede the war from taking place, a couple of platoon soldiers doing so will not. Thus, the arguments in this section will speak more directly to those lower in the chain of command.

There are different ways in which someone may refuse to participate in a war. First, there is 'surrender' in a technical sense, which is regulated by the laws of war.[25] Second, there is 'refusal', which can take place in different circumstances: there is refusal to fight in a war—by refusing conscription or by refusing deployment—refusal to participate in a mission, etc. And finally, there is desertion, which is usually a crime, punished with death in some countries: it refers to a member of the armed forces leaving or remaining absent from their unit, organization, or place of duty with the intent to remain away permanently.[26]

These differences are important, but what I will say below aims to speak to all these cases. I thus speak indistinctly of surrender, refusal, and desertion.

[24] Parfit (n 20) 70. *See* also Shelly Kagan, 'Do I Make a Difference?' (2011) 39 Philosophy & Public Affairs 105.

[25] *See* Yoram Dinstein, *The Conduct of Hostilities under the Law of International Armed Conflict* (Cambridge University Press 2004); 'How does Law Protect in War? Surrender' (International Committee of the Red Cross (ICRC) 2002) <https://casebook.icrc.org/glossary/surrender> accessed 22 August 2022.

[26] The definition is longer and more complicated than this, and it overlaps with what I call 'refusal'. *See* United States Manual for Courts-Martial, art 85 (Joint Service Committee on Military Justice 2016 edn) https://jsc.defense.gov/portals/99/documents/mcm2016.pdf?ver=2016-12-08-181411-957 accessed 6 March 2023.

While all of these have the same outcome—stopping the agent's participation in the war—I leave open the possibility that they have relevant moral differences. In particular, individuals might have reasons to accept conscription, but they might, under certain circumstances, have a duty to surrender if faced with the possibility of harming (non-liable) just combatants.

The fact that surrender, refusal, or desertion are not able to stop the war if they are not massive and collective—that they are, in a sense, useless—can be understood from the perspective of what morality demands of individuals when other individuals are not complying with their moral obligations (provided that their moral obligation is to not fight in an unjust war). In other words, it is a problem about what morality demands in non-ideal theory or under partial compliance.[27] Each combatant from the unjust side knows that if everyone complied with morality, each of them would be required to surrender. But given that they know that this will not happen, what should each individual combatant do? Is she still required to surrender? Or is she required to do something else?

This question—what to do when others do not comply with their moral obligations—can be answered in three different ways: 'do exactly your fair share, and no more than that', 'do more than your fair share', and 'do less than your fair share'.[28] It is, of course, possible for each soldier to surrender (assuming that this is, in fact, what doing their fair share requires). The question is whether they ought to do so.

Different scenarios are likely to yield different answers. There is no obvious correct choice without knowing more about the situation.[29] Laura Valentini proposes three examples, each of them with a different answer.

Doing less than your fair share might be an attractive alternative when doing one's fair share would not increase the likelihood of the desired outcome.[30] This is the case of Jason: if enough people in Jason's society protest, an unjustly imprisoned man will be released.[31] Unfortunately, none of Jason's co-citizens will do their fair share in protesting.[32] Under these circumstances, Valentini says, it is not obvious that Jason still ought to protest (ie do his fair share), since it

[27] *See* John Rawls, *A Theory of Justice* (Belknap Press of Harvard University Press 1999); Liam B Murphy, *Moral Demands in Nonideal Theory* (Oxford University Press 2003).

[28] Laura Valentini, 'Ideal vs. Non-Ideal Theory: A Conceptual Map' (2012) 7 Philosophy Compass 654, 655–56; David Miller, 'Taking Up the Slack? Responsibility and Justice in Situations of Partial Compliance' in Carl Knight and Zofia Stemplowska (eds), *Responsibility and Distributive Justice* (Oxford University Press 2011) 230.

[29] Valentini (n 28) 655.

[30] ibid 655–56.

[31] ibid.

[32] ibid.

would have no impact on the liberation of the imprisoned man, and it might generate important costs for Jason.[33]

Doing more than your fair share may be required if you are like Sandra, a rich woman who has the option of donating the amount of money that would be required to end poverty if everyone else donated their share, or donating more than that.[34] It looks like she would be required to donate more than her fair share, provided it wouldn't come at great cost to herself.[35] On the other hand, we might think it unfair that Sandra is demanded to do more simply because others are not complying with their moral obligations.[36]

Doing your fair share and nothing more might be required if you are like Hugh and dedicate five hours every week—which corresponds to his fair share—to campaigning for institutional reform.[37] If everyone else were like Hugh, institutional reform would occur faster, but they are not.[38] It is not clear whether Hugh should give more than five hours a week.[39] As Valentini notes, what is required is that everyone does five hours a week, not that Hugh does many more hours. It doesn't look like Hugh's doing more hours would make a difference.[40]

Who do combatants most resemble? Of course, the collective—the state, the armed group—might in fact be required to surrender. However, 'oughts' that apply to collectives are not immediately applicable to individuals.[41] Even if the collective ought to surrender, it doesn't follow that individuals also ought to do so.[42] This is true, argues Estlund, of essentially collective actions, that is, actions that cannot be performed by a single individual, like singing a G chord[43] or, we might add, waging war. However, members could nevertheless do their own part—surrendering or refusing to fight—which would be something else than the collective surrendering.[44]

The answer will depend on the combatant's situation. If the collective is required to surrender, soldiers who are higher up in the chain of command (and political leaders, for that matter) will have different obligations: they may be more like Sandra. They will be required not only to do their fair share

[33] ibid.
[34] ibid 655.
[35] ibid.
[36] *See* Murphy (n 27) 74–101.
[37] Valentini (n 28) 655–56.
[38] ibid.
[39] ibid.
[40] ibid.
[41] David Estlund, 'Utopophobia' (2014) 42 Philosophy & Public Affairs 113, 127.
[42] ibid.
[43] ibid.
[44] ibid.

and surrender but also to do more than their fair share and *get* people to surrender—or at least try. Now, the problem is that although refusing to fight might not be impossible or unreasonably difficult—and sometimes, it might be—it might be impossible or very difficult to *get* people to do so.[45] In some circumstances, this might make the obligation to get people to surrender supererogatory.

Other soldiers are likely to find themselves in Jason's situation: doing their fair share (surrendering) will not make the desired result (stopping the war) more likely, and it may entail great costs to the soldier and their loved ones.

Yet, some soldiers may be like Hugh, and thus required only to do their fair share (surrender or refuse to fight) and nothing more. However, when soldiers refuse conscription or surrender, someone else is likely to take their place. This situation might generate a reason not to refuse conscription in certain circumstances, as I will argue in section IV.B.

Although it is true that sometimes refusing or surrendering will not make a difference in stopping the war, it might make other kinds of differences. After all, people stand in a special relationship to what they do, and refusing to take part in an unjust war may preserve the agent's moral integrity, so to speak. When the decision not to participate harms no-one but the agent, this course of action is certainly available—perhaps not required, but admirable. But when the costs of not fighting are low or non-existent, refusing to fight or surrendering will be morally required.

In sum, it will not always be true that the best alternative available for combatants will be to surrender or refuse to fight. Sometimes, refusal will come at a high cost for soldiers and will achieve nothing or very little. And even when the costs to the soldier are low, there is the problem of someone else taking on the burden that they deflected, as we will see in section IV.B.

IV. Risking Your Life

Joining a war often comes at great personal risk to soldiers, especially to those who take part in combat:

> We sympathize with you. That's the paradox! We know that oppressed and demoralized young men were forced to take part in the war, but the fact remains

[45] ibid 126.

> that, even while you yourselves were dying, you were bringing death and destruction to another people. That was a crime rather than heroism.[46]
>
> What kind of people are we, and what right have we to ask our children to do the things they had to do there? How can we, who stayed at home, claim that our hands are cleaner than theirs? And although their suffering, their torture, has cleansed *them* of their sins, we have not yet been cleansed of ours. The machine-gunned and abandoned villages and ruined lands are not on their consciences but on ours. We were the real murderers, not they, and we murdered our own children as well as others.[47]

The fact that war is risky for those who participate in it is a particularly idiosyncratic feature of warfare—perhaps also present in some forms of policing, if to a lesser degree. And, as these passages from Alexievič's *Zinky Boys* suggest, the fact that it's risky might have an impact in shaping the reasons of combatants to fight in an unjust war.

Paul Kahn has argued that the feature of mutual risk provides a reason in favour of the symmetry thesis.[48] According to him, combatants have the licence to try to injure and kill each other as long as they stand in a relationship of mutual risk, which he takes to be a morality that constitutes a variation on the morality of individual self-defence. In other words, the soldier's privilege of self-defence is subject to the condition of reciprocity.[49] That reciprocity, which is grounded on the relationship of mutual risk, also serves to justify the principle of distinction, which does not separate the innocent from the guilty but the threatening from the non-threatening.[50] Chivalry also seems to be playing a role in the argument,[51] though it is not clear to what extent and how. Thus, if combatants are no longer threatening—and Kahn believes that the circumstances of asymmetrical warfare and killing with drones have precisely this result on the 'weaker' side—there is no longer a justification for killing, and warfare is no longer warfare: it becomes policing.[52] We do not make a moral demand, says Kahn, on police officers to risk their lives if there is no need for them to do so; in fact, policing is better to the degree that officers can accomplish their ends without risks to themselves.[53]

[46] Svetlana Aleksievič, *Zinky Boys: Soviet Voices from the Afghanistan War* (Julia Whitby and Robin Whitby trs, Norton 1992) 192.
[47] ibid 186.
[48] Paul W Kahn, 'The Paradox of Riskless Warfare' (2002) 22 Philosophy & Public Policy Quarterly 2.
[49] ibid 3.
[50] ibid 3–5.
[51] ibid 5.
[52] ibid.
[53] ibid 4.

The argument about reciprocal risk, as stated above, seems to reason in an incorrect order. Kahn's argument is that reciprocity is morally desirable in warfare, but not in policing. However, police activity is not defined by the fact that it is better to the extent that it is asymmetrical in its distribution of risks. Rather, the distribution of risks in policing is asymmetrical because there is a moral asymmetry between the police and the person who commits a crime: the first one is morally and legally justified in their actions, whereas the latter is not (or one hopes). The question about war, then, is precisely why risks should be distributed equally among combatants even in cases where their moral symmetry is lacking. We cannot begin the argument by stating that warfare is defined by reciprocal risk and that reciprocal risk justifies killing. The question is why there should be reciprocal risk at all, and if there should be, why that would justify killing.

Perhaps anticipating these issues, Kahn acknowledges that the rule of reciprocal self-defence cannot be justified by appeal to any of our moral intuitions, because 'it does not support the moral autonomy and dignity of the individual', so it stands as 'its own first principle within a circumscribed context in which individuals act in politically compelled roles'.[54] Yet people act in politically compelled roles in all sorts of situations that are not war, and it is unlikely that we would endorse a principle that failed to support the moral autonomy and dignity of the individual even in those cases, where the stakes are usually lower.

Of course, it might be highly unlikely or even impossible for there not to be reciprocal risks in war. But reciprocal risks are not *symmetrical* risks—far from it. Technological advances have been putting pressure on the feature of symmetrical risks for quite a while (at least since the invention of the machine gun), and, as a result, modern military tactics are in a large part geared towards manoeuvring the enemy into a position where they can be killed from safety.[55] The idea that symmetrical risks are constitutive of war is an illusion, in the sense that warfare aimed at winning will press towards reducing symmetrical risks—not maintaining them. This has nothing to do with whether warfare becomes policing; it is simply the way in which to fight a war with the lowest cost to one's forces. Combatants and military strategists will obviously try to avoid instances of symmetrical risks and to make them as asymmetrical as possible. This is not to say that risks are not reciprocal: they might be, but this does not entail that the risks will be equal on each side.

[54] ibid 3.
[55] Junger (n 23) 140.

Even if not symmetrical, war does involve significant risks for some combatants. And the question is whether the existence of those risks counts for anything in breaking the analogy between what unjust combatants do and murdering the innocent.

A first possibility is that they do not. The fact that war is risky for combatants can make no difference to their moral situation. Saying otherwise would be the equivalent of saying that if John decides to rob a bank, it must count for something in the assessment of the permissibility of his action that he might risk his life in doing so. This is rightly implausible: John shouldn't be robbing the bank in the first place. But this answer does not apply to all combatants—even if it does apply to combatants who are in something like John's situation.

Some combatants, as we have seen, have moral reasons for joining the war; some of them are unlike the villainous aggressor and unlike the villainous bank robber. Further, some combatants, I have argued, are not liable. Because they are not liable, they can also defend themselves and other, non-liable, individuals (section IV.A). Second, because of the existence of risks—even if not symmetrical—going to war is a heavy burden. This might have implications on the permissibility of deflecting that burden (section IV.B). A third possibility, which I will not explore here, is that combatants' facing risks does not count for permissibility, but that it may tell us something about expiation. We would be wrong, that is, to think of unjust combatants as murderers in need of punishment. Combat, we might say, is punishment enough.[56] This provides an argument in favour of maintaining combatants' immunity from prosecution—though likely in a more limited form—and against their criminal punishment, which could coexist with other arguments in favour of immunity. Ripstein, for example, believes that combatants' immunity from prosecution is demanded by Kant's idea of peace, and that states have no standing to punish other state officials—such as soldiers—for carrying out their official roles.[57]

A. Unjust Combatants, Counter-Defence, and Defence of Others

The paradigmatic case of self-defence—the villainous aggressor—is (morally) asymmetrical. When the villainous aggressor attempts to murder his innocent

[56] For a similar argument regarding the appropriateness of standard punishment, *see* Victor Tadros, *To Do, to Die, to Reason Why: Individual Ethics in War* (Oxford University Press 2020) 319.

[57] Arthur Ripstein, *Kant and the Law of War* (Oxford University Press 2021) 155.

victim, and the victim fights back, the villainous aggressor is not allowed to defend himself. Whether the villainous aggressor can fight back is usually referred to as the issue of 'counter-defence'. Most revisionists think that killing in war also has this asymmetrical nature: because unjust combatants are liable to be killed, they cannot engage in counter-defence.

The view I defended in Chapter VII qualifies this conclusion. If some unjust combatants are not liable, that is, if they haven't forfeited their rights against harm, then they are permitted to engage in counter defence and in defence of others who are not liable. That moral permission is partly explained due to the fact that the harm *wrongs* non-liable unjust combatants. It is also explained by a permission to give more weight to our lives and the lives of our loved ones than to the lives of others, given the special value that our lives have for us.[58]

That there are symmetrical cases of self-defence is controversial. Helen Frowe, for example, thinks that even those who are not liable cannot defend themselves against someone who has an objective moral justification for harming them. In war, this might mean that since just combatants have an objective justification for harming unjust ones (the just war), non-liable unjust combatants cannot defend themselves against the imposition of harm. Frowe thinks it implausible for people to become liable to harm for doing what is morally required.[59]

Frowe is right to think that it is implausible that people may forfeit their rights for doing what is morally required. But this does not provide a general objection against the possibility of there being symmetrical cases of self-defence. I have argued precisely that just combatants only have a moral permission to kill non-liable unjust combatants. I have not argued that in posing a threat to non-liable unjust combatants, just combatants forfeit for that very reason their rights. Non-liable unjust combatants may have an (agent-relative) moral permission to defend their lives and the lives of others, at least in some cases. This means that just combatants are wronged, and that they can also defend themselves.

Susanne Burri also finds agent-relative prerogatives to defend one's life against non-liable threats implausible. She argues that killing an innocent person simply because one values one's life more greatly is a gravely disrespectful thing to do; it is to make an exception of oneself in a morally

[58] *See* Jonathan Quong, 'Agent-Relative Prerogatives to Do Harm' (2016) 10 Criminal Law and Philosophy 815; Jonathan Quong, 'Killing in Self-Defense' (2009) 119 Ethics 507; Scheffler, 'Projects, Relationships, and Reasons' (n 3) 252; Cécile Fabre, *Cosmopolitan War* (Oxford University Press 2012).

[59] Helen Frowe, 'Lesser-Evil Justifications for Harming: Why We're Required to Turn the Trolley' (2018) 68 The Philosophical Quarterly 460.

objectionable way.[60] She thinks it is impermissible to depart from an impartial point of view and give more weight to one's own interests if the only way to do so is inflicting harm on someone who is not duty-bound to suffer for one's sake.[61]

But agent-relative prerogatives to inflict defensive harm are fundamental in explaining some cases that liability-based justifications cannot explain.[62] And although puzzling, they also have an intuitive pull. As Quong argues, each of us inhabits a particular perspective—our own—and as a result, our projects, our lives, are of particular importance to us.[63] In fact, our lives would be unrecognizable if we lived them through an impartial perspective; if we attached no special significance to our lives and projects.[64] Underlying these reasons, there is a claim for moderating agent-neutral demands, which arises out of understanding morality as a primarily reasonable and humane phenomenon, not entirely independent of the standpoint of the individual agent.[65]

The idea of duties to defend others in war has also been explored by Seth Lazar. He focuses on the associative duties we have towards the people we love—among which are duties of protection—and their importance for self-defence and defence of others.[66] He thinks that associative duties towards the people we love may provide a moral justification for killing non-liable people. Similar to Lazar, I think it is plausible that a father is morally required to divert a meteor so that instead of killing her daughter, it kills five other people,[67] although it is possible that he may be justified (not required) to do so, or simply, morally permitted. On the idea that morality cannot be as demanding as to require people to sacrifice the lives of their loved ones for the greater good, I also agree with Lazar.[68]

If agent-relative prerogatives can be defended successfully, they can also be employed to explain why some non-liable unjust combatants might defend

[60] Susanne Burri, 'The Toss-up between a Profiting, Innocent Threat and His Victim' (2015) 23 Journal of Political Philosophy 146, 163–64.

[61] ibid 163.

[62] Quong, *The Morality of Defensive Force* (n 5) 69–70.

[63] ibid 70.

[64] ibid.

[65] *See* Scheffler, *Human Morality* (n 8) 6.

[66] *See* Seth Lazar, 'Associative Duties and the Ethics of Killing in War' (2013) 1 Journal of Practical Ethics 3.

[67] ibid 18–19.

[68] Nevertheless, as Liam Murphy notes, it is not easy to set an uncontroversial criterion of what makes a moral demand over-demanding. In fact, it is notoriously difficult to set a baseline in comparison to which to assess the demanding nature of moral demands. Moreover, the baseline is likely to be biased in favour of the normative status quo. Here, the problem is best seen, as discussed by Nagel, as a clash between agent-neutral and agent-relative reasons or viewpoints. *See* Murphy (n 27) 34–62; Nagel (n 6) 202–05.

themselves and other non-liable individuals when they are under attack. Of course, agent-relative prerogatives to inflict defensive harm must have significant limits, and part of the challenge is coming up with them. Quong, for example, argues that the means principle limits the scope of agent-relative prerogatives.[69]

This moral permission to defend oneself and others in war is important, but it is also a modest claim. It is important because many people take part in war, and military units operate not just by individuals defending themselves, but by individuals defending each other. Less obviously, it is important because combatants do not usually think it is relevant that they are allowed to defend themselves: they are much more concerned about defending the members of their unit.[70] The study of modern war suggests that soldiers are not motivated to kill or fight in defence of themselves or in defence of some particular ideology but by other factors, among which the defence of their comrades in arms takes a central place.[71] And as we will see later, this loyalty might make even more sense or be stronger when there is a background of unjust and coercive institutions and military authorities that have put these individuals in that position.

However, that (non-liable) unjust combatants may defend themselves and others who are not liable is a very modest claim. It is modest because it covers very little of the symmetry thesis, which allows for far more than simply combatants defending themselves and individuals whose lives they hold dear. And it is modest because their permission is not, unlike the case of just combatants, accompanied by reasons related to a just cause for war. While just combatants are contributing to something that will presumably achieve some good (restoring peace, stopping genocide), unjust combatants are usually not. Nevertheless, those unjust combatants who are not liable and thus defend themselves or others are not *like* murderers. That is, the argument does explain how and why just combatants are not just like the innocent (sometimes they pose threats against those who are not liable) and why unjust combatants are not just like murderers (some of them are exercising agent-relative prerogatives to defend themselves and others who are not liable).

[69] Quong, *The Morality of Defensive Force* (n 5) 80–85.
[70] This has been documented many times. *See* Dave Grossman, *On Killing. The Psychological Cost of Learning to Kill in War and Society* (Open Road Integrated Media 2014); Junger (n 23); Finkel (n 22).
[71] *See* Grossman (n 70) section VII.

B. It's You or Me: Doing Your Share and Deflecting Harm

Soldiers sometimes explain their decisions to partake in war or in certain missions by pointing out that if they refused to do so, someone else would have to take their place. This is a particularly prevalent theme, as these quotations illustrate:

> No one ever feels like they are doing enough. If you are in a safe location, you feel guilty that your friends are getting shot at and you aren't. If you are getting shot at, you feel guilty if your buddy gets hit and you don't. If you get shot at but don't die, you feel guilty that you lived, and more guilty if you get to go home and your friends have to stay behind. I have not seen one person out here who didn't [check off] 'increased guilt' on our intake form.[72]
>
> When his mother wrote to express her considerable alarm at this attitude, [Private Ken] Webster [from the 101st Airborne] replied, 'Would you prefer for somebody else's son to die in the mud? ... Somebody has to get in and kill the enemy. Somebody has to be in the infantry and the paratroops. If the country all had your attitude, nobody would fight, everybody would be in the Quartermaster. And what kind of a country would that be?[73]
>
> How can I stay here when my friends are over there?[74]
>
> My neighbor kept getting at me—and perhaps she was right. 'Couldn't you scrape a couple of thousand of roubles together and bribe someone?' We knew a woman who did exactly that, and kept her son out. And my son had to go instead.[75]
>
> My God, what have we been reduced to? If he hadn't gone some other boy would have had to take his place, some other boy with a mother....[76]
>
> Why you? You've got a baby daughter.
>
> If I don't go, someone else will have to take my place.[77]

We can think about the argument that underlies these passages in two ways.[78] First, we might think of it as an argument for doing our share within a group or community. This argument is of limited application. Second, we might think

[72] Lisa R. Blackman, 'Alarm Red' in Andrew Carroll (ed), *Operation Homecoming: Iraq, Afghanistan, and the Home Front, in the Words of U.S. Troops and Their Families* (University of Chicago Press 2008) 307.

[73] Ambrose (n 1) 271.

[74] Aleksievič (n 46) 150.

[75] ibid 122.

[76] ibid 108.

[77] ibid 104.

[78] I am indebted to Moshe Halbertal for pointing me in the direction of this argument.

it related to the discussion on the permissibility of deflecting unjust harms or burdens towards the innocent. This strand of the argument is more promising, but works better as the foundation of an evidence-relative permission to fight.

Let's start with the idea of doing our share in our political community or in a smaller group (such as the army or the armed group the agent belongs to). The idea of doing our share as members of a group, and in particular, the idea of doing our own share as members of the state, has been the object of much discussion. Different accounts have been provided in support of the obligation to do our share within a group. Some accounts rely on the principle of fairness,[79] others on the idea of associative obligation,[80] and others on the idea of a natural duty to support just institutions.[81]

In the case of participating in an unjust war, the question is whether citizens facing conscription and soldiers facing deployment may have a duty to accept this burden, because refusing to do so would be refusing to do their share in supporting the group, contributing to its purposes, and the fulfilment of its interests (whether we justify their duty to do their fair share on the principle of fairness, associative obligations, or the existence of natural duties). In the case of soldiers, the argument might apply at several levels: the obligation to do their fair share may be owed to the political community (the state), the army or the armed group, or the combatant's unit.

The argument requires to show, first, that deployment or conscription are ways of doing our fair share, and second, that refusing to do our fair share would be a refusal to contribute to the groups' interests and purposes. Regarding the second point, arguably among the interests of a state or army, there are interests in national security, subsisting as an autonomous political community, and so on.[82] Additionally, as suggested by Benbaji, we might argue that efficient and well-ordered armies are necessary for protecting people's right to life and for minimizing the extent to which soldiers and civilians are vulnerable to threats, given that we live in states that form an anarchical society and that threats are a permanent aspect of this reality.[83]

Provided that the group is not evil or corrupt and that the norms that define what constitutes our fair share are neither gravely unjust nor irrational,[84] doing

[79] *See* Russell Hardin, The Free Rider Problem (*Stanford Encyclopedia of Philosophy*, 21 May 2013) <https://plato.stanford.edu/archives/spr2013/entries/free-rider/> accessed 22 August 2022.

[80] *See* Samuel Scheffler, 'Membership and Political Obligation' (2018) 26 Journal of Political Philosophy 3.

[81] *See* Jeremy Waldron, 'Special Ties and Natural Duties' (1993) 22 Philosophy & Public Affairs 3.

[82] *See* Yitzhak Benbaji, 'A Defense of the Traditional War Convention' (2008) 118 Ethics 464, 490.

[83] ibid.

[84] Scheffler, 'Membership and Political Obligation' (n 80) 6.

our fair share may require us to accept conscription or deployment, even in pursuit of an unjust war. For the institution itself (ie the army or the state) must be just; it is not necessary that all the actions that the institution performs are just themselves. Or so the argument might go.

It is plausible that having armies is important in deterring and minimizing threats against civilians and combatants and thus, that armies are institutions that we ought to support—although this is something that requires evidence. Moreover, the evidence should support the conclusion that having efficient and well-ordered armies is not only important in deterring threats and protecting rights, but also that it is *better* than other alternatives, which might not require such coercion. Evidence is also needed to support that blind obedience is necessary for armies to function well and efficiently. Both of these conditions are difficult to meet.

Nevertheless, beyond what might be demonstrated empirically, it is not at all clear why doing our fair share in this case would take the form of accepting conscription or deployment and thus, participating in an unjust war. Wouldn't our fair share require *precisely* refusal to fight if we believe that the war is unjust? Would just institutions collapse if people refused to fight unjust wars? And what is wrong with not doing our fair share?

Not doing our fair share might be wrong because it damages a valuable institution or practice.[85] However, sometimes not doing our fair share has little impact: your refusal to do your fair share will not harm the institution in any perceptible way. It would take many refusals to do so. Once again, this is the problem of collective harm. To the extent that states and armies are valuable institutions, and that refusing to fight in an unjust war may endanger their subsistence, there may be a *pro tanto* reason not to deflect the burden. (The burden is so demanding that we may just want to say that it will be permissible to fight and supererogatory to do so.) However, this argument relies on several conditions, most of which are exceedingly difficult to meet outside extraordinary circumstances. In fact, I do not think the argument can be successful under the present circumstances of many states and armies, where these burdens are rarely fairly distributed. Further, it is highly unlikely that just institutions would collapse if individuals refused to participate in an unjust war, and one might wonder how just those institutions truly are if they are engaged in unjust wars to begin with.

[85] *See* Rawls (n 27); HLA Hart, 'Are There Any Natural Rights?' (1955) 64 The Philosophical Review 175.

More importantly, the argument, as expressed most commonly by soldiers, is not an argument about the importance of sustaining an institution or practice. It is an argument about what is owed to other individuals who are part of the same political community and who will have to take the place of those who refuse. Thus, it is not merely—and perhaps not even primarily—an argument about sustaining valuable institutions and the imperceptible harms we may cause if we don't. It is an argument about the very perceptible harm that refusal may inflict on someone else. This strand of the argument applies not just to deflecting just or fairly imposed burdens; it may also apply to the deflection of unjust ones. It is an argument that has force, as we will see, also, and even more so, against the background of unjust institutions and evil and corrupt groups.

When applied to war, the argument of doing your fair share has different characteristics from the usual examples of disobeying the law or refusing to keep your promises, which make it quite peculiar. The first one is that if someone refuses to participate, someone else will have to take the refuser's place. This effect—of someone else having to 'take up the slack'—is unusual, although not unique to war. If I refuse to pay my taxes, for instance, no-one will be called instead of me to pay what I did not;[86] if I refuse to recycle, no-one will have to do my recycling for me. However, if I refuse to fulfil my jury duty, someone else will have to take my place and do it for me. War shares this feature with jury duty.

The second peculiar characteristic concerns the nature of the burden itself. Doing our fair share is usually burdensome or costly for us. Yet, there are few things as burdensome as going to war. After all, the person who will take the place of the refuser will not have to sit for days in the jury. That person might die and kill because she took the refuser's place.

The demandingness of the burden is a feature that can go either way in terms of the argument. If the burden is just and has been fairly imposed on you, it could make it even *more* obligatory to carry it. If you don't, someone else will die. But it might also work in the opposite direction: if you go, you might die. Can morality be as demanding as to require you to do your fair share in these circumstances? Maybe, maybe not. But in either case, accepting the burden is at the very least permissible.

If the burden is imposed unfairly, the group is corrupt, the institution to be sustained is unjust, and so on, are agents allowed to refuse it and deflect it towards somebody else? Or are they obligated to suffer it? The answer is, again,

[86] Although it is possible that if many people don't pay their taxes, taxes will need to be higher than what they would need to be if everyone paid their taxes.

far from obvious. Here, it seems like we are dealing with something that looks partly like one of the many versions of the trolley problem. The train is coming towards you. You may press a switch that will send the train towards another track, where it is certain it will kill somebody else. Can you press the switch?

The similarities with the trolley example are, of course, limited, but nonetheless illuminating. It is absurd to think that refusing to participate in an unjust war is simply like turning the trolley so that it kills somebody else. Participating in an unjust war will also entail making contributions to the achievement of something bad; it might involve causing great harm against civilians and just combatants. Yet participating in a war, whether just or unjust, comes at great cost for those who participate in it. It would also be absurd to deny this.

The argument for not deflecting an unjust burden has force even—precisely—if the group that imposes the burden is corrupt, evil, or both, and even if the norms that define what constitutes our fair share are gravely unjust and irrational. This is of particular importance in today's armed conflicts, in which terrorist groups and authoritarian states participate. The argument will also have force for soldiers belonging to armies in democratic states, especially if we think that the unjustness of the war is sufficient to render a fairly imposed burden into an unjust one.

Think about a group of terrorists that is holding five people hostages. The terrorists tell the hostages that they have to burn the bodies of those who have died but Peter refuses to do so. He says he will not contribute in any way to the terrorists' goals. The rest of the hostages complain to Peter: there are 20 bodies to move and burn. If they do not burn them, they will have to share the tiny space where they are confined with the corpses until they are rescued (if they are at all). If Peter doesn't help, the work for them is greatly increased, but Peter won't budge.

In the above situation, even if paradoxical, it seems plausible that the four hostages are right in complaining to Peter. They might, in fact, have a claim against him: that he shares in the burden of these grossly unjust duties that have been imposed on them. That he does his fair share, or that he does not deflect the unjust burden towards them. In doing so, he is harming the rest of the hostages. Nonetheless, Peter doesn't want to get his hands dirty, morally speaking. He wants to preserve his moral integrity. He might claim that he stands in a special relationship to the things that he himself does, and that he will not contribute to evil, not even by moving the bodies.[87] However, the hostages also

[87] On the idea of moral integrity, *see* Bernard Williams, 'Ethical Consistency' (1965) 103 Proceedings of the Aristotelian Society 103; JJC Smart and Bernard Williams (eds), *Utilitarianism: For and Against* (Cambridge University Press 1973).

seem to be right to complain to him. Of course, Peter is correct in claiming to stand in a special relationship to the things that he does. And he is also right about not wanting to contribute to evil. But his claim is not against the world at large, and it is not primarily against the rest of the hostages. His claim is against the terrorists. The hostages stand in no special relationship to Peter's claim for preserving his moral integrity by not contributing to evil—they are not the ones who have put his moral integrity at stake. They do, however, stand in a special relationship with Peter and his refusal to cooperate in what is a burdensome task for everyone. In this small moral world,[88] Peter might very well owe it to his fellow hostages to share in this burden—to burn the bodies. This is, of course, not plausibly owed to the terrorists or the terrorist organization. The terrorists can certainly not complain that Peter won't do his (un)fair share.

Thus, even in terrorist groups and authoritarian states some individuals—those coerced or conscripted to fight—have a *pro tanto* reason not to deflect conscription. Of course, sometimes that reason will be defeated by other, more weighty reasons. And if combatants live in a society where everyone can freely refuse to participate, then none of them should accept the burden to fight in what they reasonably believe to be an unjust war. This, however, is not how things are. It is not unusual for unjust combatants to face choices where they are certain that their refusal will harm someone else, but significantly uncertain regarding their role in the war—whether they will face combat or not—and the justness of their and the enemy's war. In those cases, they might have evidence-relative permissions to fight.

V. Conclusion: The Small Moral World of Combatants

All these reasons for fighting or killing are only *pro tanto* reasons for doing so. Reasons for fighting and killing will be at their most powerful when wars with a just cause are fought unjustly. This is far from an uncommon situation. Further, many non-liable unjust combatants will have agent-relative prerogatives to defend themselves when under attack. Regarding the rest of the reasons explored in this chapter, we should not overstate their power to make fighting in an unjust war permissible or justified. Most often, they will operate as evidence-relative prerogatives to participate in wars that are not obviously or egregiously unjust.

[88] *See* Korsgaard (n 2).

When many of the reasons explored above are present (or agents have good reason to believe that they are present), unjust combatants will be permitted to participate in an unjust war. Sometimes, killing just combatants will also be permissible, excused, and (more rarely) justified. Sometimes, refusing to fight or surrendering will be supererogatory. Other times, not surrendering will be supererogatory. As Tadros has pointed out, no simple account can be given about the permissibility of fighting and killing in an unjust war.[89]

In some ways, I have argued that agent-neutral demands cannot be excessively demanding and require the sacrifice of our lives and the lives of our loved ones for the greater good; that morality may allow that we place more weight—relative to their agent-neutral weight—on our lives and the lives of our loved ones. But even if morality demanded these sacrifices of us, it would remain an open question whether law should be similarly demanding. Even if we have our doubts about whether morality is, indeed, moderate, the argument for moderating the claims of law seems to me stronger. It is one thing to say that morality demands the ultimate sacrifice, and a very different one to say that if so, then we should set up institutions that would enforce, via criminalization or sanctions, the demand to sacrifice one's live in the name of that morality, particularly because the sacrifice of one's live will be to further someone else's goals, to save someone else's life—not one's own.

The reasons provided in this chapter can thus attenuate the fragility of the instrumentalist justification but, inevitably, to a limited extent. Although there can be reasons for participating in an unjust war and even for killing in self-defence and defence of others, participation in an unjust war will also call for actions to defeat the just side, and it will involve targeting just combatants who are innocent, who are not posing threats, or who are contributing to the achievement of peace and victory. The highest the position of power an unjust combatant finds herself in, the more important the strategic value of the place she is defending or trying to gain control over, the harder it will be to judge these actions as permissible in any relevant sense.

Yet besides criminalizing aggression at the highest levels of command, international law does not make these distinctions. The laws of war grant a legal privilege to both just and unjust combatants, in circumstances in which not *every* unjust combatant will have a moral permission to do what they are doing. This divergence between law and morality is not as deep as revisionism suggests if, as I have argued, what some unjust combatants do is not just like murdering the innocent. But this divergence is still quite troubling, for there is a

[89] Victor Tadros, 'Unjust Wars Worth Fighting For' (2016) 4 Journal of Practical Ethics 55.

legal permission in place for many individuals in high and mid-level positions of command to pursue unjust goals.

This divergence between law and morality can be justified, if only contingently, under an instrumentalist rationale, such as the one I provided in Chapter III. And if I have been able to successfully deconstruct the analogy between what unjust combatants do and murder, even if partially, the moral fragility of this instrumentalist justification is then somewhat attenuated. Nonetheless, there is more to say about that divergence.

Combatants, as things are now, inhabit a small moral world:[90] surrendering and refusing to fight are almost always exceedingly costly for agents and almost always achieve very little in terms of their ability to undermine an unjust war effort. Unjust combatants face conflicts in which agent-neutral and agent-relative moral demands pull them in opposite directions. But although some features of warfare, such as its collective nature, seem fundamental in shaping that moral world and are almost inescapable, the size and shape of the moral world that we inhabit is also determined by our political and legal institutions. As such, it is not immutable.

As Samuel Scheffler points out, the extent to which conflicts arise in our lives is not fixed once and for all: it is subject to human control.[91] Prevailing social institutions and arrangements have an important role to play in shaping the moral landscape and reducing the degree of conflict between what now seem like two unattractive alternatives. That is, the degree of conflict between agent-neutral and agent-relative demands is partly a social and political issue. As Scheffler writes:

> Human institutions can make moral decency more or less costly, and more or less difficult to achieve, and in these and other ways they can either discourage or nourish it.[92]

We should thus hope that our laws of war would not make it hard or incredibly costly for people to do what is right—or what is closer to the perpetual peace ideal—which in the case of most unjust combatants will be precisely refusing to fight or surrendering. Nonetheless, at the moment, and in many ways, our laws of war do have these effects. This is especially true of unjust combatants in high- and mid-level positions of command, who are usually in the position of

[90] As mentioned before, here I am not using the idea of a 'small moral world' in the same way as Korsgaard does. *See* Korsgaard (n 2) 46.

[91] Scheffler, *Human Morality* (n 8) 132.

[92] ibid 145.

making significant contributions to war, hold the lives of hundreds of people in their hands, and face significant costs if they decide to surrender or not follow orders. In their cases, reasons related to the unjustness of the war will weigh more heavily than for those lower in the chain of command and, thus, may counsel surrender or refusing to fight, yet the law seems to be hardly concerned at all with reducing the costs of refusal.

Indeed, our current laws of war do not concern themselves too much with how states, armies, and armed groups treat their own members. Nor are they concerned with giving guidance or reliable information to combatants about what makes a war worth fighting or with ways of making surrender less costly for them, outside of prisoner of war provisions.[93] These rules, as it should be obvious by now, are incredibly important. Soldiers who know they can surrender safely will be more inclined to do so.[94]

And yet this issue doesn't only concern the prohibition on declaring that no quarter will be given[95] and prisoner of war provisions,[96] both of which focus on how armies and states should treat enemy combatants. It is also a matter of how armies and states treat their own members who desert or refuse to fight or follow orders. At the moment, soldiers do not usually have a right to object deployment or assignment to a particular armed conflict or mission on moral or legal grounds.[97] Surprisingly, the vast majority of 8,000 Nazi soldiers who survived the Second World War with desertion convictions lived the rest of their lives as convicted felons; political clemency failed to come until the new millennium.[98] And perhaps it is no coincidence that international instruments aiming at the abolition of the death penalty include the possibility of reservation for its application 'in time of war pursuant to a conviction for a most

[93] *See Geneva Convention Relative to the Treatment of Prisoners of War (Third Geneva Convention)* (hereinafter, GCIII) 12 August 1949, 75 UNTS 135.

[94] *See* some examples in Ambrose (n 1) 342, 357.

[95] *Protocol Additional to the Geneva Conventions of 12 August 1949, and relating to the Protection of Victims of International Armed Conflicts (Protocol I)* (hereinafter, API) 8 June 1977, 1125 UNTS 3 art 40; *Protocol Additional to the Geneva Conventions of 12 August 1949, and Relating to the Protection of Victims of Non-International Armed Conflicts (Protocol II)* (hereinafter, APII) 8 June 1977, 1125 UNTS 609, art 4(1).

[96] GCIII.

[97] On the latter, *see* David Whetham, 'Do Some Soldiers Deserve to Die More Than Others?' in Bradley Jay Strawser, Ryan Jenkins, and Michael Robillard (eds), *Who Should Die?: The Ethics of Killing in War* (Oxford University Press 2018); Nathaniel Berman, 'Privileging Combat-Contemporary Conflict and the Legal Construction of War' (2004) 43 Columbia Journal of Transnational Law 1; *United States v Gillette*, 401 US 437 (1971).

[98] *See* Tom Dannenbaum, 'Why Have We Criminalized Aggressive War' (2016) 126 Yale Law Journal 1242, 1310–11. Nonetheless, I leave open the possibility that refusal to fight, surrender, and desertion might merit different treatment from both the moral and the legal standpoint.

serious crime of a military nature committed during wartime'.[99] Occasionally, this means desertion.[100]

A legal framework which would give more freedom to combatants in deciding whether to fight would certainly go beyond what the laws of war and, for that matter, what international law can do. It also concerns, more generally, the political institutions of a state, its constitution, and its laws. Such a framework is not as implausible as it sounds. For example, UK Trooper Ben Griffin served in Iraq with the elite Special Air Service and came to believe that the war was both illegal and immoral.[101] Rather than making a public protest, he explained his position to his commanding officer, expecting to be court-martialled for refusing to return to Iraq.[102] However, he was permitted to leave with an unblemished military record and a glowing testimonial.[103]

Like Dannenbaum, I think there is a powerful argument for providing combatants with a legal privilege to refuse to participate in unjust wars.[104] However, such a privilege would not be enough on its own. It would require extensive institutional arrangements so that it could be realistically exercised by combatants, hopefully even before war is initiated. These arrangements might include asylum or refugee-like protections, granting citizenship to combatants who refuse to fight, and so on. It would also require a reassessment of international laws on the resort to force and further—legal and institutional—guidance on what makes a war unjust.[105]

A framework that would effectively permit combatants to refuse to fight or to refuse certain deployments might make it more difficult for states to fight wars. Benbaji fears that a regime that obliged soldiers to ascertain whether

[99] *Second Optional Protocol to International Covenant on Civil and Political Rights, Aiming at the Abolition of the Death Penalty*, UNGA resolution 44/128, 15 December 1989, art 2. Similar language is used in *Protocol No 6 to the Convention for the Protection of Human Rights and Fundamental Freedoms (European Convention on Human Rights) Concerning the Abolition of the Death Penalty*, 1 March 1985, ETS 114 (a later protocol changed that language); *Protocol to the American Convention on Human Rights to Abolish the Death Penalty ('Pact of San Jose')*, 8 June 1990, OAS Treaty Series, No 73.

[100] Tom Dannenbaum, *The Crime of Aggression, Humanity, and the Soldier* (Cambridge University Press 2018) 123–25. Dannenbaum points out that forty-four of the forty-seven Council of Europe states have ratified a protocol that commits them not to apply the death penalty, even in war. The United States contemplates the death penalty for deserters, although it has been applied only once since the Civil War.

[101] Whetham (n 97) 72.

[102] ibid.

[103] ibid.

[104] Dannenbaum has argued in favour of providing a right to disobey to soldiers if they are asked to participate in aggressive wars, similar to the kind of privilege and related protections that soldiers currently have to refuse orders that amount to war crimes. His account would limit this privilege to aggressive war. My argument is broader in this respect, since it would also require rethinking our *jus ad bellum* norms. *See* Dannenbaum (n 98) 1306–12.

[105] McMahan, for example, has suggested the creation of an international court. *See* Jeff McMahan, 'The Sources and Status of Just War Principles' (2007) 6 Journal of Military Ethics 91, 104. I would add the need to rethink the *jus ad bellum*.

their wars are just would have the effect of making armies unable to fight just wars efficiently, to form a credible deterrent threat, to wage a just pre-emptive attack, and to provide efficient national defence.[106] This might be so.

However, the world we live in is not a world where there is too little obedience within armies. We cannot forget the atrocities that some armed forces have been guilty of in the recent past: members of the armed forces in some Latin American countries obediently carried out grossly unjust orders to torture, disappear, and murder people;[107] members of the US intelligence services and armed forces have engaged in torture;[108] and members of Myanmar's armed forces and police are dutifully engaging in the genocide of the Rohingya.[109] These examples are all too common. Anyone who has worked attempting to prosecute crimes committed by armies and other militarized institutions where obedience is paramount knows all too well how costly it is for agents to disobey even grossly unjust orders, and how strong misguided loyalty among members can be. Now I am not saying that a very entrenched culture of obedience is the only factor to blame here. The point is, rather, that there is no shortage of obedience in armies today. If anything, there is too much of it, and little is done to encourage or facilitate even what we would all agree are instances of rightful disobedience. Thus, the danger we should be concerned with in our time is emphatically not the danger of too much disobedience. Perhaps future generations will have to worry about that, if we succeed in making the current arrangements less coercive.

It is also worth repeating here that alternative institutional arrangements always seem impossible and idealistic until we have them and they work. Armies that do not require blind obedience might be as efficient as the armies we have now. And we should not underestimate the power that peer pressure has in tight-knit groups such as the army, nor the attraction that combat exercises over some individuals. In any case, making wars harder for states to fight—if it came to happen—would not be a bad thing. Senseless wars are fought all the time.

Law is helpless in countless ways. International law and institutions are also helpless in many ways. They may be helpless, particularly, in changing our

[106] Benbaji (n 82) 490.

[107] *See*, eg, Jeffrey Davis, *Seeking Human Rights Justice in Latin America: Truth, Extra-Territorial Courts, and the Process of Justice* (Cambridge University Press 2014); Elin Skaar, Jemima García-Godos, and Cath Collins (eds), *Transitional Justice in Latin America: The Uneven Road from Impunity towards Accountability* (Routledge 2017).

[108] *See*, eg, Karen J Greenberg and Joshua L Dratel (eds), *The Torture Papers: The Road to Abu Ghraib* (Cambridge University Press 2005).

[109] *See* UN Human Rights Council, Report of the Detailed Findings of the Independent International Fact-Finding Mission on Myanmar, UN Doc A/HRC/39/64 (12 September 2018).

moral convictions, at least in the short term. But law and institutions are also powerful tools in shaping the universe within which our moral lives dwell. This is why political solutions are important. They can make it less costly for agents to do what is required; that is, they can attenuate the conflict between opposing moral demands; they can effectively diminish the costs of surrender or refusal to fight. No-one should have to risk losing everything for refusing to fight in a war she believes to be unjust.

In the end, we must ask ourselves what value there is in upholding the current arrangements of most armies and states, which allow some individuals, on behalf of the state, to employ the bodies and lives of thousands of people to their (wrongful) ends. We may as well wonder why we have allowed for this, and why we have allowed for its coerciveness.

Conclusion

The Peace That Was Promised

A recurrent theme in Svetlana Alexievič's *The Unwomanly Face of War* is the expectation, and hope, about what would come after the suffering brought about by the Second World War—after victory. For example:

> We all imagined that after the war, after such oceans of tears, there would be a wonderful life. Beautiful. After the Victory ... after that day ... We imagined that all people would be very kind, would only love each other. They would all become brothers and sisters. How we waited for that day....
>
> —Olga Vasilyevna[1]

Tamara Stepanovna Umnyagina shares that hope but realizes it might not materialize:

> Do you know what thought we all had during the war? We dreamed: 'If only we survive ... People will be so happy after the war! Life will become so happy, so beautiful. People who have been through so much will feel sorry for each other. They'll love each other. They'll be changed people.' We never doubted it. Not a bit.
>
> My precious one ... People still hate each other. They go on killing. That's the most incomprehensible thing to me....[2]

This theme—the pursuit of peace—also occupies an important role in the instrumentalist justification I have provided for the symmetry thesis. In fact, in a way, the main goal of this book has been to explore ways in which the fragility of that instrumentalist justification might be reduced. I have done so by proposing an account of killing in war that differs from revisionism, and by

[1] Svetlana Alexievič, *The Unwomanly Face of War* (Random House Trade Paperbacks 2018) 157.
[2] ibid 330.

The Morality of the Laws of War. Marcela Prieto Rudolphy, Oxford University Press. © Marcela Prieto Rudolphy 2023.
DOI: 10.1093/oso/9780192855473.003.0009

denying, partially, the analogy between what unjust combatants do and the murder of the innocent.

The non-instrumentalist part of this justification relies on an assortment of moral justifications or permissions that individuals might have in the context of war. In the case of just combatants, I argued that they might have liability-based justifications for killing in self-defence, agent-relative prerogatives to defend themselves and others, and a *sui generis* permission to target non-liable unjust combatants. In the case of unjust combatants, the picture I painted was significantly more complex. I argued that some unjust combatants have a moral permission to participate in an unjust war and sometimes to kill during it based on reasons of democratic authority, duress, fairness towards members of their own group, and the fact that their choices are limited. I also argued that non-liable unjust combatants have a moral permission to defend themselves and their loved ones. The situations where these reasons fall short—as in fact, I have said, they do—can be covered by the instrumentalist justification I provided in Chapter III.

A significant omission of this book is that it does not explain why the legal privilege to kill belongs *only* to combatants who meet certain requirements (ie privileged combatants). What makes unjust combatants permissible targets, in fact, is largely independent of whether a combatant or civilian is on the other side, doing the targeting. Although I have left this very important issue aside, the justification I have provided can explain why targeting combatants is not an international crime, even if it will constitute a domestic one when it is undertaken by civilians and unprivileged combatants. This fact, perhaps, is an argument in favour of this interpretation of the legal rules: that *everyone*, both civilians and combatants, has a legal privilege to target combatants. However, the reasons I have provided pertaining to role morality and the division of labour suggest that combatants may have reasons available for participating in war that will not be often available to civilians. Nevertheless, I leave open whether this feature of our laws is justified and to what extent.

Now, even if I have successfully attenuated the fragility of the instrumentalist justification I proposed by contesting the revisionist morality of war, it is still an open question how significant this attenuation is. In fact, we may not be that much better off than where we started.

First, because by contesting the revisionist morality and arguing that most unjust combatants haven't forfeited their rights, I have effectively argued that killing in war, whether a just or unjust one, is likely to infringe the rights of both just and unjust combatants. In many cases, it will only be permissible to kill them, and so killing both just and unjust combatants will wrong them and will wrong them grievously. Thus, I have effectively increased the fragility of the *morality* of killing in war.

Someone might thus think that I have only managed to *increase* the moral fragility of the instrumentalist justification I provided in Chapter III. Under revisionism, most unjust combatants have forfeited their rights. As a result, when just combatants kill unjust combatants, they do not wrong them; rights are not violated nor infringed. Just combatants' rights, of course, are violated when unjust combatants kill them. If revisionism is the correct moral theory about killing in war, the symmetrical nature of the legal privilege of combatants means that our laws of war permit people (unjust combatants) to murder the innocent (just combatants).

Under my account, our laws still permit *some* unjust combatants to do what is the equivalent of murdering the innocent. But some killings in war, undertaken by either just or unjust combatants, are not like this; they are not just like the murder of the innocent. They infringe rights, but they don't violate them.

The objection, then, would be that in my account of the morality of war, the number of rights infringements is higher than the number of rights violations that take place under the revisionist morality, and that either a higher number of rights infringements is always morally worse than a lower number of rights violations, or that there is a point where the number of rights infringements becomes high enough to be morally worse than a lower number of rights violations. Even if we accepted these two controversial propositions—that any higher number of rights infringements is always morally worse than any lower number of rights violations or that there is a point where the number of rights infringements becomes high enough to be morally worse than a lower number of rights violations—it is unclear what follows from them. On the one hand, and most obviously, they do not present any kind of objection against the morality of war I have proposed: it is not a failure of the view that rights on both sides of a war are infringed. On the other hand, it does put into question whether I have reduced the fragility of the instrumentalist justification at all. I may have done the opposite, at least if we accept that there is a point where rights infringements are high enough to constitute a state of affairs that is morally worse than a state of affairs where fewer rights violations occur. Note, however, that this is a strange claim to make, for it would require arguing that there is a point where *morally permissible* rights infringements are high enough to be worse than fewer *morally prohibited* rights violations.

However, the issue here is not best understood as a problem of how many rights violations or infringements are allowed under a given set of laws—at least not in this case. For the laws of war do not really allow for a given number of rights violations/infringements to occur; they only allow for certain conduct (ie targeting combatants) and the conduct allowed may, if I am correct about the morality of war, violate or infringe rights, depending on the circumstances. In other words, the *number* of rights violations or infringements that occur in any given armed conflict are largely contingent; they will depend on certain features of the conflict, and they may, within certain circumstances, affect the permissibility of resorting to war if those numbers can be reasonably predicted *ex ante*. By contrast, under the revisionist morality, the laws of war decidedly allow for the massive violation of rights by allowing unjust combatants not only to do what is morally prohibited, but by allowing them to do what in effect is the moral equivalent of murdering the innocent. And the point about reducing the fragility of the instrumentalist justification of the laws of war is best understood as a point about how laws that permit the murder of the innocent are morally worse than laws that permit for people to infringe each other's rights by doing what is morally permissible.

The second reason why it is still an open question how significantly I have managed to attenuate the fragility of the instrumentalist justification—and why we might not be much better off, morally speaking—relates to certain features of the instrumentalist justification itself. For both the appeal of the justification I have provided and its fragility lie precisely in the ability of the laws of armed conflict to lead us in the direction of moral progress, in the direction of peace. They have to *work*. But have they?

Between 480,000 and 507,000 people were killed from 2001 to 2018 in the armed conflicts taking place in Iraq, Afghanistan, and Pakistan.[3] More than 500,000 people have died in Syria since 2011.[4] As a result of these wars, around 21 million people have become refugees or have been internally displaced from their homes.[5] In February 2022, Russia began an invasion of Ukraine that has displaced millions of people and killed thousands. Fourteen million people remain at risk of starvation and death in Yemen, largely as a result of the armed conflict.[6] We don't really know the number of drone strike casualties.[7] More

[3] *See* Neta C Crawford, 'Human Cost of the Post-9/11 Wars: Lethality and the Need for Transparency' (Watson Institute, Brown University, November 2018) <https://watson.brown.edu/costsofwar/files/cow/imce/papers/2018/Human%20Costs%2C%20Nov%208%202018%20CoW.pdf> accessed 22 August 2022.

[4] ibid.

[5] ibid.

[6] 'Yemen' (*Human Rights Watch World Report*, 2019) <https://www.hrw.org/world-report/2019/country-chapters/yemen> accessed 22 August 2022.

[7] Naureen Shah, 'Drone Strike Casualties Estimates Likely Understated' <https://hri.law.columbia.edu/sites/default/files/publications/CountingDroneDeathsPresserFINAL.pdf> accessed 22 August 2022.

than 240 million children live today in countries affected by armed conflict.[8] At least 19,000 children under the age of 18 are believed to be participating in the conflict in South Sudan.[9] Within the past five years, the number of child soldiers involved in conflicts globally has increased 159 per cent, with almost 30,000 verified recruitment cases since 2012.[10] From 2008 to 2016, more than 6,000 US veterans committed suicide each year.[11] Veterans' suicide rate is 1.5 times greater than the rate for non-veteran adults, even after adjusting for age and gender.[12] This list could easily go on for pages.

Even if we cannot attribute the current state of affairs to our laws of war, and even if we do not know what the current state of affairs would look like if our laws of war did not exist, the fact that war keeps happening, the fact that peace seems so distant, might make us wonder whether our laws help at all in achieving peace—we might wonder whether we have achieved any moral progress and at what cost. Perhaps the moral costs of regulating war, the moral fragility of the enterprise, are larger than what seems at first sight—even if it is attenuated in the ways I have suggested.

Moreover, modern conflicts show worrying trends. The nature of most of today's conflicts makes it almost impossible to draw a sharp line between times of war and times of peace (or the cessation of hostilities), and between law enforcement and war. As a result, we seem to be facing the prospect of war without end.[13]

Whether these facts—the deterioration of the boundaries between warfare and policing, between war and peace, between civilians and combatants; the increasing number of civilian casualties; the sheer number of armed conflicts and their impact—can be attributed to some sort of failure of our laws of war in effectively constructing moral progress and achieving peace is hard to say. But they might pose a problem for the instrumentalist justification I proposed.

[8] 'Child Soldiers International Annual Report 2017–18' (*OCHA*, 24 September 2018) <https://reliefweb.int/report/world/child-soldiers-international-annual-report-2017-18> accessed 22 August 2022.

[9] ibid.

[10] Anna Varfolomeeva, 'Number of Child Soldiers Involved in Conflicts Worldwide Jumps 159% in 5 Years' (*The Defense Post*, 11 February 2019) <https://thedefensepost.com/2019/02/11/child-soldiers-global-increase/> 22 August 2022.

[11] 'VA National Suicide Data Report 2005–2016' (*U.S. Department of Veterans Affairs*, 2018) 2 <https://www.mentalhealth.va.gov/docs/data-sheets/OMHSP_National_Suicide_Data_Report_2005-2016_508.pdf> accessed 22 August 2022.

[12] ibid.

[13] *See* Rosa Ehrenreich Brooks, 'War Everywhere: Rights, National Security Law, and the Law of Armed Conflict in the Age of Terror' (2004) 153 University of Pennsylvania Law Review 675, 725–28; Claire O. Finkelstein, 'The Equality of Combatants in Asymmetric War' in Jens David Ohlin, Larry May, and Claire Finkelstein (eds), *Weighing Lives in War* (Oxford University Press 2017).

We took a risk—a moral risk—in deciding to regulate warfare. Even if the account defended in Chapters VII and VIII is successful, the *content* of the symmetry thesis is not fully justified. Its *adoption* might be justified—and only to the extent that we wouldn't be able to agree on better alternatives. But settling for what is second best, settling for regulating war when what there should be in ideal circumstances is a prohibition and *no wars*, can have unexpected consequences, and some of those consequences will be morally troubling, to say the least. Perhaps, it shouldn't surprise us that some things have turned out the way they have. That war has lost its territory, its time frame, and its distinctive actors;[14] that there are so many meaningless wars; that so many people still die and kill in them. After all, if we have legal rules that allow combatants to kill each other during armed conflict, that tell people and states how to conduct warfare, institutions will take their place in this arrangement. There will be armies trained for fighting, scientists creating more powerful weapons, commanders refining military strategies. Would these things exist today anyway? Would they be less or more powerful and pervasive? We can only imagine.

In managing duels, there was eventually a choice to make: allowing duels to continue regulated by the rules of honour that in fact distinguished them from regular crime and assault[15] or prohibiting them. And when duelling in fact was prohibited and criminalized, even with all the associated enforcement difficulties,[16] the rules that regulated duelling became pointless. It was a crime either way.

Yet war is prohibited, there should be no wars, and still we regulate it—if hopefully temporarily. I have tried to answer why, and, in particular, I have tried to answer why we allow combatants to kill each other. And while some killings might be morally permitted, their permissibility is contingent on there being no other way of dealing with collective threats. Further, the instrumentalist justification I have provided is also contingent upon there being no other alternatives available. As a result, there most likely will come a point where permissive regulation is no longer good enough—and we may be closer to that point than we would like to think.

Be that as it may, we are where we are, and where we are is not very good. And we have to face the possibility that it might be a tragic consequence of war's

[14] Finkelstein (n 13).

[15] *See* David S Parker, 'Law, Honor, and Impunity in Spanish America: The Debate over Dueling, 1870–1920' (2001) 19 Law and History Review 311, 328.

[16] This is regardless of the discussion that preceded the criminalization of duels and the actual enforcement of the law once it was approved. *See* Parker (n 15); Robert B Shoemaker, 'The Taming of the Duel: Masculinity, Honour and Ritual Violence in London, 1660–1800' (2002) 45 The Historical Journal 525.

regulation that it has entrenched war itself. We have tried to humanize war, and in doing so, we may have made war longer; we have tried to humanize war and in doing so, we may have made it more palatable, legitimate; we may have forgotten to question war itself.[17] Given the way things have developed, it might seem unthinkable to get rid of any of it: of armies, of the stock-piled weapons, of our laws of war. Some might react with fear: if we don't have standing armies, if we don't have weapons of mass destruction, if we don't invest in technology, then someone else will, and they might not have good intentions (as we presumably do). And if we don't have any of these, or if we got rid of them, the world might come to be ruled by those who, less morally inclined than us, would keep them and use them against us. And if there are no rules, then there would be chaos—it would be so much worse.

Nonetheless, there is also—or there should be—the nagging thought that *it didn't have to be this way*. That we could have come up with something else. After all, although war is an activity that can exist without and prior to any laws,[18] war today is at least in part the result of our institutions,[19] of our laws. And our laws, even if they don't create the possibility of warfare, are in some ways constitutive of it:[20] they make it the case that 'X counts as Y in context C'; that killing B counts as targeting a combatant during war.

If it didn't have to be this way, then it seems that something has indeed gone very wrong. And not only that, but it is difficult to imagine how we can get out of the current situation. We seem to be tied to the current practice by the daunting prospect of trying to dismantle it and replace it with something else, and the fear that doing so will be a moral catastrophe.

More troubling is the fact that, despite all I have said, pacifism may be true. Already the targeting of unjust combatants is hard to justify. In fact, I have only provided the structure of the argument, through the introduction of the idea of permissibility, and have left open many parts of the substantive argument in favour of the permissibility of targeting non-liable unjust combatants. But there is also one area of our laws that I have barely touched on: IHL allows for civilian casualties when they are the foreseeable result of an attack against a legitimate military target, provided that the expected casualties are proportional to the expected military advantage. But civilians, I have argued, are innocent. There is no moral permission to kill them.

[17] See Samuel Moyn, *Humane: How the United States Abandoned Peace and Reinvented War* (1st edn, Farrar, Straus and Giroux 2021).

[18] John R Searle, *The Construction of Social Reality* (Free Press 1995) 27–28.

[19] ibid 27.

[20] *See* ibid 27–28.

If that is so, then whatever fragility might have been attenuated by my attempts at deconstructing the analogy between murder and what unjust combatants do pales in comparison to the moral fragility behind this legal permission, and as a result, to the moral fragility of regulating war. And although the distinction between intentionally and foreseeably causing someone's death—the well-known doctrine of double effect—might have some role in diminishing this fragility,[21] this potential diminution could not be so great as to plausibly give us much (moral) comfort. We might as well wonder if waging war is permissible at all if this is the cost to pay.

If choosing to regulate war is in fact not leading us to peace, and if pacifism is true, then the instrumentalist justification I have provided is even more fragile than what the past pages suggest. It begins to look like we have regulated murder at a vast scale. It might be heavily organized; it might be disciplined. There might be rules that states (sometimes) follow; there might be (some) enforcement mechanisms. But it is still, largely, regulated murder; it is still an abomination. We might be glad that at least it is a *regulated* abomination,[22] the way in which we might be glad that something like capital punishment is a regulated abomination. It may be that some features of both practices do, indeed, have value; that some features make them preferable over other alternatives: total war, executions without trials.

It starts to become less clear whether we would want to dismantle them—it is not even clear that we *could* dismantle them. Perhaps this is precisely why we feel uneasy about the ongoing deterioration of the boundaries between war and peace, combatants and civilians, warfare and policing. All of these are boundaries made relatively bright by our laws of war. We do not want to dismantle these features; these features are precisely what make our laws valuable, even if they are an abomination, even if they regulate murder at a vast scale.

Ultimately, while there might be cause for celebration that our laws of war exist, it should also be a cause for sorrow. The content of the legal permission to target just combatants is, at least sometimes, very close to murder, even if its

[21] *See* Kai Draper, 'Double Effect and the Laws of War' in Saba Bazargan and Samuel C Rickless (eds), *The Ethics of War* (Oxford University Press 2017); Philippa Foot, 'The Problem of Abortion and the Doctrine of Double Effect' (1967) 5 Oxford Review 5; Alison McIntyre, 'Doing Away with Double Effect' (2001) 111 Ethics 219.

[22] The most famous member of the fast-growing peace movement of the 1830s and 1840s, Charles Sumner, insisted that regulating warfare was a moral abomination. *See* John Fabian Witt, 'The Dismal History of the Laws of War' (2011) 1 University of California at Irvine Law Review 895, 896. Tolstoy held a similar position and even went further: he worried that making war more humane would bring more rather than fewer wars; that it would entrench it. *See* Samuel Moyn, 'Tolstoy's Case Against Humane War' (*Plough Quarterly Magazine*, 25 March 2020) <https://www.plough.com/en/topics/justice/politics/tolstoys-case-against-humane-war> accessed 22 August 2022; Moyn (n 17).

adoption can be contingently justified as the preferable alternative to no laws of war. And the *sui generis* moral permission to target non-liable unjust combatants is also often quite close to murder, even if it can be contingently explained by the impossibility of dealing with collective threats by only targeting those who are liable. But contingent justifications or permissions are only that—contingent. As a result, the moral pressure to revise the permissive dimensions of our laws, and of morality, is enormous.

We have done the best we could, perhaps, but it remains unclear whether the best we could do in the past is the best we could do now. And it might be that the best we could do was not, after all, very good at all.

The conclusion might then be that humanity made a leap of faith in choosing to regulate war—and yet, we may have risked too much. Or, in the words of Tamara Stepanovna Umnyagina: 'My precious one ... People still hate each other. They go on killing. That's the most incomprehensible thing to me ....'[23]

[23] Alexievič (n 1) 330.

Bibliography

Adams DM, 'Hohfeld on Rights and Privileges' [1985] ARSP: Archiv für Rechts-und Sozialphilosophie/Archives for Philosophy of Law and Social Philosophy 84

Aleksievič SA, *Zinky Boys: Soviet Voices from the Afghanistan War* (Julia Whitby and Robin Whitby trs, Norton 1992)

Aleksievič SA, *The Unwomanly Face of War* (Random House Trade Paperbacks 2018)

Alexander LA, 'Self-Defense and the Killing of Noncombatants: A Reply to Fullinwider' (1976) 5 Philosophy & Public Affairs 408

Alexander LA, 'Deontology at the Threshold' (2000) 37 San Diego Law Review 893

Alexandra A, 'Political Pacifism' (2003) 29 Social Theory and Practice 589

Ambrose SE, *Citizen Soldiers: The U.S. Army from the Normandy Beaches to the Bulge to the Surrender of Germany, June 7, 1944–May 7, 1945* (Simon & Schuster 1998)

Applbaum AI, *Ethics for Adversaries: The Morality of Roles in Public and Professional Life* (Princeton University Press 2001)

Aquinas T, *Summa Theologica* (Fathers of the English Dominican Providence tr, Burns Oats & Washbourne 1947)

Arendt H, *The Origins of Totalitarianism* (7th edn, Meridian Books, The World Publishing Company 1962)

Arneson RJ, 'Just Warfare Theory and Noncombatant Immunity' (2006) 39 Cornell International Law Journal 663

Arneson RJ, 'Resolving the Responsibility Dilemma' in Saba Bazargan and Samuel C Rickless (eds), *The Ethics of War* (Oxford University Press 2017)

Bald W, 'Rejoining Moral Culpability with Criminal Liability: Reconsideration of the Felony Murder Doctrine for the Current Time' (2017) 44 Journal of Legislation 239

Baron M, 'Self-Defense: The Imminence Requirement' in Leslie Green and Brian Leiter (eds), *Oxford Studies in Philosophy of Law*, vol 1 (Oxford University Press 2011)

Barry C and Christie L, 'The Moral Equality of Combatants' in Seth Lazar and Helen Frowe (eds), *The Oxford Handbook of Ethics of War* (Oxford University Press 2015)

Bazargan S, 'Complicitous Liability in War' (2013) 165 Philosophical Studies 177

Bazargan S, 'Morally Heterogeneous Wars' (2013) 41 Philosophia 959

Bazargan S, 'Killing Minimally Responsible Threats' (2014) 125 Ethics 114

Bazargan S, 'Defensive Liability Without Culpability' in Christian Coons and Michael Weber (eds), *The Ethics of Self-Defense* (Oxford University Press 2016)

Beer Y, 'Humanity Considerations Cannot Reduce War's Hazards Alone: Revitalizing the Concept of Military Necessity' (2015) 26 European Journal of International Law 801

Beer Y, *Military Professionalism and Humanitarian Law: The Struggle to Reduce the Hazards of War* (Oxford University Press 2018)

Benbaji Y, 'The Responsibility of Soldiers and the Ethics of Killing in War' (2007) 57 The Philosophical Quarterly 558

Benbaji Y, 'A Defense of the Traditional War Convention' (2008) 118 Ethics 464

Benbaji Y, 'The War Convention and the Moral Division of Labour' (2009) 59 The Philosophical Quarterly 593

Benbaji Y and Statman D, *War by Agreement: A Contractarian Ethics of War* (Oxford University Press 2019)

Benvenisti E, 'Legal Battle to Define the Law on Transnational Asymmetric Warfare, The' (2009) 20 Duke Journal of Comparative & International Law 339

Berman MN, 'Punishment and Justification' (2008) 118 Ethics 258

Berman N, 'Privileging Combat-Contemporary Conflict and the Legal Construction of War' (2004) 43 Columbia Journal of Transnational Law 1

Biggar N, *In Defence of War* (Oxford University Press 2013)

Blackman LR, 'Alarm Red' in Andrew Carroll (ed), *Operation Homecoming: Iraq, Afghanistan, and the Home Front, in the Words of U.S. Troops and Their Families* (Updated edn, University of Chicago Press 2008)

Blum G, 'The Dispensable Lives of Soldiers' (2010) 2 Journal of Legal Analysis 115

Bolinger RJ, 'The Moral Grounds of Reasonably Mistaken Self-Defense' (2021) 103 Philosophy and Phenomenological Research 140

Brandt RB, 'Utilitarianism and the Rules of War' (1972) 1 Philosophy & Public Affairs145

Brooks RE, 'War Everywhere: Rights, National Security Law, and the Law of Armed Conflict in the Age of Terror' (2004) 153 University of Pennsylvania Law Review 675

Burri S, 'The Toss-up between a Profiting, Innocent Threat and His Victim' (2015) 23 Journal of Political Philosophy 146

Burri S, 'Morally Permissible Risk Imposition and Liability to Defensive Harm' (2020) 39 Law and Philosophy 381

Burri S, 'Defensive Liability: A Matter of Rights Enforcement, Not Distributive Justice' [2021] Criminal Law and Philosophy 1

Burri S, 'If You Care About a Rule, Why Weaken Its Enforcement Dimension? On a Tension in the War Convention' [2022] Law and Philosophy 1

Carnahan BM, 'Lincoln, Lieber and the Laws of War: The Origins and Limits of the Principle of Military Necessity' [1998] American Journal of International Law 213

Carroll A (ed), *Operation Homecoming: Iraq, Afghanistan, and the Home Front, in the Words of U.S. Troops and Their Families* (Updated edn, University of Chicago Press 2008)

Chan DK, *Beyond Just War: A Virtue Ethics Approach* (Palgrave Macmillan 2012)

Chehtman, A, *A Theory of Asymmetric Warfare* (Oxford University Press, forthcoming 2023)

Clausewitz C von, *On War* (Michael Eliot Howard and Peter Paret eds, Princeton University Press 1989)

Coll S, *Ghost Wars: The Secret History of the CIA, Afghanistan, and Bin Laden, from the Soviet Invasion to September 10, 2001* (Penguin Books 2005)

Cook WW, 'Hohfeld's Contributions to the Science of Law' (1919) 28 Yale Law Journal 721

Coons C and Weber M, 'The Ethics of Self-Defense' in Christian Coons and Michael Weber (eds), *The Ethics of Self-Defense* (Oxford University Press 2016)

Corbin AL, 'Legal Analysis and Terminology' (1919) 29 The Yale Law Journal 163

Cover RM, *Justice Accused: Antislavery and the Judicial Process* (Yale University Press 1975)

Christiano T, 'Authority' (*Stanford Encyclopedia of Philosophy*, 2 July 2004) <https://plato.stanford.edu/archives/spr2013/entries/authority/> accessed 17 August 2022

Christiano T, 'Democracy' (*Stanford Encyclopedia of Philosophy*, 27 July 2006) <https://plato.stanford.edu/archives/fall2018/entries/democracy/> accessed 17 August 2022

Crawford NC, 'Human Cost of the Post-9/11 Wars: Lethality and the Need for Transparency,' (Watson Institute, Brown University, November 2018) <https://watson.brown.edu/costsofwar/files/cow/imce/papers/2018/Human%20Costs%2C%20Nov%208%202018%20CoW.pdf> accessed 22 August 2022

Cullison AD, 'Review of Hohfeld's Fundamental Legal Concepts, A' (1967) 16 Cleveland-Marshall Law Review 559

Dan-Cohen M, 'Decision Rules and Conduct Rules: On Acoustic Separation in Criminal Law' (1984) 97 Harvard Law Review 625

Dannenbaum T, 'Why Have We Criminalized Aggressive War' (2016) 126 Yale Law Journal 1242

Dannenbaum T, *The Crime of Aggression, Humanity, and the Soldier* (Cambridge University Press 2018)

Davis J, *Seeking Human Rights Justice in Latin America: Truth, Extra-Territorial Courts, and the Process of Justice* (Cambridge University Press 2014)

Dempsey MM, 'Against Liability' in Christian Coons and Michael Weber (eds), *The Ethics of Self-Defense* (Oxford University Press 2016)

Dill J and Shue H, 'Limiting the Killing in War: Military Necessity and the St. Petersburg Assumption' (2012) 26 Ethics & International Affairs 311

Dill J, 'Towards a Moral Division of Labour between IHL and IHRL during the Conduct of Hostilities' in Ziv Bohrer, Helen Duffy, and Janina Dill (eds), *Law Applicable to Armed Conflict* (Cambridge University Press 2020)

Dinstein Y, *The Conduct of Hostilities under the Law of International Armed Conflict* (Cambridge University Press 2004)

Dinstein Y, *War, Aggression, and Self-Defence* (5th edn, Cambridge University Press 2012)

Dinstein Y, *Non-International Armed Conflicts in International Law* (Cambridge University Press 2014)

Draper K, *War and Individual Rights: The Foundations of Just War Theory* (Oxford University Press 2016)

Draper K, 'Double Effect and the Laws of War' in Saba Bazargan and Samuel C Rickless (eds), *The Ethics of War* (Oxford University Press 2017)

Draper K, 'Who Should Die?: The Ethics of Killing in War' in Bradley Jay Strawser, Ryan Jenkins, and Michael Robillard (eds), *Defensive Liability: Four Common Mistakes* (Oxford University Press 2018)

Duff RA, 'I Might Be Guilty, but You Can't Try Me: Estoppel and Other Bars to Trial' (2003) 1 Ohio State Journal of Criminal Law 245

Dworkin R, *Taking Rights Seriously* (Harvard Univ Press 1978)

Dworkin R, *Law's Empire* (Hart 2010)

Estlund D, 'On Following Orders in an Unjust War' (2007) 15 Journal of Political Philosophy 213

Estlund D, *Democratic Authority: A Philosophical Framework* (Princeton University Press 2009)

Estlund D, 'Human Nature and the Limits (If Any) of Political Philosophy' (2011) 39 Philosophy & Public Affairs 207

Estlund D, 'Utopophobia' (2014) 42 Philosophy & Public Affairs 113

Fabre C, *Cosmopolitan War* (Oxford University Press 2012)

Fabre C, 'On Jan Narveson's "Pacifism: A Philosophical Analysis"' (2015) 125 Ethics 823

Fabre C, *Cosmopolitan Peace* (Oxford University Press 2016)

Feinberg J and Narveson J, 'The Nature and Value of Rights' (1970) 4 The Journal of Value Inquiry 243

Ferzan KK, 'Culpable Aggression: The Basis for Moral Liability to Defensive Killing' (2011) 9 Ohio State Journal of Criminal Law 669

Ferzan KK, 'Forfeiture and Self-Defense' in Christian Coons and Michael Weber (eds), *The Ethics of Self-Defense* (Oxford University Press 2016)

Filkins D, *The Forever War* (Vintage Books 2009)
Finkel D, *The Good Soldiers* (Third edition, Picador USA 2010)
Finkelstein CO, 'Duress: A Philosophical Account of the Defense in Law' (1995) 37 Arizona Law Review 251
Finkelstein CO, 'The Equality of Combatants in Asymmetric War' in Jens David Ohlin, Larry May, and Claire Finkelstein (eds), *Weighing Lives in War* (Oxford University Press 2017)
Finlay CJ, 'Justification and Legitimacy at War: On the Sources of Moral Guidance for Soldiers' (2019) 129 Ethics 576
Finnis J, 'Some Professorial Fallacies about Rights' (1971) 4 Adelaide Law Review 377
Finnis J, *Natural Law and Natural Rights* (2nd edn, Oxford University Press 2011)
Fishback I, 'Necessity and Institutions in the Self Defense and War' in Christian Coons and Michael Weber (eds), *The Ethics of Self-Defense* (Oxford University Press 2016)
Fletcher GP and Ohlin JD, *Defending Humanity: When Force Is Justified and Why* (Oxford University Press 2008)
Foot P, 'The Problem of Abortion and the Doctrine of Double Effect' (1967) 5 Oxford Review 5
Frowe H, 'A Practical Account of Self-Defence' (2010) 29 Law and Philosophy 245
Frowe H, 'Self-Defence and the Principle of Non-Combatant Immunity' (2011) 8 Journal of Moral Philosophy 530
Frowe H, *Defensive Killing* (Oxford University Press 2014)
Frowe H, 'Can Reductive Individualists Allow Defense Against Political Aggression?' (2015) 1 Oxford Studies in Political Philosophy 173
Frowe H, 'The Just War Framework' in Seth Lazar and Helen Frowe (eds), *The Oxford Handbook of Ethics of War* (Oxford University Press 2015)
Frowe H, 'Lesser-Evil Justifications for Harming: Why We're Required to Turn the Trolley' (2018) 68 The Philosophical Quarterly 460
Frowe H, 'Introduction' (2019) 129 Ethics 544
Frowe H, 'The Moral Irrelevance of Moral Coercion' (2021) 178 Philosophical Studies 3465
Fullinwider RK, 'War and Innocence' (1975) 5 Philosophy & Public Affairs 90
Gardner J and Macklem T, 'Reasons' in Jules Coleman and Scott Shapiro (eds), *The Oxford Handbook of Jurisprudence and Philosophy of Law* (Oxford University Press 2002)
Gardner J and Tanguay-Renaud F, 'Desert and Avoidability in Self-Defense' (2011) 122 Ethics 111
Gardner J, *Offences and Defences: Selected Essays in the Philosophy of Criminal Law* (Oxford University Press 2007)
Gardner J, 'Justification under Authority' (2010) 23 Canadian Journal of Law & Jurisprudence 71
Gardner J, 'What Is Tort Law for? Part 1. The Place of Corrective Justice' (2011) 30 Law and Philosophy 1
Gardner J, 'Law and Morality' (*Social Science Research Network* 2008) SSRN Scholarly Paper ID 2344794 <http://papers.ssrn.com/abstract=2344794> accessed 21 December 2015
Gardner J, 'Tort Law and Its Theory' in John Tasioulas (ed), *The Cambridge Companion to Philosophy of Law* (Cambridge University Press 2018)
Gentili A, *De Iure Belli Libri Tres* (John C Rolfe tr, The Clarendon Press 1933)
Gert J, 'The Distinction between Justifying and Requiring: Nothing to Fear' in Errol Lord and Barry Maguire (eds), *Weighing Reasons* (Oxford University Press 2016)
Gibson K, 'Contrasting Role Morality and Professional Morality: Implications for Practice' (2003) 20 Journal of Applied Philosophy 17
Glover J, 'It Makes No Difference Whether or Not I Do It' (1975) 49 Proceedings of the Aristotelian Society 171

Glover J, *Humanity: A Moral History of the Twentieth Century* (2nd edn, Yale University Press 2012)
Goodman R, 'The Power to Kill or Capture Enemy Combatants' (2013) 24 European Journal of International Law 819
Gordon-Solmon K, 'Self-Defence Against Multiple Threats' (2017) 14 Journal of Moral Philosophy 125
Gordon-Solmon K, 'How (and How Not) to Defend Lesser-Evil Options' (2022) 1 Journal of Moral Philosophy 1
Greenberg KJ and Dratel JL (eds), *The Torture Papers: The Road to Abu Ghraib* (Cambridge University Press 2005)
Greenberg M, 'The Moral Impact Theory of Law' (2013) 123 Yale Law Journal 1288
Gregory E, 'What Do We Want from the Just War Tradition? New Challenges of Surveillance and the Security State' (2014) 27 Studies in Christian Ethics 50
Grossman LtColD, *On Killing. The Psychological Cost of Learning to Kill in War and Society* (Revised, Open Road Integrated Media 2014)
Grotius H, *Hugo Grotius on the law of war and peace* (Stephen C Neff ed, Cambridge University Press 2012)
Gunneflo M, *Targeted Killing: A Legal and Political History* (Cambridge University Press 2016)
Halbertal M, 'The Goldstone Illusion' (*New Republic*, 5 November 2009) <https://newrepublic.com/article/70956/the-goldstone-illusion> accessed 18 August 2022
Halbertal M, *On Sacrifice* (Princeton University Press 2012)
Halpin AK, 'Hohfeld's Conceptions: From Eight to Two' (1985) 44 The Cambridge Law Journal 435
Hamlin A and Stemplowska Z, 'Theory, Ideal Theory and the Theory of Ideals' (2012) 10 Political Studies Review 48
Haque AA, 'Law and Morality at War' (2014) 8 Criminal Law and Philosophy 79
Haque AA, *Law and Morality at War* (Oxford University Press 2017)
Haque AA, 'Indeterminacy in the Law of Armed Conflict' (2019) 95 International Law Studies 119
Hardin R, The Free Rider Problem (*Stanford Encyclopedia of Philosophy*, 21 May 2013), <https://plato.stanford.edu/archives/spr2013/entries/free-rider/> accessed 22 August 2022
Hart HLA, 'Are There Any Natural Rights?' (1955) 64 The Philosophical Review 175
Hart HLA, *Essays on Bentham. Jurisprudence and Political Philosophy* (Oxford University Press 1982)
Hart HLA, *The Concept of Law* (Penelope A Bulloch and Joseph Raz eds, 2nd edn, Clarendon Press 1998)
Hart HLA, 'Bentham on Legal Rights' in Carl Wellman (ed), *Rights and duties* (Routledge 2002)
Hippo A, *City of God* (Gerald G Walsh and Daniel J Honan tr, Catholic University of America Press 2008)
Hohfeld WN, 'Some Fundamental Legal Conceptions as Applied in Judicial Reasoning' (1913) 23 Yale Law Journal 16
Hohfeld WN, 'Fundamental Legal Conceptions as Applied in Judicial Reasoning' (1917) 26 Yale Law Journal 710
Honoré T, 'The Dependence of Morality on Law' (1993) 13 Oxford Journal of Legal Studies 1
Hosein A, 'Are Justified Aggressors a Threat to the Rights Theory of Self-Defense?' in Helen Frowe and Gerald Lang (eds), *How We Fight* (Oxford University Press 2014)

Hurka T, 'Proportionality and Necessity' in Larry May and Emily Crookston (eds), *War: Essays in Political Philosophy* (Cambridge University Press 2008)
Ihara CK, 'In Defense of a Version of Pacifism' (1978) 88 Ethics 369
Jenkins RC, Robillard M, and Strawser BJ, *Who Should Die?: The Ethics of Killing in War* (Oxford University Press 2018)
Johnson JT, 'The Broken Tradition' (1996) 45 The National Interest 27
Junger S, *War* (Reprint edition, Twelve 2011)
Kadish SH, 'Torture, the State and the Individual' (1989) 23 Israel Law Review 345
Kagan S, 'Do I Make a Difference?' (2011) 39 Philosophy & Public Affairs 105
Kahn PW, 'Lessons for International Law from the Gulf War' (1992) 45 Stanford Law Review 425
Kahn PW, 'War and Sacrifice in Kosovo' (1999) 13 Philosophical Dimensions of Public Policy 201
Kahn PW, 'War Powers and the Millennium' (2000) 34 Loyola of Los Angeles Law Review 11
Kahn PW, 'The Paradox of Riskless Warfare' (2002) 22 Philosophy & Public Policy Quarterly 2
Kamm FM, 'Non-Consequentialism, the Person as an End-in-Itself, and the Significance of Status'(1992) 21 Philosophy & Public Affairs 354
Kamm FM, *Ethics for Enemies: Terror, Torture, and War* (Oxford University Press, 2011)
Kant I, 'To Perpetual Peace. A Philosophical Sketch' *in Perpetual Peace and Other Essays* (Hackett Publishing Company 1983)
Kaufman W, 'What Is the Scope of Civilian Immunity in Wartime?' (2003) 2 Journal of Military Ethics 186
Kaufman W, 'Is There a "Right" to Self-Defense?' (2004) 23 Criminal Justice Ethics 20
Kaufman W, 'What's Wrong with Preventive War? The Moral and Legal Basis for the Preventive Use of Force' (2005) 19 Ethics & International Affairs 23
Kaufman W, 'Torture and the "Distributive Justice" Theory of Self-Defense: An Assessment' (2008) 22 Ethics & International Affairs 93
Kaufman W, 'Self-Defense, Innocent Aggressors, and The Duty of Martyrdom' (2010) 91 Pacific Philosophical Quarterly 78
Kelly R, 'This is Not a Game,' in Andrew Carroll (ed), *Operation Homecoming: Iraq, Afghanistan, and the Home Front, in the Words of U.S. Troops and Their Families* (Updated ed, University of Chicago Press 2008) 279
Kelsen H, 'Collective and Individual Responsibility in International Law with Particular Regard to the Punishment of War Criminals' (1942) 31 Califprnia Law Review 530
Kelsen H, *Pure Theory of Law* (University of California Press 1967)
Kelsen H, *General Theory of Norms* (Clarendon Press 1991)
Kershaw I, *The End: The Defiance and Destruction of Hitler's Germany, 1944–1945* (Penguin Press 2011)
Klay P, *Redeployment* (Penguin Books 2015)
Klay P, 'The Citizen-Soldier: Moral Risk and the Modern Military' (*The Brookings Institution* 24 May 2016) <http://csweb.brookings.edu/content/research/essays/2016/the-citizen-soldier.html> accessed 17 August 2022
Kocourek A, 'Hohfeld System of Fundamental Legal Concepts' (1920) 15 Illinois Law Review 24
Korsgaard CM, 'The Reasons We Can Share: An Attack on the Distinction between Agent-Relative and Agent-Neutral Values' (1993) 10 Social Philosophy and Policy 24
Koskenniemi M, '"The Lady Doth Protest Too Much" Kosovo, and the Turn to Ethics in International Law' (2002) 65 The Modern Law Review 159

Kramer MH and Steiner H, 'Theories of Rights: Is There a Third Way?' (2007) 27 Oxford Journal of Legal Studies 281

Kutz C, 'Acting Together' (2000) 61 Philosophical and Phenomenological Research 1

Kutz C, 'The Difference Uniforms Make: Collective Violence in Criminal Law and War' (2005) 33 Philosophy & Public Affairs 148

Kutz C, *On War and Democracy* (Princeton University Press 2016)

Lamb A, *Ethics and the Laws of War: The Moral Justification of Legal Norms* (Routledge 2013)

Lang G, 'Why Not Forfeiture?' in Helen Frowe and Gerald Lang (eds), *How We Fight* (Oxford University Press 2014)

Lang G, 'What Follows from Defensive Non-Liability?' (2017) 117 *Proceedings of the Aristotelian Society* 231

Lazar S and Valentini L, 'Proxy Battles in Just War Theory' in Peter Vallentyne, David Sobel, and Steven Wall (eds), *Oxford Studies in Political Philosophy*, vol 3 (Oxford University Press 2017)

Lazar S, 'Responsibility, Risk, and Killing in Self-Defense' (2009) 119 Ethics 699

Lazar S, 'The Responsibility Dilemma for Killing in War: A Review Essay' (2010) 38 Philosophy & Public Affairs 180

Lazar S, 'Necessity in Self-Defense and War' (2012) 40 Philosophy & Public Affairs 3

Lazar S, "The Morality and Law of War' in Andrei Marmor (ed), *Routledge Companion to the Philosophy of Law* (Routledge 2012)

Lazar S, 'Associative Duties and the Ethics of Killing in War' (2013) 1 Journal of Practical Ethics 3

Lazar S, 'National Defence, Self Defence, and the Problem of Political Aggression' in Cécile Fabre and Seth Lazar (eds), *The Morality of Defensive War* (Oxford University Press 2014)

Lazar S, 'Method in the Morality of War' in Seth Lazar and Helen Frowe (eds), *The Oxford Handbook of Ethics of War* (Oxford University Press 2015)

Lazar S, *Sparing Civilians* (Oxford University Press 2015)

Lazar S, 'War' (*The Stanford Encyclopedia of Philosophy*, 3 May 2016) <https://plato.stanford.edu/archives/spr2017/entries/war/> accessed 18 August 2022

Lazar S, 'Just War Theory: Revisionists versus Traditionalists' (2017) 20 Annual Review of Political Science 37

Lazar S, Bazargan S, and Rickless SC, 'War's Endings and the Structure of Just War Theory' in *The Ethics of War: Essays* (Oxford University Press 2017)

Lazar S, 'Strengthening Moral Distinction' (2018) 37 Law and Philosophy 327

Lieblich E, 'The Facilitative Function of Jus in Bello' (2019) 30 European Journal of International Law 321

Locke J, *Two Treatises of Government: And a Letter Concerning Toleration* (Ian Shapiro ed, Yale University Press 2003)

Luban D, *Lawyers and Justice: An Ethical Study* (Princeton University Press 1988)

Luban D, 'War Crimes. The Law of Hell' in Larry May and Emily Crookston (eds), *War: Essays in Political Philosophy* (Cambridge University Press 2008)

Luban D, 'Military Necessity and the Cultures of Military Law' (2013) 26 Leiden Journal of International Law 315

Luban D, 'Just War Theory and the Laws of War as Nonidentical Twins' (2017) 31 Ethics & International Affairs 433

Lyons D, 'The Correlativity of Rights and Duties' [1970] Nous 45

Macias A, 'From Aruba to Iceland, These 36 Nations Have No Standing Military' (*CNBC* 3 April 2018) <https://www.cnbc.com/2018/04/03/countries-that-do-not-have-a-standing-army-according-to-cia-world-factbook.html> accessed 19 August 2022

Maginnis VL, 'Limiting Diplomatic Immunity: Lessons Learned From the 1946 Convention on the Privileges and Immunities of the United Nations' (2002) 28 Brookings Journal of International Law 989

Mapel DR, 'Moral Liability to Defensive Killing and Symmetrical Self-Defense' (2010) 18 Journal of Political Philosophy 198

McDonough MB, 'Privileged Outlaws: Diplomats, Crime and Immunity' (1996) 20 Suffolk Transnational Law Review 475

Mavrodes GI, 'Conventions and the Morality of War' (1975) 4 Philosophy & Public Affairs 117

May L, *Aggression and Crimes Against Peace* (Cambridge University Press 2008)

May L, 'Contingent Pacifism and the Moral Risks of Participating in War' (2011) 25 Public Affairs Quarterly 95

May L, 'Humanity, Necessity, and the Rights of Soldiers' in Jens David Ohlin, Larry May, and Claire Finkelstein (eds), *Weighing Lives in War* (Oxford University Press 2017)

McCary J, 'To the Fallen, E-mail' in Andrew Carroll (ed), *Operation Homecoming: Iraq, Afghanistan, and the Home Front, in the Words of U.S. Troops and Their Families* (Updated ed, University of Chicago Press 2008)

McConnell T, 'Moral Dilemmas' (*Stanford Encyclopedia of Philosophy*, 15 April 2002) <https://plato.stanford.edu/archives/fall2018/entries/moral-dilemmas/> accessed 19 August 2022

McIntyre A, 'Doing Away with Double Effect' (2001) 111 Ethics 219

McMahan J, 'Self-Defense and the Problem of the Innocent Attacker' (1994) 104 Ethics 252

McMahan J, 'The Ethics of Killing in War' (2004) 114 Ethics 693

McMahan J, 'Just Cause for War' (2005) 19 Ethics & International Affairs 1

McMahan J, 'The Basis of Moral Liability to Defensive Killing' (2005) 15 Philosophical Issues 386

McMahan J, 'On the Moral Equality of Combatants' (2006) 14 Journal of Political Philosophy 377

McMahan J, 'The Sources and Status of Just War Principles' (2007) 6 Journal of Military Ethics 91

McMahan J, 'Debate: Justification and Liability in War' (2008) 16 Journal of Political Philosophy 227

McMahan J, 'The Morality of War and The Law of War' in David Rodin and Henry Shue (eds), *Just and Unjust Warriors: The Moral and Legal Status of Soldiers* (Oxford University Press 2008)

McMahan J, *Killing in War* (Oxford University Press 2009)

McMahan J, 'Duty, Obedience, Desert, and Proportionality in War: A Response' (2011) 122 Ethics 135

McMahan J, 'Who Is Morally Liable to Be Killed in War' (2011) 71 Analysis 544

McMahan J, 'Individual Liability in War: A Response to Fabre, Leveringhaus and Tadros' (2012) 24 Utilitas 278

McMahan J, 'Self-Defense against Justified Threateners' in Gerald Lang and Helen Frowe (eds), *How We Fight: Ethics in War* (Oxford University Press 2014)

McMahan J, 'What Rights May Be Defended by Means of War?' in Cécile Fabre and Seth Lazar (eds), *The Morality of Defensive War* (Oxford University Press 2014)

McMahan J, 'Liability, Proportionality, and the Number of Aggressors' in Saba Bazargan and Samuel C Rickless (eds), *The Ethics of War* (Oxford University Press 2017)

McMahan J, 'Necessity and Proportionality in Morality and Law' in Claus Kress and Robert Lawless (eds), *Necessity and Proportionality in International Peace and Security Law* (Oxford University Press 2020)

Meisels T and Waldron J, *Debating Targeted Killing: Counter-terrorism or Extrajudicial execution?* (Oxford University Press, 2020)
Melzer N, *Targeted Killing in International Law* (Oxford University Press 2008)
Meron T, 'The Humanization of Humanitarian Law' (2000) 94 American Journal of International Law 239
Mill JS, *Utilitarianism* (George Sher ed, 2nd edn, Hackett Publishing Company 2001)
Miller D, 'Taking Up the Slack? Responsibility and Justice in Situations of Partial Compliance' in Carl Knight and Zofia Stemplowska (eds), *Responsibility and Distributive Justice* (Oxford University Press 2011)
Moir L, *The Law of Internal Armed Conflict* (Cambridge University Press 2002)
Moorer T, 'JAG in the Sandbox' in Andrew Carroll (ed), *Operation Homecoming: Iraq, Afghanistan, and the Home Front, in the Words of U.S. Troops and Their Families* (Updated ed, University of Chicago Press 2008)
Montague P, 'Self-Defense and Choosing between Lives' (1981) 40 Philosophical Studies 207
Moyn S, 'Tolstoy's Case Against Humane War' (*Plough Quarterly Magazine,* 25 March 2020) <https://www.plough.com/en/topics/justice/politics/tolstoys-case-against-humane-war> accessed 22 August 2022
Moyn S, *Humane: How the United States Abandoned Peace and Reinvented War* (1st edn, Farrar, Straus and Giroux 2021)
Murphy LB, 'Institutions and the Demands of Justice' (1998) 27 Philosophy & Public Affairs 251
Murphy LB, *Moral Demands in Nonideal Theory* (Oxford University Press 2000)
Nagel T, 'War and Massacre' (1972) 1 Philosophy & Public Affairs 123
Nagel T, *The View from Nowhere* (Oxford University Press 1989)
Nagel T, *Equality and Partiality* (Oxford University Press 1995)
Nagel T, 'The Fragmentation of Value' in *Mortal Questions* (Cambridge University Press 2012)
Narveson J, 'Pacifism: A Philosophical Analysis' (1965) 75 Ethics 259
Narveson J, 'Is Pacifism Consistent?' (1968) 78 Ethics 148
Ninh, B *The Sorrow of War: A Novel of North Vietnam* (Frank Palmos tr, Riverhead Books 1996)
Nozick R, *Anarchy, Utopia and State* (Basil Blackwell 1974)
Nyquist C, 'Teaching Wesley Hohfeld's Theory of Legal Relations' (2002) 52 Journal of Legal Education 238
O'Brien T, *The Things They Carried* (1st Mariner Books edn, Houghton Mifflin Harcourt 2009)
O'Connell ME, 'The Legal Case against the Global War on Terror' (2004) 36 Case Western Reserve Journal of International Law 349
O'Connell ME, 'Combatants and the Combat Zone' (2008) 43 University of Richmond Law Review 845
O'Donovan O, *The Just War Revisited* (Cambridge University Press 2003)
Orend B, 'Kant's Just War Theory' (1999) 37 Journal of the History of Philosophy 323
Otsuka M, 'Killing the Innocent in Self-Defense' (1994) 23 Philosophy & Public Affairs 74
Otsuka M, 'The Moral-Responsibility Account of Liability to Defensive Killing' in Christian Coons and Michael Weber (eds), *The Ethics of Self-Defense* (Oxford University Press 2016)
Parfit D, *Reasons and Persons* (Clarendon Press 1992)
Parfit D, *On What Matters*, vol 1 (Oxford University Press 2013)
Parker DS, 'Law, Honor, and Impunity in Spanish America: The Debate over Dueling, 1870–1920' (2001) 19 Law and History Review 311

Philips M, 'Weighing Moral Reasons' (1987) 96 Mind 367

Prieto Rudolphy M, 'Who is at War? On the Question of Co-belligerency' (forthcoming) Yearbook of International Humanitarian Law

Quong J and Stone R, 'Rules and Rights' in David Sobel, Peter Vallentyne, and Steven Wall (eds), *Oxford Studies in Political Philosophy, Volume 1* (Oxford University Press 2015)

Quong J, 'Killing in Self-Defense' (2009) 119 Ethics 507

Quong J, 'Liability to Defensive Harm' (2012) 40 Philosophy & Public Affairs 45

Quong J, 'Agent-Relative Prerogatives to Do Harm' (2016) 10 Criminal Law and Philosophy 815

Quong J, *The Morality of Defensive Force* (Oxford University Press 2020)

Rawls J, *A Theory of Justice* (Rev, Belknap Press of Harvard University Press 1999)

Rawls J, *The Law of Peoples* (Harvard University Press 1999)

Rawls J, *Justice as Fairness: A Restatement* (Erin Kelly ed, Harvard University Press 2001)

Raz J, 'On the Nature of Rights' (1984) 93 Mind 194

Raz J, *Practical Reason and Norms* (Oxford Univ Press 2002)

Raz, 'Personal Practical Conflicts' in Peter Baumann and Monika Betzler (eds), *Practical Conflicts: New Philosophical Essays* (Cambridge University Press 2004)

Raz J, 'Value and the Weight of Practical Reasons' in Errol Lord and Barry Maguire (eds), *Weighing Reasons* (Oxford University Press 2016)

Renzo M, 'Rights Forfeiture and Liability to Harm' (2017) 25 Journal of Political Philosophy 324

Renzo M, 'Political Authority and Unjust Wars' (2019) 99 Philosophy and Phenomenological Research 336

Ripstein A, *Kant and the Law of War* (Oxford University Press 2021)

Roberts A, 'The Principle of Equal Application of the Laws of War' in David Rodin and Henry Shue (eds), *Just and Unjust Warriors: The Moral and Legal Status of Soldiers* (Oxford University Press 2008)

Robin GD, 'Justifiable Homicide by Police Officers' (1963) 54 The Journal of Criminal Law, Criminology, and Police Science 225

Robinson PH and Williams TS, *Mapping American Criminal Law: Variations across the 50 States* (Praeger 2018)

Robinson PH, 'A Theory of Justification: Societal Harm as a Prerequisite for Criminal Liability' (1975) 23 UCLA Law Review 266

Rodin D, *War and Self-Defense* (Clarendon Press; Oxford University Press 2002)

Rodin D, 'Justifying Harm' (2011) 122 Ethics 74

Rodin D, 'Morality and Law in War' in Hew Strachan and Sibylle Scheipers (eds), *The Changing Character of War* (Oxford University Press 2011)

Rodin D, 'The Lesser Evil Obligation' in Saba Bazargan and Samuel C Rickless (eds), *The Ethics of War* (Oxford University Press 2017)

Roser M, 'Homicides' *(Our World in Data* 2019) <https://ourworldindata.org/homicides#sources-homicide-data-in-the-post-war-time> accessed 17 August 2022

Ryan CC, 'Self-Defense and the Obligations to Kill and to Die' (2004) 18 Ethics & International Affairs 69

Ryan CC, 'The Hard Hand of War' (2018) 37 Law and Philosophy 269

Scanlon T, *What We Owe to Each Other* (Belknap Press of Harvard University Press 1998)

Shah N, 'Drone Strike Casualties Estimates Likely Understated' <https://hri.law.columbia.edu/sites/default/files/publications/CountingDroneDeathsPresserFINAL.pdf> accessed 22 August 2022

Scheffler S, *Human Morality* (Oxford University Press 1993)

Scheffler S, *The Rejection of Consequentialism: A Philosophical Investigation of the Considerations Underlying Rival Moral Conceptions* (Clarendon Press 1995)
Scheffler S, 'Projects, Relationships, and Reasons' in R Jay Wallace (ed), *Reason and Value: Themes from the Moral Philosophy of Joseph Raz* (Oxford University Press 2004)
Scheffler S, *Death and the Afterlife* (Kolodny N ed, Oxford University Press 2013)
Scheffler S, 'Membership and Political Obligation' (2018) 26 Journal of Political Philosophy 3
Schmitt MN, 'Asymmetrical Warfare and International Humanitarian Law' in WH von Heinegg and V Epping (eds), *International Humanitarian Law Facing New Challenges* (Springer 2007)
Schmitt MN, *The Interpretive Guidance on the Notion of Direct Participation in Hostilities: A Critical Analysis,* Essays on Law and War at the Fault Lines (Springer 2011)
Schmitt MN, 'Wound, Capture, or Kill: A Reply to Ryan Goodman's "The Power to Kill or Capture Enemy Combatants"' (2013) 24 European Journal of International Law 855
Searle JR, *The Construction of Social Reality* (Free Press 1995)
Sherman N, *The Untold War: Inside the Hearts, Minds, and Souls of Our Soldiers* (1st edn, W W Norton 2010)
Shoemaker RB, 'The Taming of the Duel: Masculinity, Honour and Ritual Violence in London, 1660–1800' (2002) 45 The Historical Journal 525
Shue H, 'Do We Need a "Morality of War"?' in David Rodin and Henry Shue (eds), *Just and Unjust Warriors: The Moral and Legal Status of Soldiers* (Oxford University Press 2008)
Shue H, 'Laws of War' in Samantha Besson and John Tasioulas (eds), *The Philosophy of International Law* (Oxford University Press 2010)
Simmons AJ, 'Ideal and Nonideal Theory' (2010) 38 Philosophy & Public Affairs 5
Skaar E, García-Godos J, and Collins C (eds), *Transitional Justice in Latin America: The Uneven Road from Impunity towards Accountability* (Routledge 2017)
Smart JJC and Williams B, *Utilitarianism: For and Against* (Cambridge University Press 1973)
Smart JJC, 'An Outline of a System of Utilitarian Ethics' in JJC Smart and Bernard Williams (eds), *Utilitarianism: For and Against* (Cambridge University Press 1973)
Smith JC, 'Liberties and Choice' (1974) 19 American Journal of Jurisprudence 87
Stahl L, 'What the Last Nuremberg Prosecutor Alive Wants the World to Know' (*CBS News*, 30 June 2019) <https://www.cbsnews.com/news/what-the-last-nuremberg-prosecutor-alive-ben-ferencz-wants-the-world-to-know-60-minutes-2019-06-30/> accessed 17 August 2022
Statman D, 'On the Success Condition for Legitimate Self-Defense' (2008) 118 Ethics 659
Steinhoff U, 'Debate: Jeff McMahan on the Moral Inequality of Combatants' (2008) 16 Journal of Political Philosophy 220
Steinhoff U, 'Rights, Liability, and the Moral Equality of Combatants' (2012) 16 The Journal of Ethics 339
Steinhoff U, 'The Liability of Justified Attackers' (2016) 19 Ethical Theory and Moral Practice 1015
Steinhoff U, 'Against a "Combined Liability-Lesser-Evil Justification"' (2019) 47 Philosophia 533
Stemplowska Z and Swift A, 'Ideal and Nonideal Theory' in David Estlund (ed), *The Oxford Handbook of Political Philosophy* (Oxford University Press 2012)
Stemplowska Z, 'What's Ideal about Ideal Theory?' (2008) 34 Social Theory and Practice 319
Tadros V, *The Ends of Harm: The Moral Foundations of Criminal Law* (Oxford University Press 2011)
Tadros V, 'Duty and Liability' (2012) 24 Utilitas 259

Tadros V, 'Orwell's Battle with Brittain: Vicarious Liability for Unjust Aggression' (2014) 42 Philosophy & Public Affairs 42

Tadros V, 'Unjust Wars Worth Fighting For' (2016) 4 Journal of Practical Ethics 52

Tadros V, 'Duress and Duty' in Saba Bazargan-Forward and Samuel C Rickless (eds), *The Ethics of War: Essays* (Oxford University Press 2017)

Tadros V, 'The Moral Distinction Between Combatants and Noncombatants: Vulnerable and Defenceless' (2018) 37 Law and Philosophy 289

Tadros V, *To Do, to Die, to Reason Why: Individual Ethics in War* (Oxford University Press 2020)

Tanguay-Renaud F, 'Can States Be Corporately Liable to Attack in War?' in Saba Bazargan and Samuel C Rickless (eds), *The Ethics of War: Essays* (Oxford University Press 2017)

Thomson JJ, 'Self-Defense and Rights' (The Lindley Lecture, University of Kansas, 1976) <https://kuscholarworks.ku.edu/bitstream/handle/1808/12392/Self-Defense%20and%20Rights-1976.pdf;sequence=> accessed 18 August 2022

Thomson JJ, *Rights, Restitution, and Risk: Essays, in Moral Theory* (William Parent ed, Harvard University Press 1986)

Thomson JJ, *The Realm of Rights* (Harvard University Press 1990)

Thomson JJ, 'Self-Defense' (1991) 10 Philosophy & Public Affairs 283

Thorburn M, 'Soldiers as Public Officials: A Moral Justification for Combatant Immunity' (2019) 32 Ratio Juris 395

Tomlin P, 'Innocence Lost: A Problem for Punishment as Duty' (2017) 36 Law and Philosophy 225

Valentini L, 'Ideal vs. Non-Ideal Theory: A Conceptual Map' (2012) 7 Philosophy Compass 654

Varfolomeeva A, 'Number of Child Soldiers Involved in Conflicts Worldwide Jumps 159% in 5 Years,' (*The Defense Post*, 11 February 2019) <https://thedefensepost.com/2019/02/11/child-soldiers-global-increase/> 22 August 2022

Vattel E de, *The Law of Nations, or, Principles of the Law of Nature, Applied to the Conduct and Affairs of Nations and Sovereigns, with Three Early Essays on the Origin and Nature of Natural Law and on Luxury* (Bela Kapossy and Richard Whatmore eds, Thomas Nugent tr, Liberty Fund 2008)

Viehoff D, 'Democratic Equality and Political Authority' (2014) 42 Philosophy & Public Affairs 337

Vitoria F de, *De Indis et de Iure Belli Relectiones* (John Pawley tr, Carnegie Institution of Washington 1917)

Vitoria F de, *Reflection on Homicide & Commentary on Summa Theologiae* (John P Doyle tr, Marquette University Press 1997)

Von Elbe J, 'The Evolution of the Concept of the Just War in International Law' (1939) 33 American Journal of International Law 665

Waldron J, 'Special Ties and Natural Duties' (1993) 22 Philosophy & Public Affairs 3

Waldron J, *Law and Disagreement* (Clarendon Press 2004)

Waldron J, 'Torture, Suicide, and Determinatio' (2010) 55 American Journal of Jurisprudence 10

Waldron J, *Torture, Terror, and Trade-Offs: Philosophy for the White House* (Oxford University Press 2010)

Waldron J, 'Responses to Zedner, Haque and Mendus' (2014) 8 Criminal Law and Philosophy 135

Waldron J, 'Deep Morality and The Laws of War' in Seth Lazar and Helen Frowe (eds), *The Oxford Handbook of Ethics of War* (2015)

Walen A, *The Mechanics of Claims and Permissible Killing in War* (Oxford University Press 2019)

Walzer M and Margalit A, 'Israel: Civilians & Combatants' (*The New York Review of Books*, 2009) <http://www.nybooks.com/articles/archives/2009/may/14/israel-civilians-combatants/> accessed 28 October 2015

Walzer M, 'World War II: Why Was This War Different?' (1971) 1 Philosophy & Public Affairs 3

Walzer M, 'Political Action: The Problem of Dirty Hands' (1973) 2 Philosophy & Public Affairs 160

Walzer M, *Just and Unjust Wars: A Moral Argument with Historical Illustrations* (Fifth edition Basic Books 2006)

Walzer M, 'Response to McMahan's Paper' (2006) 34 Philosophia 43

Wellman C, *An Approach to Rights: Studies in the Philosophy of Law and Morals* (Kluwer Academic Publishers 1997)

Wellman C, 'Legal Rights' in Carl Wellman (ed), *Rights and Duties* (Routledge 2002)

Wenar L, 'Rights' in Edward N Zalta (ed), *The Stanford Encyclopedia of Philosophy* (Fall 2015, 2015) <http://plato.stanford.edu/archives/fall2015/entries/rights/> accessed 6 June 2016

Whetham D, 'Do Some Soldiers Deserve to Die More Than Others?' in Bradley Jay Strawser, Ryan Jenkins, and Michael Robillard (eds), *Who Should Die?: The Ethics of Killing in War* (Oxford University Press 2018)

Whitman MJ, 'Is Pacifism Self-Contradictory?' (1966) 76 Ethics 307

Wiens D, 'Political Ideals and the Feasibility Frontier' (2015) 31 Economics & Philosophy 447

Williams B, 'Ethical Consistency' (1965) 103 Proceedings of the Aristotelian Society 103

Williams B, 'A Critique of Utilitarianism' in JJC Smart and Bernard Williams (eds), *Utilitarianism: For and Against* (Cambridge University Press 1973)

Williams B, 'Politics and Moral Character' in Stuart Hampshire (ed), *Public & Private Morality* (Cambridge University Press 1978)

Williams G, 'The Concept of Legal Liberty' (1956) 56 Columbia Law Review 1129

Witt JF, 'The Dismal History of the Laws of War' (2011) 1 University of California Irvine Law Review 895

Zohar NJ, 'Innocence and Complex Threats: Upholding the War Ethic and the Condemnation of Terrorism' (2004) 114 Ethics 734

Legal Materials

African Charter on Human and Peoples' Rights. 21 June 1986, OAU Doc. CAB/LEG/67/3 rev.5, 21 ILM 58 (entered into force 21 October 1986)

Commentary of 1987 to Protocol Additional to the Geneva Conventions of 12 August 1949, and relating to the Protection of Victims of International Armed Conflicts (Protocol I), 8 June 1977 <https://ihl-databases.icrc.org/applic/ihl/ihl.nsf/Treaty.xsp?documentId=D9E6B6264D7723C3C12563CD002D6CE4&action=openDocument> accessed 18 August 2022

Commentary of 2020 to Convention (III) Relative to the Treatment of Prisoners of War (Geneva, 12 August 1949) (hereinafter, 2020 Commentary to GCIII), Article 85: Offences Committed Before Capture, para 3634. <https://ihl-databases.icrc.org/applic/ihl/ihl.nsf/Comment.xsp?action=openDocument&documentId=3A23C27AF17D2326C1258585005366D3>.

Convention (IV) respecting the Laws and Customs of War on Land and its annex: Regulations concerning the Laws and Customs of War on Land (Hague Convention IV), The Hague, 18 October 1907

Declaration Renouncing the Use, in Time of War, of Explosive Projectiles Under 400 Grammes Weight (hereinafter, 1868 St. Petersburg Convention) 11 December 1868

European Convention on Human Rights. Convention for the Protection of Human Rights and Fundamental Freedoms, 4 November 1950, 213 UNTS 222

Geneva Convention for the Amelioration of the Condition of the Wounded, Sick and Shipwrecked Members of Armed Forces at Sea (Second Geneva Convention), 12 August 1949, 75 UNTS 85

Geneva Convention Relative to the Protection of Civilian Persons in Time of War (Fourth Geneva Convention) (hereinafter, GCIV) 12 August 1949, 75 UNTS 287

Geneva Convention Relative to the Treatment of Prisoners of War (Third Geneva Convention) (hereinafter, GCIII) 12 August 1949, 75 UNTS 135

Hague Convention (IV) Concerning the Laws and Customs of War on Land and the Regulations annexed thereto, preamble, 18 October 1907, 36 Stat. 2277, 1 Bevans 631 (entered into force 26 January 1910)

Human Rights Committee General Comment No. 36 (2018)

ICC, Amendments to the Rome Statute of the International Criminal Court on the crime of aggression, 11 June 2010, Res. RC/Res.6/Annex 1

International Covenant on Civil and Political Rights, 1966, 999 UNTS 171 (entered into force 23 March 1976)

Model Penal Code (American Law Institute, Tentative Draft No. 4 2016)

Protocol Additional to the Geneva Conventions of 12 August 1949, and relating to the Protection of Victims of International Armed Conflicts (Protocol I), 8 June 1977, 1125 UNTS 3

Protocol Additional to the Geneva Conventions of 12 August 1949, and relating to the Protection of Victims of Non-International Armed Conflicts (Protocol II), 8 June 1977, 1125 UNTS 609

Protocol No. 6 to the Convention for the Protection of Human Rights and Fundamental Freedoms (European Convention on Human Rights) concerning the abolition of the death penalty

Protocol to the American Convention on Human Rights to Abolish the Death Penalty ('Pact of San Jose', 8 June 1990, OAS Treaty Series, No. 73

Rome Statute of the International Criminal Court ISBN No. 92-9227-227-6, 17 July 1998

Second optional protocol to International Covenant on Civil and Political Rights, aiming at the abolition of the death penalty, GA resolution 44/128, 15 December 1989

UN Human Rights Council, Report of the Detailed Findings of the Independent International Fact-Finding Mission on Myanmar, UN Doc. A/HRC/39/64 (12 September 2018)

UN Security Council. Charter of the United Nations 24 October 1945, 1 UNTS XVI

UNGA, International Convention against the Recruitment, Use, Financing and Training of Mercenaries, 4 December 1989, A/RES/44/34

Uniform Code of Military Justice, 10 USC (2016)

United States Manual for Courts Martial (2016 Edition)

US Army Field Manual FM-03, Department of the Army, United States of America, ADP 3-0 Operations, Army Publishing Directorate, <https://atiam.train.army.mil/catalog/dashboard> accessed 18 August 2022

Vienna Convention on Diplomatic Relations, 18 April 1961, 500 UNTS 95

Case Law

ICJ, On Legality of the Threat or Use of Nuclear Weapons, Advisory Opinion, 8 July 1996

ICTY, Prosecutor v Dusko Tadić, IT-94-1-A, Appeals Chamber, Decision on the Defense Motion for Interlocutory Appeal on Jurisdiction, 2 October 1995

United States v Gillette, 401 US 437 (1971)

Websites

Advisory Serv. on Int'l Humanitarian Law, Int'l Comm. of the Red Cross, What Is International Humanitarian Law? (2004) https://www.icrc.org/en/war-and-law accessed 18 August 2022

'Child Recruitment and Use' (*Office of the Special Representative of the Secretary-General of the United Nations for Children and Armed Conflict*) <https://childrenandarmedconflict.un.org/six-grave-violations/child-soldiers/> accessed 18 August 2022

'Child Soldiers International Annual Report 2017–18' (*OCHA*, 24 September 2018) <https://reliefweb.int/report/world/child-soldiers-international-annual-report-2017-18> accessed 22 August 2022

'Global Conflict Tracker' (*Council on Foreign Relations*) <https://www.cfr.org/interactive/global-conflict-tracker/?category=us> accessed 22 August 2022

'How Does Law Protect in War? Surrender' (*International Committee of the Red Cross (ICRC)* 2002) <https://casebook.icrc.org/glossary/surrender> accessed 22 August 2022

ICRC. IHL Database, Customary IHL, Rule 1 <https://ihl-databases.icrc.org/customary-ihl/eng/docs/v1_rul_rule1> accessed 18 August 2022

Prisoners of War and Detainees Protected Under International Humanitarian Law *in* ICRC <https://www.icrc.org/eng/war-and-law/protected-persons/prisoners-war/overview-detainees-protected-persons.htm> accessed 18 August 2022

'VA National Suicide Data Report 2005-2016,' (*US Department of Veterans Affairs*, 2018) 2, <https://www.mentalhealth.va.gov/suicide_prevention/data.asp> accessed 22 August 2022

'Yemen' (*Human Rights Watch World Report*, 2019) <https://www.hrw.org/world-report/2019/country-chapters/yemen> accessed 22 August 2022

'Yemenis Are Left so Poor They Kill Themselves Before Hunger Does' (*The Guardian*, 11 December 2018) <http://theguardian.com/global-development/2018/dec/11/yemen-war-norwegian-refugee-council> accessed 18 August 2022

Index

For the benefit of digital users, indexed terms that span two pages (e.g., 52–53) may, on occasion, appear on only one of those pages.

Additional Protocol I 17, 18–19, 33–34, 43
Additional Protocol II 17
agent-relative reasons or prerogatives 11, 94, 179–80, 239
aggression, crime of 14–15, 49, 74–75, 76, 81, 191, 194, 263–64
Arneson, Richard 245

Bazargan-Forward, Saba 160–61, 169–70, 171, 173–74, 243
Beer, Yishai 194, 225, 226
Benbaji, Yitzhak 95, 97, 98–99, 100–2, 266–67
Blum, Gabriella 220, 225
Burri, Susanne 105n.58, 130n.90, 131, 134n.108, 254–55

chain of command issue 168–69, 181–85, 216–17, 233–34
civilians
  principle of distinction 1–2, 4, 17–20, 55–56, 61, 113, 195, 196–98, 231–32, 275
Collective threats 10, 156, 157–58, 187–89, 191, 195–98, 212–13, 230, 232–33, 243–50
  macro-threat and micro-threat distinction 127–28, 133, 140, 143–44, 146–47, 150, 156, 165, 173–74, 175, 181–82, 185, 187, 196
collectivism 114–15, 117–18, 213
  collective liability 150–51
combatants
  privileged combatants 16–17, 36–40, 270
  just combatants 2–3, 14, 55–56
  unjust combatants
    liability of 2–3, 55–56, 111, 118–23
    non-liability of 111–77

conscientious driver 131–33, 135, 144–45, 152–53, 169–70
conscientious refusal 247–48
  see also surrender
conscription 138
  duty to accept conscription 258–62
consequentialism
  instrumentalism 6–7, 56–60, 63–67, 73–74
  instrumentalism and perpetual peace ideal *see* perpetual peace ideal
  and non-consequentialism 58–59, 63, 66–67, 69–72, 189–90
  non-instrumentalism 93–94, 95, 238, 270
  rule-consequentialism 189
  weak instrumentalist requirement 68–69
    instrumentalism and moral fragility 12, 85–86, 87, 271–77

Dan-Cohen, Meir 49–50
Dannenbaum, Tom 266
democratic authority 94, 105–10
desertion 247–48, 265–66
Dill, Janina 60, 62–63
diplomatic immunity 46–48, 50, 53
doing our share (fair share) 248–50, 258–62
Draper, Kaila 59
duress 141, 154, 159–60, 241–43
duty
  canceling condition 207–8
  exclusionary reasons 205–6

eliminative killing 175–76
Estlund, David 97–98, 99, 101–7, 249–50

evidence-relative permissibility 14, 147, 167–69, 181–84, 195–96, 231–32, 233–34, 240, 242–43, 257–58, 262
excuse(s) 53, 199, 200, 239, 242–43, 263

Fabre, Cécile 112–13
fact-relative permissibility 14, 53, 167–68, 184, 196, 232, 233
forfeiture *see* liability
Frowe, Helen 112–13, 115, 137–38, 160–61, 171–72, 244–45, 254

Glover, Jonathan 157–58
Gordon-Solmon, Kerah 223

Halbertal, Moshe 194n.38, 195n.42, 257n.78
Haque, Adil 6, 27–29, 40–53, 59, 112, 138–39, 147, 225
Hohfeld, Wesley
  Hohfeldian privilege 28–29, 30–36, 40–41, 42–43, 44–46, 50, 52, 54
  and liability to defensive harm 119–21
humanitarian intervention 14–15, 191, 194
humanitarian view 56, 57–60, 61–64, 65–67, 68–74, 81–82, 84–85, 189–90

ideal theory
  and non-ideal theory (as non-compliance) 57–58, 75–77, 78–85, 187, 191, 248–50
immunity from prosecution 17, 27, 29, 36, 40–53, 55, 84–85, 253
individualism 111–15, 118, 127–28, 133, 135–36, 150–51
innocent bystander 125–26, 131, 220
instrumentalism *see* consequentialism
international armed conflict (IAC) 16–17, 18–19, 20–22, 24, 34–35, 42, 43, 56

jus ad bellum 1–2, 4, 14–15, 16–17, 48–51, 52, 81
jus in bello 1–2, 4, 48–51, 93–94
just cause for war 14, 49, 118, 145, 190–93, 194–98, 201, 216–17, 223–24, 242–43, 244–45, 256, 262
just war 2–3, 8, 14, 55–56, 118, 127–28, 186, 187–89, 190–93, 243–45
  moral fragility of 7, 77, 86, 182, 228–29
just war theory 4–5, 76, 78–79, 80, 191–92
justification 199–200

Kahn, Paul 22, 66, 251–52
Kant, Immanuel 71
Kelsen, Hans 28–29, 37–39, 41–42, 50
killing and letting die, distinction 72–73, 162–63, 185
Korsgaard, Christine 240–41, 264
Kutz, Christopher 64–65

Lamb, Anthony 59
Lazar, Seth 67, 77n.100, 84n.135, 114–16, 117n.36, 130–31, 147, 164, 175, 186, 190n.23, 255
  legal equality of combatants 1–3, 4, 6, 7, 12–13, 16–17, 42, 52–53, 54, 60, 67, 68, 73–74, 84–85, 90–91, 94, 113, 118–19, 184, 238, 251–52, 256, 274
lesser-evil justification 60, 72–73, 130–31, 145, 160–63, 182, 183–84, 185, 220–23, 224, 229, 232–33, 242
  the combined view 169–76
  infringement model of rights 162, 229
liability 119–21, 139–47
  causal contributions 140, 141–42, 144, 149, 151, 155–57, 160, 213
  comparative account of liability 134–35
  complex account 169–70
  the moral responsibility account 122–23, 125–39
  narrow proportionality 140–43, 144, 148, 149, 155, 164, 169, 172, 173
  necessity 140, 143–47, 148, 149, 164, 173, 192–93
  stringency principle 142–43
  unjust combatants non-liability *see* unjust combatants
Lieblich, Eliav 51–52, 53
Luban, David 68, 104

McMahan, Jeff 59, 60, 61–62, 84–85, 112–13, 119–20, 131–32, 134–37, 140–41, 142, 144–45, 147, 160–61, 169, 170–71, 172, 173–74, 221–23
means principle 199, 255–56
Moyn, Samuel 275n.17, 276n.22
Murphy, Liam 96, 248

Nagel, Thomas 95, 103–4, 109, 194–95
necessity *see* liability

NIAC 17, 19, 20–21, 23, 24, 45
non-threatening combatants *see* civilians: micro and macro threat
Nozick, Robert 227–28

opportunistic killing 175, 183–84, 185, 220–21, 233

pacifism; pacifist 10–11, 12, 75, 76, 77–78, 124, 125, 182, 184–85, 186–87, 195–96, 200–1, 212–13, 215, 223, 228–29, 234–35, 275, 276
Parfit, Derek 246–47
partially excused threats 147
permissibility 199–234
  comparative and non-comparative account 195–96, 231–32
  moral residue 201–3, 219, 224
  morally indifferent actions 200
  proportionality 220–27
  and reductivism 187–89, 232–33
  small difference in weight of reasons 181, 203–5, 209–16, 219, 227–30
perpetual peace ideal *see* instrumentalism
proportionality
  narrow proportionality *see* liability
  proportionality and permissibility *see* permissibility
  proportionality in the aggregate 222–23
  wide proportionality 171, 220

Quong, Jonathan 9, 142–43, 199, 255–56

Rawls, John 75
Raz, Joseph 206, 218–19
reasons
  exclusionary reasons 205–6
  first-order reasons 102–3, 205, 206
rights
  violations/infringements 59, 72, 119, 162, 209, 224, 227–28, 229–30, 271–72
reductivism 111, 112, 113, 115–18, 133
  exceptionalism or non-reductivism 113
  see also revisionism
reductivist individualism 124
  see also revisionism
Renzo, Massimo 105
the responsibility dilemma 130
revisionism 113–23, 148
  methodological revisionism 114–18, 148
  substantive revisionism 118–23
Ripstein, Arthur 46, 93–94, 253
Rodin, David 70, 71, 122, 221–22, 223
role morality 7, 93–105, 109–10, 270
  moral division of labor 96–97, 102–4
Rome Statute 14–15, 22, 74–75

Scheffler, Samuel 77n.102, 180, 217n.123, 264–65
self-defence *see* liability
Shue, Henry 59, 60, 62–63
small moral world 240–41, 261–68
Statman, Daniel 97, 100–1
surrender
  incentives to surrender 61
  obligation to surrender 240–41, 245–50, 263, 264–68
  opportunity to surrender 219
symmetry thesis *see* legal equality of combatants

Tadros, Victor 112, 114, 171–72, 263
Thomson, Judith Jarvis 126
threshold deontology 166–67
traditionalism *see* revisionism

uncertainty 11, 117–18, 147, 176, 180, 183–84, 198, 214, 215–17, 231–32, 242–43

Valentini, Laura 248–49
villainous aggressor 125–30, 132–33, 149, 150, 152, 155–56, 157, 158–59, 176, 179, 186, 202–3, 221–22, 238–39, 244, 253–54

Waldron, Jeremy 19, 29, 63–64, 67–68, 75, 83–84, 198
Walen, Alec 162, 172, 229
Walzer, Michael 2, 4–5, 113, 117–18, 134–35, 218, 227–28
Williams, Bernard 82–83, 176–77, 209, 218

Zohar, Noam 127–28, 133